Whole Body Vibrations

Physical and Biological Effects on the Human Body

Whole Body Vibrations

Physical and Biological Effects on the Human Body

Edited by

Redha Taiar, Christiano Bittencourt Machado, Xavier Chiementin, and Mario Bernardo-Filho

CRC Press
Taylor & Francis Group
Boca Raton London New York

CRC Press is an imprint of the
Taylor & Francis Group, an **informa** business

CRC Press
Taylor & Francis Group
6000 Broken Sound Parkway NW, Suite 300
Boca Raton, FL 33487-2742

Printed on acid-free paper

International Standard Book Number-13: 978-1-138-50001-3 (Hardback)

Library of Congress Cataloging-in-Publication Data

Names: Taiar, Redha, editor. | Machado, Christiano B., editor. | Chiementin, Xavier, editor. | Bernardo-Filho, Mario, editor.
Title: Whole body vibrations : physical and biological effects on the human body / [edited by] Redha Taiar, Christiano Bittencourt Machado, Xavier Chiementin, and Mario Bernardo-Filho.
Description: Boca Raton : CRC Press/Taylor & Francis, 2018. | Includes bibliographical references.
Identifiers: LCCN 2018034178 | ISBN 9781138500013 (hardback : alk. paper)
Subjects: | MESH: Vibration--adverse effects | Biomechanical Phenomena--physiology
Classification: LCC R856 | NLM WD 640 | DDC 610.28--dc23
LC record available at https://lccn.loc.gov/2018034178

Visit the Taylor & Francis Web site at
http://www.taylorandfrancis.com

and the CRC Press Web site at
http://www.crcpress.com

Contents

Preface

Writing a scientific book, as it is known today, depends on a series of major factors such as regrouping researchers, informing and exchanging with them, and at least, to keep them motivated to achieve this objective. This book comes after many years of collaboration with most of the participant writers (thirty-eight individuals from eight countries). We have learned much from each other and decided to transfer our knowledge to others interested in the impact of whole body vibration on healthy subjects and on the daily life of patients.

Mechanical vibrations are present in our lives in different situations. There are natural vibrations in the tissues and organs that are responsible for various and important functions, such as heart rate, peristaltic movement of the digestive tract, contractions of the blood vessels, and skeletal muscle contractions. In addition, during various daily activities our bodies are exposed to mechanical vibrations from the external environment that can transmit energy to the organs and tissues. These absorbed energies can aid the organism to perform metabolic functions efficiently.

In a healthy subject, the transmission of mechanical vibrations from the environment to the body can occur in simple activities, as walking or running due to the impact of the feet with the ground. This book allows an understanding about the qualities and disadvantages of vibration exposure on the human body with a biomechanical and medical perspective. It offers a comprehensive range of principles, methods, techniques, and tools to provide the reader with a clear knowledge of the impact of vibration on human tissues and physiological processes. The text considers physical, mechanical, and biomechanical aspects and it is illustrated by application in key domains such as sports and medicine. The first three chapters provide useful tools for measuring, generating, simulating, and processing vibration signals. The following six chapters are "applications" in different fields of expertise, from performance to health, with localized or global effects. Since, unfortunately, there can be undesirable effects to being exposed to mechanical vibrations, a final chapter is dedicated to this issue. Engineers, researchers, and students from biomedical engineering and health sciences, as well as industry professionals can profit from this compendium of knowledge about mechanical vibration applied to the human body.

Acknowledgments

We express our thanks to all the authors for their successful collaboration on this book. We recognize and appreciate everything you have done to achieve this hard objective. Thank you colleagues and friends for your kind attention.

Editors

Redha Taiar, PhD, is a professor at the University of Reims Champagne-Ardenne, France. He is a specialist in the biomechanics of health disease and rehabilitation. His research focuses on industry engineering for medicine and high-level sport. He is an engineer for different industries like Arena for high-level sport, and Sidas and Medicapteur for medical development. For industry workers, his lastest research focused on the conception and validation of anti-fatigue mats for Notrax manufacturing group. His most recent research for the sports industry focused on the development of swimsuits for the triathlon, swimming, and skiing for the 2016 Olympic Games hosted in Rio de Janeiro, Brazil and the 2014 Winter Olympic Games held in Sochi, Russia. Prof. Taiar is head of the RTBE (Redha Taiar Biomechanical Engineering) society, which was developed to provide sport and medical advice for industry professionals. He has served as the vice head of the congress of Physical Medicine and Rehabilitation Society (SOFMER) since October 2013 in Reims city. He is on the editorial board for the *Biomechanics Journal* series and is a reviewer for the *Journal of Biomechanics*, the *Journal of Human Movement Science*, the *International Journal of Sports Medicine*, *The Sport Technology Journal*, and the *Journal of Applied Biomechanics*. He has developed patents for new kinds of knee rehabilitation tools for high-level athletes and pathological subjects and the concept of a new shoe for high-level athletes and pathological subjects.

Further information about Prof. Taiar can be found at the following address: www.redha taiar.com

Xavier Chiementin, PhD, is an associate professor at the University of Reims Champagne-Ardenne, France. He received his PhD in mechanical engineering from the University of Reimsin in 2007. His research interests include dynamic structure and sports engineering. Dr. Chiementin focuses on human exposure to vibration. He studies the mechanical and vibrational behavior of equipment by considering the interaction between man and machine. This research has applications in numerous industrial sectors like transport and sport. He is the principal investigator involved in several national and international research activities relating to the study of vibration.

He has published nearly 70 technical and scientific papers in international peer-reviewed journals and conference proceedings. He is a member of editorial boards for *The Open Mechanical Engineering Journal*, *Vibration*, and *Shock and Vibration* and he is associate editor of the *International Robotics & Automation Journal*. He was chair and a participating member of numerous international congresses in the field of vibration research. He is a member of the research working group "Vibration/ Acoustic" as part of the French Society of Mechanics and a member of the World Association of Vibration Exercises Experts – WAVEX.

Further information about Dr. Chiementin can be find at the following address: www.grespi.fr

Christiano Bittencourt Machado, PhD, has a bachelor's degree in physical therapy (2002) and a licentiate degree in mathematics (2018) from Estácio de Sá University (Brazil), a master's degree in biomedical engineering from the Federal University of Rio de Janeiro, UFRJ (2005), and a PhD in mechanical, acoustic, electronic, and robotic sciences from the University of Paris Sorbonne (France) as well as in biomedical engineering from UFRJ (2011). Prof. Machado has a formation on human anatomy from the University of Michigan. He is currently head of the Biomedical Ultrasound Laboratory (LabusBio) at Estácio de Sá University, developing research on quantitative ultrasound to characterize bone tissue, infrared thermography applied to medicine, and whole-body vibration. He is also a professor at the same university on undergraduate health programs (human anatomy and physiology, biomechanics, biochemistry and electrotherapy) and on the master's program of family health, with a focus on the application of technology for primary care and public health. Currently, he is on several international review boards (for instance, Ultrasonics, IEEE Transactions on Ultrasonics, Ferroelectrics, and Frequency Control, Biomedical Signal Processing and Control, Medical & Biological Engineering and Computing, Materials, PAHCE – Pan American Health Care Exchanges), which work in collaboration with several countries.

More information about the Prof. Christiano can be found at the following address: christiano-machado.webnode.com/

Mario Bernardo-Filho, PhD, is a professor at Rio de Janeiro State University, Brazil. He is a physiotherapist and his research involves integrative and complementary medicine (auriculotherapy and acupuncture) and mechanical vibrations generated in an oscillating/vibratory platform that produces whole body vibration exercises (WBVE) when a subject is in contact with that platform. His research includes ongoing studies to evaluate the consequences of WBVE and extracts of medicinal plants in rats and investigations about the effects of the WBVE in individuals with different diseases (metabolic syndrome, chronic obstructive pulmonary disease, arthrosis, and osteoporosis) and healthy people. He is supervisor to various professionals that are preparing their Master of Science or PhD thesis. He teaches courses in graduate and undergraduate courses at Rio de Janeiro State University. He was the head of the First International Congress on Mechanical Vibration and Integrative and Complementary Practices in Cabo Frio, Brazil (2016). At this meeting, the World

Association of Vibration Exercise Experts (WAVEX) was created and in 2018, the First Congress of the WAVEX was held in Groningen, the Netherlands. He has more than 130 publications indexed in PubMed (https://www.ncbi.nlm.nih.gov/pubmed/?term=bernardo-filho). Prof. Bernardo-Filho is also on the editorial board for the *Biomechanics Journal,* the *Brazilian Archives and Biology and Technology*, and *Fisioterapia Brasil.* He is also part of the peer review process for the *American Journal of Physical Medicine & Rehabilitation*, *Anais da Academia Brasileira de Ciências*, *Annual Research & Review in Biology*, *Brazilian Archives of Biology and Technology*, *Brazilian Journal of Medical and Biological Research*, *Brazilian Journal of Microbiology*, *British Journal of Applied Science and Technology*, *British Journal of Medicine and Medical Research*, *Cancer Biotherapy & Radiopharmaceuticals, Clinics, Evidence-Based Complementary and Alternative Medicine* (Online), *International Journal of Therapy and Rehabilitation*, *Journal of Radioanalytical and Nuclear Chemistry (JRNC)*, *Journal of Exercise, Sports & Orthopedics*, *Journal of Scientific Research and Reports*, *Journal of Steroid Biochemistry and Molecular Biology*, *Memórias do Instituto Oswaldo Cruz*, *Pharmaceutical Nanotechnology*, *Brazilian Journal of Geriatrics and Gerontology*, *Stem cell research & therapy*, *The Scandinavian of Clinical and Laboratory Investigation*, *Physiotherapy Theory and Practice*, *Physiotherapy Research International*, and *The Physician and Sportsmedicine*.

More information about Prof. Mario Bernardo-Filho can be found at the following address: http://lattes.cnpq.br/9941440001544010.

List of Contributors

Armèle Dornelas de Andrade
Laboratório de Fisiologia e Fisioterapia Cardiopulmonar
Universidade Federal de Pernambuco
Recife, Brazil

Frédéric Bonnardot
Université de Lyon
UJM Saint Etienne
LASPI
Roanne, France

Christophe Corbier
Université de Lyon
UJM Saint Etienne
LASPI
Roanne, France

Samuel Crequy
MSMP, Arts et Métiers ParisTech
Paris, France

D. da Cunha de Sá-Caputo
Laboratório de Vibrações Mecânicas e Práticas Integrativas
Departamento de Biofísica e Biometria
Instituto de Biologia Roberto Alcantara Gomes, Universidade do Estado do Rio de Janeiro, Brazil

Helga C. Muniz de Souza
Laboratório de Fisiologia e Fisioterapia Cardiopulmonar
Universidade Federal de Pernambuco
Recife, Brazil

Anselm B. M. Fuermaier
Department of Clinical and Developmental Neuropsychology
University of Groningen
Groningen, the Netherlands

Helen K. Bastos Fuzari
Laboratório de Fisiologia e Fisioterapia Cardiopulmonar
Universidade Federal de Pernambuco
Recife, Brazil

Maria Teresa García-Gutiérrez
CyMO Research Institute
Valladolid, Spain

Marelle Heesterbeek
Department of Molecular Neurobiology
GELIFES, University of Groningen
Groningen, the Netherlands

Marciele S. Hopp
Department of Physical Education and Health
Universidade de Santa Cruz do Sul
Santa Cruz do Sul, Brazil

Georges Kouroussis
Faculty of Engineering, Department of Theoretical Mechanics, Dynamics and Vibrations
University of Mons — UMONS
Mons, Belgium

Patrícia E. M. Marinho
Laboratório de Fisiologia e Fisioterapia Cardiopulmonar
Universidade Federal de Pernambuco
Recife, Brazil

Pedro J. Marín
CyMO Research Institute
Valladolid, Spain

Marcela Múnera
Department of Biomedical Engineering
Colombian School of Engineering Julio Garavito
Bogotá, Colombia

Sébastien Murer
Université de Reims Champagne-Ardenne
Reims, France

M. Fritsch Neves
Departamento de Clínica Médica, Faculdade de Ciências Médicas
Universidade do Estado do Rio de Janeiro
Rio de Janeiro, Brazil

Dulciane N. Paiva
Department of Physical Education and Health
Universidade de Santa Cruz do Sul
Santa Cruz do Sul, Brazil

Maíra Florentino Pessoa
Laboratório de Fisiologia e Fisioterapia Cardiopulmonar
Universidade Federal de Pernambuco
Recife, Brazil

Thomas Provot
EPF – Graduate School of Engineering
Sceaux, France
and
Institut de Biomécanique Humaine Georges Charpak, Arts et Métiers ParisTech
Paris, France

Lanto Rasolofondraibe
CReSTIC, Université de Reims Champagne-Ardenne
Reims, France

Matthew R. Rhea
A.T. Still University
Mesa, Arizona

Caroline Cabral Robinson
Universidade Federal de Ciências da Saúde de Porto Alegre
Hospital Moinhos de Vento de Porto Alegre
Porto Alegre, Brazil

Borja Sañudo
Departamento de Educación Física y Deporte
Universidad de Sevilla
Seville, Spain

Roger Serra
INSA Centre Val de Loire, Laboratoire de Mécanique Gabriel Lamé
Blois, France

Eckhard Schoenau
University Hospital Cologne, Children's and Adolescent's Hospital
Cologne, Germany

Christina Stark
University Hospital Cologne, Children's and Adolescent's Hospital
Cologne, Germany

Oliver Tucha
Department of Clinical and Developmental Neuropsychology
University of Groningen
Groningen, the Netherlands

Marieke J. G. van Heuvelen
Center for Human Movement Sciences
University Medical Center Groningen/University of Groningen
Groningen, the Netherlands

Litiele E. Wagner
Department of Physical Education and Health
Universidade de Santa Cruz do Sul
Santa Cruz do Sul, Brazil

Eddy A. van der Zee
Department of Molecular Neurobiology
GELIFES, University of Groningen
Groningen, the Netherlands

1 Instrumentation for Mechanical Vibration Analysis

Thomas Provot, Roger Serra, and Samuel Crequy

CONTENTS

1.1 INTRODUCTION

Vibration is an oscillation motion around an equilibrium point. It is defined according to the amplitude of these oscillations but also by their frequencies (Griffin, 1990). Today, the study of human vibrations has been associated with the notion of health. Indeed, when the vibrations are controlled, it can show a positive effect, such as an improvement in sports performance (Chen et al., 2014) or as a useful treatment for some diseases (Ruck et al., 2010). On the other hand, when vibrations are not controlled, it can show many negative aspects such as performance decreases (Samuelson et al., 1989), or acute or chronic pathologies (Boshuizen et al., 1990). The quantification of the vibrations perceived by the human body is therefore a very important subject, studied by an experimental phase requiring appropriate instrumentation.

Experimental testing and associated techniques play a significant role in how a wide range of practical vibration problems are characterized. Maximum benefits could be obtained when the instrumentation and analysis techniques used are consistent with the desired test objectives for a given machine and/or structure (in this case the human body). In fact, the quality of the measured data is fundamental. In vibration practice, perhaps the most important task of instrumentation is to measure vibration. Thus, it is very important to understand the fundamental principles that are involved in mechanical vibration testing.

The mechanical vibration testing of a specific item is composed of different parts. Figure 1.1 shows a generic test item (De Silva, 2007). The tested object (in the case of this chapter, the human body) is submitted to vibrations by the way of an exciter. This last one could be a simulating exciter such as a shaker or a vibration platform, or a natural exciter such as a vibration tool or the natural operating vibration of a system. The output motion of the tested object is measured by a response sensor and the input measured by a control sensor (usually accelerometers). The electrical signals from these sensors are amplified electronically and analyzed. The frequency multi-analyzer associated with a computer recorder is commonly employed for analysis purposes. Thus, an experimental vibration system generally consists of four main subsystems: (i) test item (object), (ii) excitation system (vibration exciter), (iii) control and response systems and (iv) signal acquisition and modification system.

This chapter will present the different tools, techniques and characteristics useful for the acquisition of vibration in the case of measurement on the human body. The dynamic signal analysis will be described in Chapter 2.

1.2 CONTROL SYSTEM: SENSORS AND TRANSDUCERS

In order to assess and process acceleration and vibration measurement, it is necessary to deploy all the conventional elements of an acquisition chain (Figure 1.2).

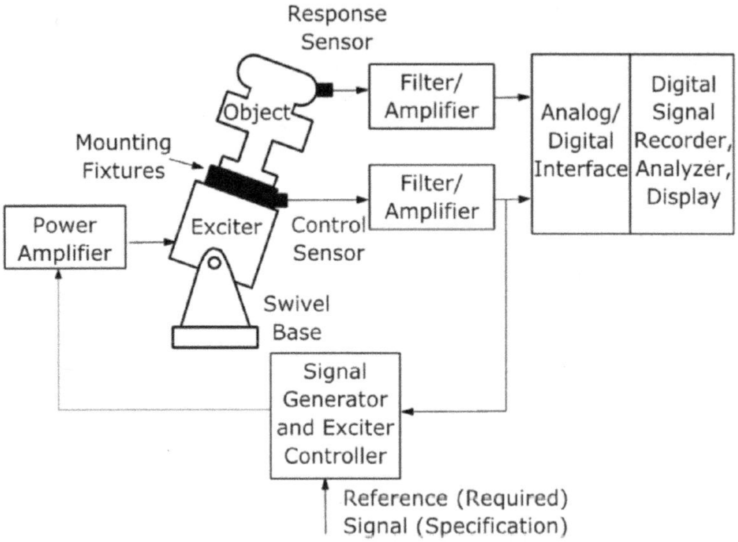

FIGURE 1.1 Example of test flowchart and instrumentation in vibration testing (De Silva, 2007).

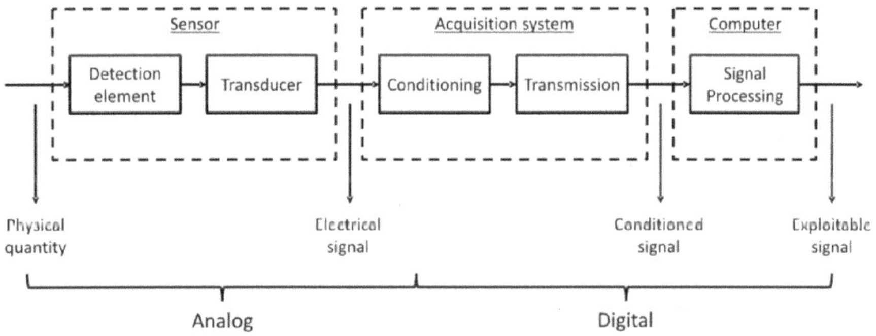

FIGURE 1.2 Synoptic of an acquisition chain.

The physical quantity to assess occurs at the input of this chain, in this case the vibration. At high amplitudes and low frequencies, vibration could be expressed naturally as a displacement over time. This vibration can then be seen by an observer. In the case of high frequencies and smaller amplitudes, the vibration could be then expressed as an acceleration or a speed. These terms are related to the energy involved in the oscillation movement.

The first elements of the acquisition chain are associated to the sensor. It is composed of a detection element, representing the body that will be subjected to physical magnitude. In the case of the vibration measurement, the detection element is generally a seismic mass associated with a biasing means (spring). This mass will therefore be submitted to a movement whose oscillation will be constrained by the biasing means. The detection element is followed by the transducer. Transduction is a process that transforms an input physical quantity into another quantifiable physical quantity. The transducer, is based on the existence of different physical or chemical effects allowing transformation of the variation of the movement into a quantity generally associated with an electrical effect. The second elements of the acquisition chain are associated generally to an acquisition system or acquisition card. The first part is the conditioning. Its role may be to amplify and pre-filter the electrical signal or to convert the analog signal into a digital signal. The second part is the transmission element which will transmit the conditioned signal to a storage or data processing device. The last element is generally associated with a computer. The conditioned signal will be processed in order to be transformed as an exploitable signal. This section will especially focus on the transduction elements used.

1.2.1 Transduction

In the case of accelerometers, it is essentially four transduction principles which are used to transform the mechanical input signal into an electrical output signal. These four principles are the potentiometric, capacitive, piezoelectric and piezoresistive.

1.2.1.1 Potentiometric

A potentiometer is a three-terminal resistive element (Figure 1.3). If two fixed terminals are connected to both ends of the resistor, the third represents a mobile cursor that can slide along the resistance track. It is then possible to measure between a fixed terminal and the mobile terminal a voltage V which will depend on the position l of the cursor. Thus, a potentiometric transducer allows the creation of a simple

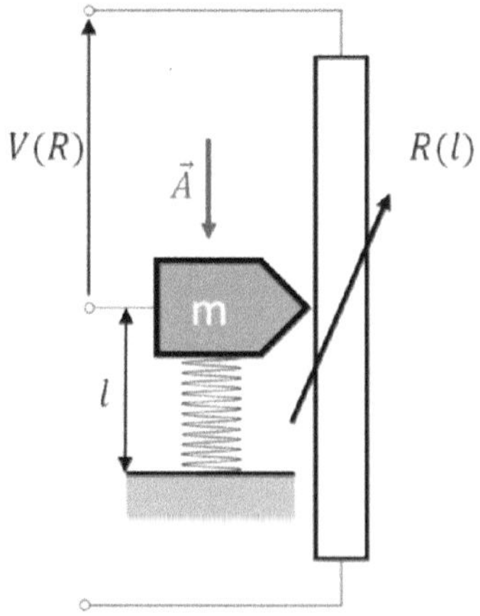

FIGURE 1.3 Principle of potentiometric accelerometer. $\vec{A}$ is the acceleration applied to the seismic mass m. The potentiometer presents a resistance R which depends on the distance l of the cursor. V is the electrical voltage measured between the cursor and the fixed terminal of the potentiometer.

sensor transforming a mechanical displacement into a variation in resistance *R* and therefore in tension. In the case of the vibration measurement, the seismic mass *m* is directly connected to the cursor of the potentiometer. The displacement *l*, or the acceleration of the mass $\vec{A}$, representative of the vibration, will therefore be measured by means of the variation of the resistance *R*.

1.2.1.2 Variable Capacitance Transducers

A capacitive body is a body that stores an electrical charge (Figure 1.4). The concept of mutual capacitance occurs when two parallel conductors (often associated with plates) are isolated from each other. The capacity of this system is therefore, a function of the distance between these two plates. In the case of a capacitive transducer, the measured displacement is directly related to the distance between the two conductive bodies. A variation in capacitance, and thus, a variation in voltage in the electrical system is observed. In the case of the vibration measurement, the seismic mass *m* is associated with a first mobile conductive plate while the second fixed plate is associated with the sensor frame. When the seismic mass is subjected to an acceleration $\vec{A}$, the distance *l* between the two plates varies, it is therefore, possible to determine the value of the reaction force of the spring and the acceleration of the mass as a function of the variation in capacity *C*.

1.2.1.3 Piezoelectric Transducers

A piezoelectric material is a material that is electrically charged under the action of mechanical stress (Figure 1.5). A piezoelectric transducer will therefore observe the stresses which are imposed by the detection element by means of a voltage variation. In the case of the vibration measurement, the spring element is therefore a piezoelectric element (crystal) which emits an electrical signal *V* as a function of the stress σ applied to it by the seismic mass *m*. In this case, the sensor measures the inertial force induced by the seismic mass which is a function of the acceleration undergone by this mass and which is expressed as a voltage variation.

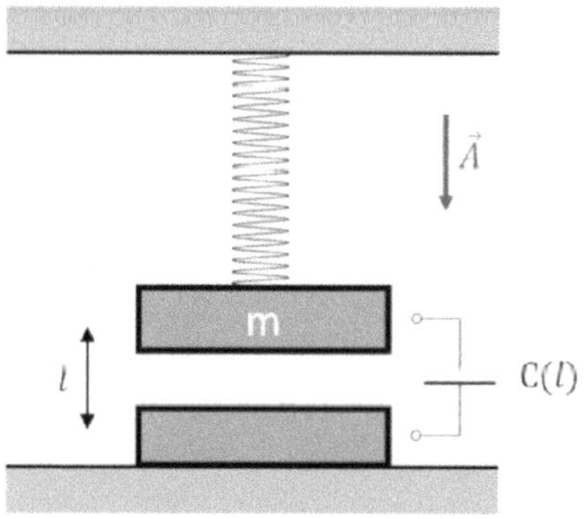

FIGURE 1.4 Principle of capacitive accelerometer. $\vec{A}$ is the acceleration applied to the seismic mass *m*. The capacity *C* between depends on the distance l between the two plates.

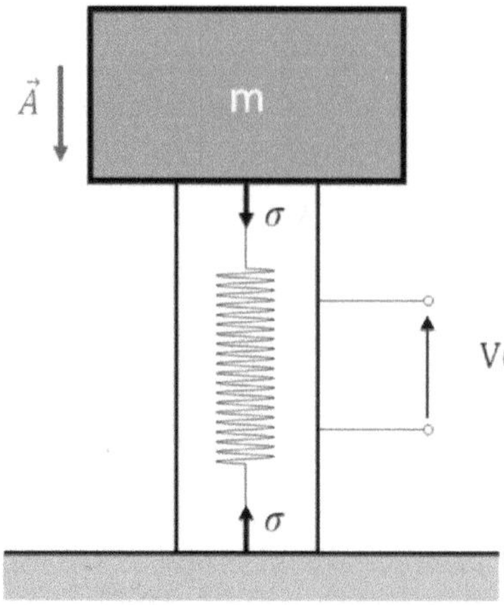

FIGURE 1.5 Principle of piezoelectric accelerometer. $\vec{A}$ is the acceleration applied to the seismic mass *m*. The electrical voltage *V* produced by the piezoelectrical element depends on the stresses σ applied by the seismic mass.

1.2.1.4 Piezoresistive Transducers

A piezoresistive material is a material whose electrical resistance changes under the action of mechanical stress (Figure 1.6). In a piezoresistive transducer, the effect of stresses σ on the piezoresistive element will cause a modification of its geometry and thus of its resistance. In the case of the vibration measurement, the spring element represents a beam to which the seismic mass *m* is suspended. Therefore, the action of the acceleration causes a deformation of the beam ε which is measured by means of the variation in resistance of a piezoresistive element. The displacement of the seismic mass is thus measured with the aid of a typical strain gauge element.

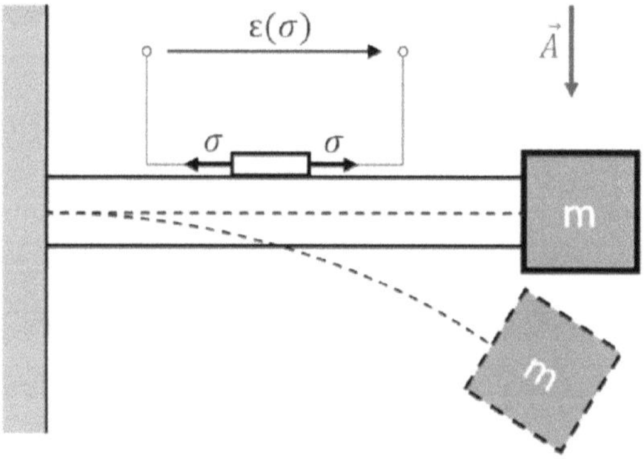

FIGURE 1.6 Principle of piezoresistive accelerometer. $\vec{A}$ is the acceleration applied to the seismic mass *m*. The measured deformation ε depends on the stresses σ applied by the seismic mass.

1.2.2 Other Transduction Techniques

There are also other transduction techniques such as inductive transduction and optical transduction. In the case of inductive transduction, the displacement of a ferromagnetic core (seismic mass) through one or more coils will cause a variation of the magnetic flux and therefore a variation in voltage. The optical transduction is commonly used for laser vibrometers. This tool is based on the Doppler effect to measure the variations of distance and therefore the vibrations of a body. It is the detection of the variation in the frequency of the emitted and then returned light to determine the amplitude of the acceleration.

1.2.3 Actual Accelerometers Technology

Currently, the evolution of micro-fabrication techniques based on silicon micromachining and microelectronics have allowed the production of miniature systems. These micro electro mechanical systems (MEMS) come in various forms, from simple structures without moving elements to highly complex electromechanical structures. MEMS are typically assigned to sensor or actuator functions (Amendola et al., 2011). The use of MEMS technology for accelerometers is quickly expanding, thanks in particular to the automotive industry, which carries these small electronic components in different parts of the vehicles (airbag, belt, seat, etc.).

The use of MEMS accelerometers can be found in many other application areas such as mobile telephony, machine monitoring or video game consoles. In addition to presenting the advantages of the transduction principles presented above, MEMS accelerometers offer other advantages. First, their extreme miniaturization allows them to take innovative measures at strategic points of the system studied. Moreover, these components benefit from a very low cost due to the silicon collective machining technologies used in the microelectronics industry. Several studies have shown that MEMS accelerometers can be a suitable solution for human vibration exposure (Stein et al., 2011; De Capua et al., 2009; Aiello et al., 2012; Tarabini et al., 2012; Provot et al., 2017).

These systems allow the development of on-board instrumentation at low cost without the need for specialized equipment. These electronic components are used on measuring devices for the study of vibrations on the human body, such as the Svantek SV38 accelerometers (for seat measurement) and SV105 (for hand measurement).

1.3 PERFORMANCE AND SPECIFICATION

1.3.1 Sensor Specifications

In order to properly select the sensor allowing vibration measurement, it is necessary to know the different technical characteristics that represent it. These different characteristics are generally found on the data sheet provided by the sensor's manufacturer and are detailed in this section as well as qualifiers of the sensor's performance.

1.3.1.1 Sensitivity

The sensor's sensitivity simply represents the ratio between the physical quantity collected at the sensor output and the desired to measure physical quantity of input. In the case of an accelerometric sensor with an electrical transduction, the sensitivity can therefore be expressed in $V/(m.s^2)$. The sensitivity depends on the calibration conditions of the sensor, so it is only valid for a given frequency and temperature. It is also valid only for the amplitude range for which the sensor is specified. If a tolerance expressed as a percentage is specified with the sensitivity, this means that it will remain within this range over the frequency range for which the sensor is specified.

1.3.1.2 Amplitude Response

The amplitude response (sometimes referred to as "frequency response") describes the amplitude sensitivity of the sensor over the entire frequency range that the accelerometer allows to measure (Figure 1.7). More specifically the amplitude response represents the tolerance associated with the sensitivity. The amplitude response can be expressed as a percentage or in decibels. This characteristic therefore makes it possible to express tolerance of the sensitivity of the sensor for any measured frequency.

1.3.1.3 Resonance

The resonance frequency is a frequency at which a system exhibits high sensitivity (Figure 1.8). In the case of a vibration sensor the measurement of nearby frequencies will cause distorted measurements and corrupted data. It is therefore essential for the user to ensure that the measured frequencies do not reach these extreme frequencies.

1.3.1.4 Linearity

In an ideal case, a sensor has an output value that increases linearly with the input value. However, in reality, there is a difference between the measured curve and this linear ideal curve. Thus, linearity represents the maximum error between reality and the ideal case. Linearity represents a tolerance expressed as a percentage in which the sensitivity of the sensor may vary.

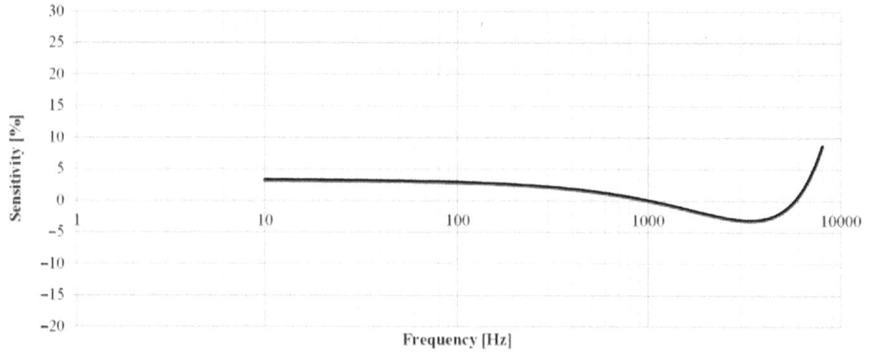

FIGURE 1.7 Amplitude response function representing the sensitivity as a function of frequency.

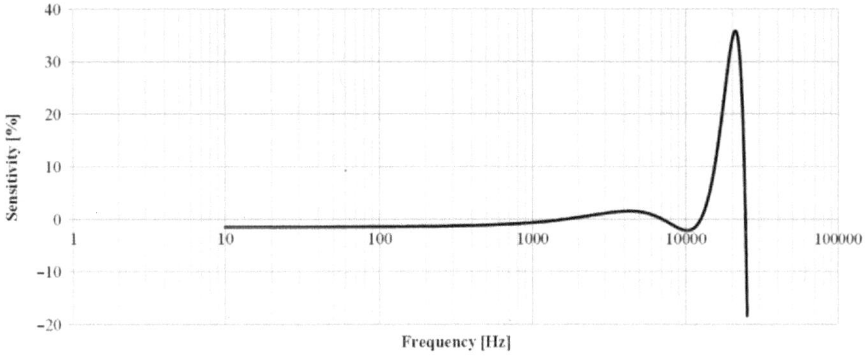

FIGURE 1.8 Resonance frequency at 20 kHz.

1.3.1.5 Accuracy and Precision

In the case where a lack of information exists for a sensor, it may be necessary to qualify it. In a first step, two parameters can be defined: accuracy and precision (Figure 1.9). Accuracy represents the proximity between the average of the measured values by the sensor and the actual reference value. Precision represents the dispersion of the measured values around the mean value. The precision can also be presented as the repeatability of a sensor to give the same value under constant conditions.

1.3.2 Acquisition System Specifications

The quality of the acquisition will also depend on the acquisition system's setting. It is a preliminary study of the characteristics of the structure that will make it possible to define the adapted configuration for the acquisition system.

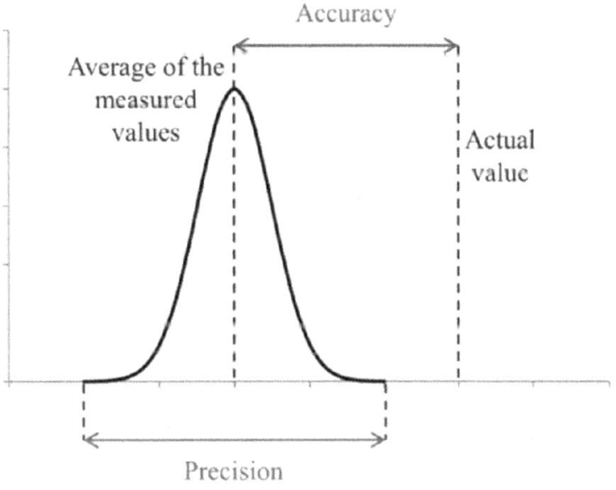

FIGURE 1.9 Accuracy and precision.

1.3.2.1 Sampling Frequency

The set of values of an analog signal taken at successive instants constitutes a discrete sample (Asch, 2006). The discretization of a signal (during the conditioning process) induces a loss of information which can be detrimental to the vibratory information. This loss of information is related to the number of high frequency components introduced by the process. The discretization operation is performed periodically on the time signal at a period T_s, as shown in Figure 1.10. The equivalent sampling frequency f_s depends of the discretization (Equation 1.1). The choice of the sampling frequency will condition the quality of the recorded signal. A too small sampling frequency may provide a low spectral resolution while a high sampling frequency may provide an over-sampled time signal (Asch, 2006).

$$f_s = \frac{1}{T_s} \tag{1.1}$$

The preliminary study of the structure's characteristics allows us to define a maximum frequency $f_{\max}$ for the observation of vibratory phenomena. In practice, the application of Shannon's theorem allows us to avoid phenomena of spectral aliasing. To this, it is necessary to choose $f_s \geq 2 \cdot f_{\max}$. In practice, the value chosen for the sampling frequency is $f_s = 2.56 \cdot f_{\max}$.

1.3.2.2 Spectral Resolution

To perform quality measurements, it is necessary to choose a good spectral resolution in the working frequency band. A resolution, Δf, is satisfactory when it allows us to dissociate neighboring frequencies. The spectral resolution is linked to the acquisition time T_{acq} (Equation 1.2).

$$\Delta f = \frac{1}{T_{\text{acq}}} \tag{1.2}$$

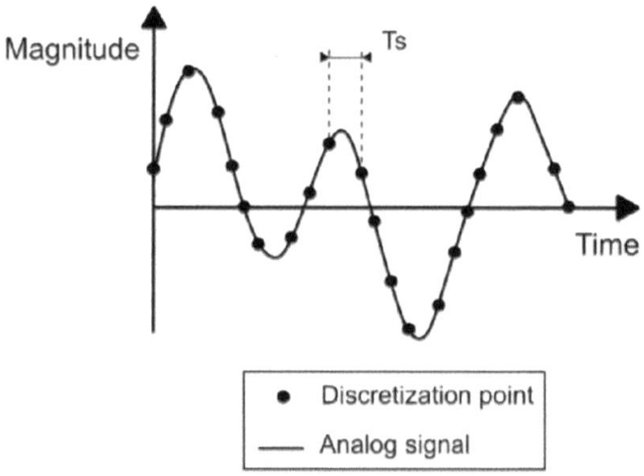

FIGURE 1.10 Signal sampling.

1.3.3 New Specifications

The emergence of investigations in the field of sport and the use of new measurement technologies have revealed new specifications which can influence the measurement of vibrations. Studies of vibratory effects in sports have often been conducted in controlled laboratory environments, based on the use of fixed sensors. However, these studies in simulated conditions tend to influence the athletes technique (Lee et al., 2010). Measurement of the sport therefore requires a continuous and embedded measurement in order to follow the athlete's movement during the activity. In order to meet the embedded measurement constraints, the measurement systems must therefore present an energy autonomy in order to follow the athlete from several minutes to several hours. In addition, the measurement tool must also present a memory autonomy in the case of data storage, or data transfer based on wireless technologies in order to not disturb the athlete. The energy and memory autonomy will directly impact the volume and the mass of the sensor which must be minimized in order to not influence the technique and not present any discomfort to the practice of the activity.

Depending on the constraints present in the sports field, it is difficult to choose a sensor perfectly adapted to the activity. Most of the industrial sensors used for vibration measurement also require an acquisition system. The latter presented an additional embedded constraint during several studies of the vibrations of athletes (Bonnardot and El Badaoui, 2010; Chiementin et al., 2012; Tarabini et al., 2015) as shown in Figure 1.11 (Crequy, 2015).

1.4 REGULATIONS AND STANDARDS

Several questions arise when choosing suitable instrumentation for measuring vibration on the human body. What to measure? Which ranges are suitable for my measurement? But also, what characteristics should my sensor have to ensure correct measurement of vibration? Answers to these questions are presented in different

(a)

(b)

FIGURE 1.11 Vibration studies during sport activity: Equipment of acquisition systems, Crequy (2015).

standards that describe the assessment of human exposure to vibration. Standards ISO 5349 (ISO-5349 2002) and ISO 2631 (ISO-2631 1997) allow us to set measurement ranges and indicators to measure and quantify the vibration. Sensor requirements are described in the ISO 8041 standard (ISO-8041 2005). According to the standard, the measurement must be carried out in three dimensions at the interface between the vibrating source and the body. The standards presented are focused on professional activities and are in general limited in size, or not adapted to sport activities.

1.4.1 Range and Evaluation

1.4.1.1 ISO 2631

The main objective of ISO 2631 standard (ISO-2631 1997) is to quantify the vibrations quantity transmitted to the whole body. This standard is applied to human health and comfort, the probability of vibration perception, and the incidence of motion sickness. For the study of vibrations exposure transmitted to the body, the frequency bands considered are: 0.1 to 0.5 Hz for transport sickness and 0.5 to 80 Hz for health, comfort and perception. The measurement must be carried out along the axes of a system of coordinates having as origin the point of entry of the vibrations. The coordinate systems are described in the standard and considered for different configurations (sitting, standing, lying). To ensure good accuracy and a measure of exposure, the measurement time must be long enough.

Since the impact of various frequencies constituting the accelerometric measurement on health and comfort is different, the standard proposes to weight the raw signals $a(t)$ in order to determine weighted signals $a_w(t)$. Several weighting filters are then proposed, two main filters are used essentially to differentiate the configurations in terms of positions, but other filters exist for more particular cases: vibrations in the backrest, motion sickness, rotational vibrations or vibrations under the head. Depending on the condition, the acceleration $a_w(t)$ is determined and the standard proposes to compute the weighted effective value a_w (Equation 1.3) representing a quantitative indicator of acceleration. This value is calculated as a function of the measurement time T.

$$a_w = \left[\frac{1}{T} \int_0^T a_w^2(t)\,dt \right]^2 \quad (1.3)$$

Thus, it is possible to determine the total value of the effective values of the weighted accelerations a_v (Equation 1.4). This total value is calculated from the weighted effective values a_{wx}, a_{wy} and a_{wz} of the three measurement axes and therefore represents the exposure indicator of an individual vibration. The values of the factors k_x, k_y and k_z depend on the position of the subject and the contact point between the vibrating surface and the body.

$$a_v = \sqrt{k_x^2.a_{wx}^2 + k_y^2.a_{wy}^2 + k_z^2.a_{wz}^2} \quad (1.4)$$

1.4.1.2 ISO 5349

The ISO 5349 standard (ISO-5349 2002) sets out the general requirements for measuring and evaluating the human exposure to hand-transmitted vibrations. The aim is to provide information that guarantees protection for the majority of exposed workers. For the study of exposure to vibrations transmitted to the hand–arm system, the frequency range considered is from 8 to 1000 Hz. The measurement must be based on a coordinate system whose origin is at the source of the vibrations.

The effects of vibrations on the body depend on the coupling between hand and source of vibration, so it is important to perform the measurement under normal and conventional conditions (gripping force, positioning…). The standard considers different factors that may affect an individual's health: frequency of vibration, amplitude of vibration and duration of daily exposure.

As for the case of transmitted whole body vibrations, the impact of the various frequencies composing the signal on health is different. The standard therefore proposes the use of a weighting filter. The acceleration measurement is therefore frequency weighted and then presented as a root mean square value representing a quantitative acceleration indicator. Since the measurement is generally tri-axial, it is assumed that each axis has an equivalent detrimental potential. The values of the weighted accelerations for the three axes x, y, and z are therefore denoted a_{hwx}, a_{hwy} and a_{hwz}. The total vibration value a_{hv} is thus determined by Equation 1.5.

$$a_{hv} = \sqrt{a_{hwx}^2 + a_{hwy}^2 + a_{hwz}^2} \tag{1.5}$$

The a_{hv} indicator is used to quantify the effects of vibration amplitude frequencies, but does not consider daily exposure duration. To take this parameter into account it is necessary to determine the daily vibration exposure $A(8)$ which is defined for an activity related to a reference period of eight hours of labor by Equation 1.6, where T is the duration of exposure to the activity having the total vibration value a_{hv} and T_0 the reference time of eight hours.

$$A(8) = a_{hv}\sqrt{\frac{T}{T_0}} \tag{1.6}$$

1.4.2 Instrumentation ISO 8041

It is the ISO 8041 standard (ISO-8041 2005) which will describe performance specifications and tolerance limits for instruments dedicated to vibration measurement on the human body. This standard provides all the necessary information to design a system to monitor vibrations and to comply with standards for the exposure of individuals to vibrations. It details the requirements for weighting filters, exposure indicators, and measurement display (refresh, sensor overload indication, etc.). The standard also describes the verification of the measuring equipment in the form of conformity tests, but also of maintenance control of the tool.

If the standard presents the characteristics of a complete measuring system in order to totally control the measurement chain, it specifies particularly the limit of technical specifications relating to the sensor of this system. These technical characteristics which are necessary for the sensor dedicated to the measurement of the individuals' response to vibrations are presented in Table 1.1.

1.4.3 Measurement Reality

Several studies show limitations in the application of these standards and that these are not always very well adapted to the observed activity. The work of Griffin (1997) shows that the frequency range imposed by the ISO 5349 (ISO-5349 2002) is too specific. Indeed, many tools do not produce frequencies above 250 Hz. On the other hand, since frequencies above 500 Hz are weighted only by a coefficient less than 5%, their effects on health are very minimal. This study concludes that the effects of vibration would not be overestimated if the standard imposed only a 500 Hz range. The study by Tarabini et al. (2012) emphasizes that this range is difficult to respect for some MEMS accelerometers. It is concluded that the ISO 8041 standard (ISO-8041 2005) has been drafted by imagining an evolution of industrial piezoelectric sensors widely used for vibration measurement without considering new technologies, which are valid for measurement of vibration, with a minimum of processing.

Other studies have shown that the world of sport is also very impacted by vibrations. However, if athletes are subjected to close or even greater amplitude as well as similar pathologies, the link between vibration and health is rarely made. The work of Chiementin et al. (2012) followed by that of Arpinar-Asvar et al. (2013) are interested in the vibration exposure for the hand–arm system in cyclists. These two studies emphasize the importance of changing standards to the athlete who is subjected to the same vibrations as the worker. Similar work is presented by Tarabini et al. (2015) for the case of athletes exposed to whole body vibration.

TABLE 1.1
Limit Specifications for the Sensor for Measuring Vibrations on the Human Body

Specifications	Hand–arm Vibrations	Whole-Body Vibrations
Mass (g)	5	5
Dimension (mm)	25×25×25	200×200×50
Mounting height (mm)	10	0
Temperature range (°C)	−10 à 50	−10 à 50
Electromagnetic sensibility	< 30 m/s^2 /T	< 2 m/s^2 /T
Acoustic sensibility	< 0.01 m/s^2 /kPa	< 0.05 m/s^2 /kPa
Transverse sensibility (%)	<5	<5
Shock (m/s^2)	30000	500
Phase response	No fast change	No fast change
Minimum resonance frequency	10 kHz	800 Hz
Ingress Protection (IP)	IP55	IP55
Frequency range (Hz)	8 à 1000	0.1 à 80

Moreover, if athletes are subjected to vibrations, the measurement and the control of these vibrations do not necessarily require the same specifications as in the field of work. The practice of sport imposes large and rapid individual movements, which are not considered by the ISO 8041 standard for the development of measuring equipment. It is therefore necessary to include for these measurement systems specifications for autonomy, space requirement and data recording as it could be proposed in Section 3.3.

1.5 HUMAN MEASUREMENTS

1.5.1 Attachment and Placement Techniques

The attachment of the sensor to the studied structure must be made from a perfect connection. The fixation method plays an important role in the quality and repeatability of the measurements. Two types of fixation for measurement of human activities should be distinguished. First, the indirect measurement, through the structure on which the subject evolves, then the direct measurement, on a member of the body of the subject studied.

In the first case, for indirect measurement, the rules dedicated to vibrations analysis of a mechanical structure are generally applied. There are several techniques for maintaining the position of a sensor: by adhesion (with a cyanoacrylate glue, an epoxy adhesive, or a "beeswax" type adhesive), by the use of a threaded stud which requires the drilling of the instrumented structure or by the use of a magnetic base that requires a steel base and above all does not allow displacement. The fixing method plays an important role on the measured frequency range and on the repeatability of the measurement. Only the screwed stud and the use of hard glue as cyanoacrylate offer real guarantees of the reliability of the measurements. The use of cyanoacrylate glue allows to work up to 5000 Hz (Boulenger and Pachaud, 1998).

In the second case, for direct human measurement, the instrumented area will depend of the aim of the study. For the study of human exposure to vibration (ISO-2631 1997) (ISO-5349 2002), the measuring point should be closest to the interface between the body and the vibration source, in order to determine exactly the vibratory quantity undergone by the body. For the study of vibration transmissibility (Munera et al., 2015; Crequy, 2015), the aim could be to avoid areas with soft tissues whose the viscoelastic properties have been emphasized in numerous studies to justify the absorption or amplification of some frequency ranges (Griffin, 1990; Lafortune et al., 1995; Kucharova et al., 2007). The sensor is then located closest to the bone in non-adipose areas. Finally, for the study of the vibratory response of the muscular system and its activity, the measurement point is generally associated with the bulk of the muscle (Friesenbichler et al., 2011; Giandolini et al., 2017). Generally, the attachment of the sensor is carried out by means of adhesive tape and medical elastic band. This method of fixing is equivalent to the hand-held method presenting a linearity range up to 1000 Hz, which largely covers the useful range for measurement on humans. However, many parameters can affect the frequency band. The work of Nokes et al. (1984) reported the effect of preloading using skin and bone mounted accelerometers in a cadaveric study. The results of this study allow to demonstrate that a preload on the skin-mounted sensor between 3.8 and 5.4 N (depending on the thickness of adipose tissue) was necessary to present similar information between both sensors. A too weak preloading presented the damping effect of the skin, while a too high preloading showed a distortion of the signal by the appearance of high frequencies. However, in the case of the skin-mounted accelerometer using an elastic strap, it is always difficult to evaluate the preloading. In the case of running, it is advisable to preload sensors using the elastic strap to the limit of subject comfort in order to minimize the effect of soft tissues (Shorten and Winslow, 1992). The study by Ziegert and Lewis (1979) showed the importance of the mass of the sensors. Indeed, a sensor with a low mass would avoid the movements of the soft tissues. The reproducibility of the hand-held method has also been studied several times (Decker et al., 2011; Provot et al., 2016).

1.5.2 Example of Human Vibration Measurement

The study by Adewusi et al. (2010) is interested in the vibratory transfer of the arm according to different parameters such as the position, the pushing force, and the gripping force of the hand. In this study, the arm was excited by a shaker. The vibration was measured at four points of the body: wrist, close of the elbow (on the arm and the forearm), and shoulder. The four accelerometers were attached to a Velcro strap tightly fastened near the joints so as to minimize the contributions due to skin artifacts. The strap was tightened so that it was neither loose nor too tight.

The works of Chiementin et al. (2012) followed by those of Crequy (2015) were based on the study of cyclist exposure to vibration during paved races (Figure 1.12). Sensors were positioned on the right arm on three non-adipose areas near the base of the ulna for the wrist, the epicondyle at the elbow, and the acromion at the shoulder. Sensors were then attached by a medical elastic adhesive tape by surrounding the limb. As for the study of Adewusi et al. (2010) all the sensors were positioned close to bone portion in order to limit the vibration response of soft tissues.

Several works used accelerometers during running in order to study impact shock and attenuation (Purcell et al., 2005; Giandolini et al., 2013; Giandolini et al., 2015; Provot et al., 2017). Sensors are generally positioned on the antero-medial aspect of

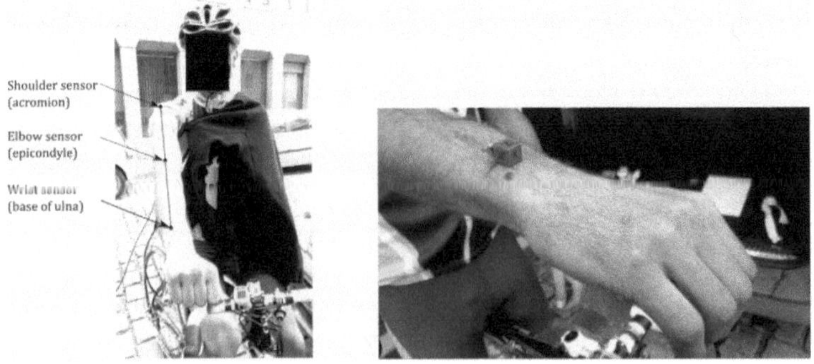

FIGURE 1.12 Experimental set-up of the study of Crequy (2015) during cycling.

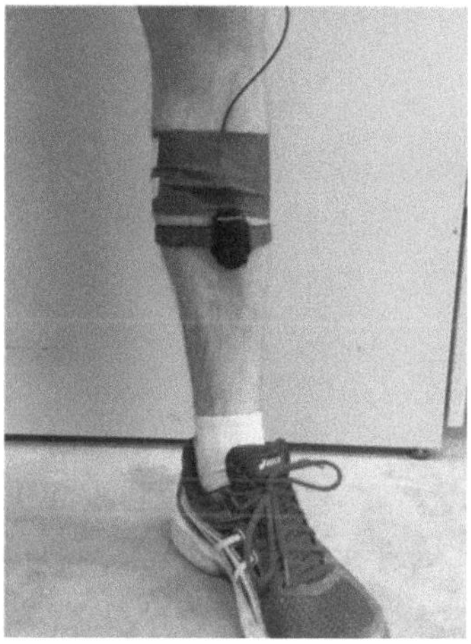

FIGURE 1.13 Experimental set-up for running of Provot et al. (2017).

the distal part of the tibia representing an area with a minimum influence of soft tissues. The accelerometers are fixed by the use of an elastic strap tightened to the limit of the subject's comfort (Figure 1.13).

1.6 EXCITATION SYSTEMS

Several vibration experiments may require an external exciter to generate the necessary vibration. It is rather well kept that these exciters significantly interact with the structure under test, and hence, can significantly influence test results (often either overlooked or not understood by the user). In the context of the human body, the vibration exciters will be used to calibrate the instrumentation used in operational conditions or to study the structural human response to vibration for several conditions of test (coupled with equipment, position, movement, etc.).

1.6.1 Instrumentation Calibration

Many different types of vibration exciter schemes are employed in vibration testing with different capabilities and principles of operation. Vibration exciters can take any number of forms and here the more common excitation techniques are presented in order to describe their characteristics. Three basic types of vibration exciters (shakers) are widely used: electrodynamic shakers with high bandwidth, moderate power, complex and multi frequency excitations; hydraulic shakers with moderate to high bandwidth, high power, complex, and multi-frequency excitations; and inertial shakers with low bandwidth, low power, and single-frequency harmonic excitations. The operation-capability ranges of typical exciters in these three categories are summarized in De Silvia's book (2007) (Table 1.2).

As presented in Section 2, vibration could be expressed as a displacement, a speed or an acceleration over time according to the amplitude but above all frequency.

TABLE 1.2
Typical Operation-Capability Ranges for Various Shaker Types (De Silva, 2007)

	Typical Operational Capabilities					
Shaker Type	**Frequency**	**Maximum Displacement**	**Maximum Velocity**	**Minimum Acceleration**	**Maximum Force**	**Excitation Waveform**
Hydraulic	Low (0.1–500 Hz)	High (50 cm)	Intermediate (125 cm/s)	Intermediate (20 g)	High (450000 N)	Average Flexibility
Inertial	Intermediate (2–50 Hz)	Low (2.5 cm)	Intermediate (125 cm/s)	Intermediate (20 g)	Intermediate (4500 N)	Sinusoidal
Electromagnetic	High (2–10000 Hz)	Low (2.5 cm)	Intermediate (125 cm/s)	High (100 g)	Low to intermediate (2000 N)	High flexibility and accuracy

Thus an idealized performance curve of an exciter should have the following frequency-velocity plane: a constant displacement-amplitude in low frequency domain, a constant velocity-amplitude in intermediate frequency domain and a constant acceleration-amplitude in high frequency domain (Figure 1.14).

From Figure 1.14, several observations could be made. First, in the constant-peak displacement region, the peak velocity increases proportionally with the excitation frequency and the peak acceleration increases proportionally with the square of the excitation frequency. Then, in the constant-peak velocity region, the peak displacement varies inversely with the excitation frequency and the peak acceleration increases proportionally with the excitation frequency. Finally, in the constant-peak acceleration region, the peak displacement varies inversely with the square of the excitation frequency and the peak velocity varies inversely with the excitation frequency. This further explains why rated stroke, maximum velocity and maximum acceleration values could not be simultaneously realized. Moreover, as the mass increases, the shaker performance curve decreases. The acceleration limit of a shaker depends on the mass of the test object (load), the full load supported by the shaker corresponds to the heaviest object that could be tested.

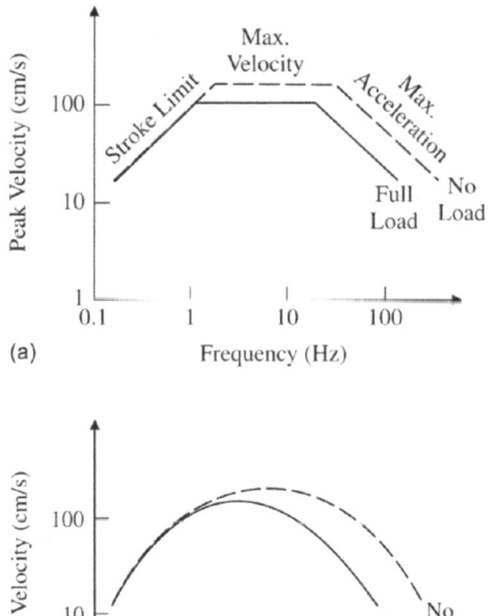

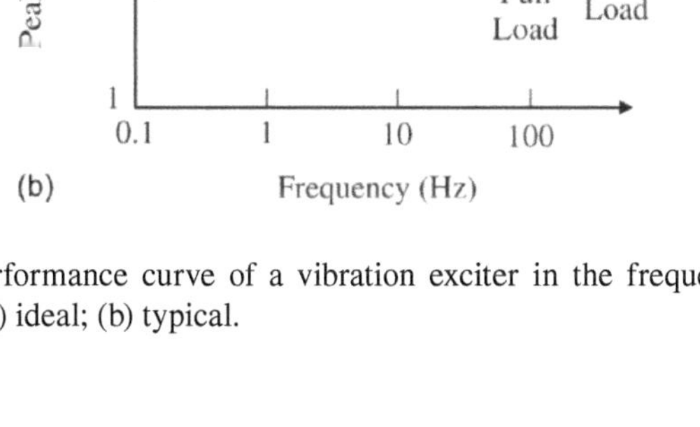

FIGURE 1.14 Performance curve of a vibration exciter in the frequency-velocity plane (De Silva, 2007): (a) ideal; (b) typical.

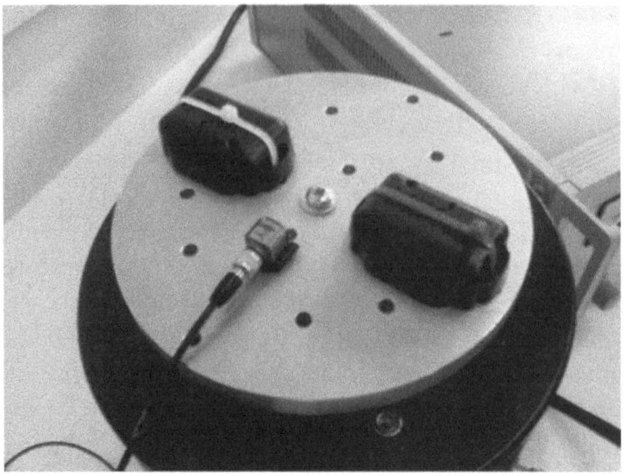

FIGURE 1.15 Comparison on shaker between a reference industrial accelerometer (gold standard) and two accelerometers embedded on an inertial measurement unit.

Thus, vibration exciters are very useful to calibrate instrumentations which will be used for studying human vibration exposure (Figure 1.15). The work of Tarabini et al. (2012) studies the potential of MEMS accelerometers for human exposure to vibration. By comparing, on shaker, the measured signal of the tested sensor and a reference sensor (gold standard), this study validates the use of MEMS accelerometers for vibration measurement. Thus, thanks to the particularities of the MEMS sensors (particularly the size) new vibration measurement approaches have been developed. A similar approach on shaker has been proposed by Provot et al. (2017) to validate the accelerometer of an inertial measurement unit for the study of vibration in the case of running.

1.6.2 Test Bench Examples

The literature mentions various test benches in order to excite the whole-body or the hand–arm system. Many of these benches have been developed in the field of sports engineering. This review of literature is non-exhaustive and allows the reader to project himself into the design of benches in relation to the planned studies.

In the context of whole body vibration, several studies used an upgraded vibrating platform (Munera et al., 2016; Kiiski et al., 2008). In the study by Munera et al. (2016) the movement of the platforms is governed by unbalanced motors. The technology of these platforms provides the possibility to automate, optimize test times and to set random cycles of excitation. The tests are carried out by varying the working frequencies of the platform from 16 to 80 Hz by defined increments and 2 or 4 mm amplitudes.

Another ergometers have been developed with excitations just applied only to the lower limbs and oriented in the vertical (Sperlich and Kleinoeder, 2009) and sagittal direction (Filingeri et al., 2012). For example, on the bench of Sperlich and Kleinoeder, the frame is connected to a rigid frame without any vibration, the ergometer, consists of an aluminum frame, a steel bicycle frame (D), a vibrating platform

(B) and a resistance unit (A). The bottom bracket (C) is physically disconnected from the chassis, and connected to which the crankset is in turn connected to a vibrating lower part with frequencies between 20 Hz and 4 mm amplitude.

But for a global excitation in cycling and in order to include the whole bike and subject, Champoux and al. (2007) have used a treadmill for exciting in a controlled environment. It is the use of a cylinder glued to the carpet which created an impact on the bicycle wheels. The size of this cylinder could vary to modify the impact force. However, this excitation technique does not reproduce faithfully the entire frequency spectrum of the excitation of the road. Thus, for this study, a hydraulic or electrodynamic exciters placed under the wheels of a bicycle as sources of excitation was used. Different excitations signals could be used, deterministic signals such as sinusoidal waves, reproductions of road excitations or stochastic signals as a random signal. The advantage of this technique is that it is possible to excite each wheel independently (Lépine et al., 2013). This is a very costly technique since this solution can amount to several hundred thousand euro.

1.7 CONCLUSION

In this chapter, we have presented the essential instrumentation, acquisition chain as well as its necessary characteristics to the determination of vibrations transmitted to the global human system composed of the body, its equipment and its activity. The principal sensors with their different transduction principles and the specifications of the acquisition systems were discussed. Then, the normative context that governs the instrumentation and the measurement for human exposure to vibrations was introduced to fix a framework of the experimental context. Various examples of activity involving the human body subjected to vibration measurement and implementation of the measurement system were presented. Finally, human body vibratory experimentation was illustrated by some examples of test benches. This chapter gives to the reader the basics concepts in order to use the instrumentation for human mechanical vibration analysis. Chapter 2 will present the signal processing following the acquisition chain and the measurement system.

REFERENCES

Adewusi, S. A. et al. (2010). Vibration transmissibility characteristics of the human hand–arm system under different postures, hand forces and excitation levels. *Journal of Sound and Vibration*, 329(14):2953–2971.

Aiello, G. et al. (2012). Real time assessment of hand–arm vibration system based on capacitive MEMS accelerometers. *Computers and Electronics in Agriculture*, 85:45–52.

Amendola, G. et al. (2011). Capteurs MEMS - Techniques de mesures. *Techniques de l'Ingénieur.*

Arpinar-Avsar, P. et al. (2013). The effects of surface-induced loads on forearm muscle activity during steering a bicycle. *Journal of Sports Science and Medicine*, 12(3):512–520.

Asch, G., ed. (2006). *Acquisition de données - Du capteur à l'ordinateur.* Dunod.

Bonnardot, F. and El Badaoui, M. (2010). Etude de la fatigue d'un coureur, de l'instrumentation à l'analyse vibratoire. In *10ème Congrès Français d'Acoustique.*

Boshuizen, H. C., Bongers, P. M., and Hulshof, C. T. J. (1990). Back disorders and occupational exposure to whole-body vibration. *International Journal of Industrial Ergonomics*, 6(1):55–59.

Boulenger, A. and Pachaud, C., eds. (1998). *Diagnostic vibratoire en maintenance préventive*. Dunod.

Champoux, Y., Richard, S., and Drouet, J.-M. (2007). Bicycle structural dynamics. *Sound and Vibration*, 41(7):16–25.

Chen, C.-H. et al. (2014). Chronic effects of whole-body vibration on jumping performance and body balance using different frequencies and amplitudes with identical acceleration load. *Journal of Science and Medicine in Sport*, 17(1):107–112.

Chiementin, X. et al. (2012). Hand–arm vibration in cycling. *Journal of Vibration and Control*, 19(16):2551–2560.

Crequy, S. (2015). *Analyse accélérométrique pour l'optimisation de la performance et la prévention des risques en cyclisme*. PhD thesis, Université de Reims Champagne-Ardenne.

De Capua, C., Meduri, A., and Morello, R. (2009). A hand–arm vibration meter monitoring the percussion exposure for health risk prevention applications. In *2009 IEEE International Workshop on Medical Measurements and Applications*, 45–50.

De Silva, C. (2007). *Vibration Monitoring, Testing, and Instrumentation*. CRC Press.

Decker, C., Prasad, N., and Kawchuk, G. N. (2011). The reproducibility of signals from skin-mounted accelerometers following removal and replacement. *Gait and Posture*, 34(3):432–434.

Filingeri, D. et al. (2012). The effects of vibration during maximal graded cycling exercise: A pilot study. *Journal of Sports Science and Medicine*, 11(3): 423–429.

Friesenbichler, B. et al. (2011). Tissue vibration in prolonged running. *Journal of Biomechanics*, 44(1):116–120.

Giandolini, M. et al. (2013). Impact reduction through long-term intervention in recreational runners: Midfoot strike pattern versus low-drop/low-heel height footwear. *European Journal of Applied Physiology*, 113(8):2077–2090.

Giandolini, M. et al. (2015). Foot strike pattern and impact continuous measurements during a trail running race: Proof of concept in a world-class athlete. *Footwear Science*, 1–11.

Giandolini, M. et al. (2017). Footwear influences soft-tissue vibrations in rearfoot strike runners. *Footwear Science*, 9(May):S25–S27.

Griffin, M. J. (1990). *Handbook of Human Vibration*.

Griffin, M. J. (1997). Measurement, evaluation, and assessment of occupational exposures to hand-transmitted vibration. *Occupational and Environmental Medicine*, 54(2):73–89.

ISO-2631 (1997). Mechanical vibration and shock – Evaluation of human exposure to whole-body vibration.

ISO-5349 (2002). Mechanical vibration – Measurement and evaluation of human exposure to hand-transmitted vibration.

ISO-8041 (2005). Human response to vibration – Measuring instrumentation.

Kiiski, J. et al. (2008). Transmission of vertical whole body vibration to the human body. *Journal of Bone and Mineral Research*, 23(8):1318–1325.

Kucharová, M. et al. (2007). Viscoelasticity of biological materials – Measurement and practical impact on biomedicine. *Physiological Research*, 56(Suppl. 1):33–37.

Lafortune, M. A., Henning, E., and Valiant, G. A. (1995). Tibial shock measured with bone and skin mounted transducers. *Journal of Biomechanics*, 28(8):989–993.

Lee, J. B., Mellifont, R. B., and Burkett, B. J. (2010). The use of a single inertial sensor to identify stride, step, and stance durations of running gait. *Journal of Science and Medicine in Sport*, 13(2):270–273.

Lépine, J., Champoux, Y., and Drouet, J.-M. (2013). *A Laboratory Excitation Technique to Test Road Bike Vibration Transmission*. PhD thesis.

Munera, M. et al. (2016). Transmission of whole body vibration to the lower body in static and dynamic half-squat exercises. *Sports Biomechanics*, 15(4):409–428.

Munera, M. et al. (2015). Physiological and dynamic response to vibration in cycling: A feasibility study. *Mechanics and Industry*, 16(5):503.

Nokes, L. et al. (1984). Vibration analysis of human tibia: The effect of soft tissue on the output from skin-mounted accelerometers. *Journal of Biomedical Engineering*, 6(3):223–226.

Provot, T. et al. (2016). Intra and inter test repeatability of accelerometric indicators measured while running. *Procedia Engineering*, 147:573–577.

Provot, T. et al. (2017). Validation of a high sampling rate inertial measurement unit for acceleration during running. *Sensors*, 17(9):1958.

Purcell, B. et al. (2005). Use of accelerometers for detecting foot-ground contact time during running. In *Proceedings of SPIE – The International Society for Optical Engineering*.

Ruck, J., Chabot, G., and Rauch, F. (2010). Vibration treatment in cerebral palsy: A randomized controlled pilot study. *Journal of Musculoskeletal and Neuronal Interactions*, 10(1):77–83.

Samuelson, B., Jorfeldt, L., and Ahlborg, B. (1989). Influence of vibration on work performance during ergometer cycling. *Upsala Journal of Medical Sciences*, 94(1):73–79.

Shorten, M. R. and Winslow, D. S. (1992). Spectral analysis of impact shock during running. *Journal of Biomechanics*, 8:288–304.

Sperlich, B. and Kleinoeder, H. (2009). Physiological and perceptual responses of adding vibration to cycling. *Journal of Exercise Physiology*, 12(2):40–46.

Stein, G. J., Chmúrny, R., and Rosík, V. (2011). Compact vibration measuring system for in-vehicle applications. *Measurement Science Review*, 11(5):154–159.

Tarabini, M., Saggin, B., and Scaccabarozzi, D. (2015). Whole-body vibration exposure in sport: Four relevant cases. *Ergonomics*, 58(7):1143–1150.

Tarabini, M. et al. (2012). The potential of micro-electro-mechanical accelerometers in human vibration measurements. *Journal of Sound and Vibration*, 331(2):487–499.

Ziegert, J. C. and Lewis, J. L. (1979). The effect of soft tissue on measurements of vibrational bone motion by skin-mounted accelerometers. *Journal of Bio-mechanical Engineering*, 101(August):218–220.

2 Signal Processing Applied to the Analysis of Vibrations Transmitted in the Human Body

Frederic Bonnardot, Christophe Corbier, and Lanto Rasolofondraibe

CONTENTS

2.1 INTRODUCTION

Exposure to whole body vibration (WBV) causes movement and effects in the body, which can cause discomfort, negatively affect performance for athletes and present a risk to health and safety. The assessment of health risks involves the analysis of recorded vibration signals. This analysis can be done in the time domain, in the frequency domain and/or in the time-frequency domain. For this purpose, it is necessary firstly to choose the characteristic parameters of the disorder, and secondly to monitor the evolution of these indicators over time in order to predict the change in the person's behavior it front of the vibrations. Several parameters can be used both in the time domain and in the frequency domain. The scientific and technological difficulties lie in the choice of the indicators to better follow the behavior of the person subjected to vibrations and to better diagnose the disturbances related to vibrations. These selected parameters, extracted from the signal (temporal and/or frequency), will make it possible to evaluate and predict the deleterious effects of vibrations on the body (visual disturbances, cardiac, pulmonary or osteoarticular disorders, etc.).

We can classify the vibrations as follows:

- Periodic vibrations of sinusoidal or harmonic type.
- Periodic vibrations of impulse type are thus called by reference to the forces that generate them and their brutal, brief and periodic nature. These shocks can be produced by normal events or abnormal events.
- Random vibrations of impulse type, etc.

All the characteristics of the vibrations (amplitude, distribution of amplitudes, frequency, etc.) can be exploited and analyzed to reveal precise details on the risks to the health and the safety of a person exposed to vibrations. In this chapter, we will present the classical signal processing tools that can be used to extract characteristic parameters of a vibratory signal, as well as advanced signal characterization techniques to better track the behavior of the person subjected to vibrations. Two case studies are presented at the end of this chapter.

2.2 CLASSICAL METHODS OF TREATMENT OF VIBRATORY SIGNALS

2.2.1 Temporal Analysis

A time indicator is a magnitude that characterizes the power, amplitude or the distribution of amplitudes of the vibratory signal. Many indicators exist in the literature and some are the result of the combination of several of them. The most commonly used parameters are rms value (x_{rms}), kurtosis (*ku*), skewness (*skew*), crest factor (*CF*) and impulse factor (*IF*) (Table 2.1).

The rms value is used to measure the average energy of the signal, it is used to detect abnormally high energy dissipations in the raw signal. Unlike the rms value, specific indicators like the crest factor is better suited to represent a signal induced by impulse forces. Kurtosis (*ku*) is a temporal analysis method based on the examination of the amplitude distribution of a vibratory signal (Pachaud, 1997) In fact, the appearance of a transient and periodic signal in the raw signal modifies the shape of the distribution.

Statistically, kurtosis (*ku*) is a coefficient that determines the degree of flattening of a distribution. The dissymmetry coefficient (*skew*) corresponds to a measure of the asymmetry of the distribution of a real random variable.

2.2.2 Spectral Analysis

Frequency analysis has become the fundamental tool for the processing of vibratory signals (Debbal and Bereksi-Reguig, 2007). It relies on the Fourier transform. This approach makes it possible to know the spectral content of the signal and to locate the characteristic frequencies of the different periodic phenomena contained in the raw signal. The Fourier transform is defined by the following equation:

$$X(f)=\int_{-\infty}^{+\infty} x(t)e^{-j2\pi ft}dt \tag{2.1}$$

Frequency domain analysis is the essential counterpart of time domain analysis of the signal. As it consists decomposing the energy of the signal analyzed by frequency bands, this results in a finer analysis than the temporal analysis. The "classical" frequency analysis is obtained from a Fourier transform of the time signal (periodogram power spectral density).

The periodogram method is based on the calculation of the discrete Fourier transform of the weighted or non-weighted data and the spectral power density (the square of the magnitude of the discrete Fourier transform):

$$X(k)=\sum_{n=0}^{N-1}\left[w(n)\times x(n)\right]e^{-j2\pi\frac{nk}{N}} \tag{2.2}$$

$$S(k)=A\left|X(k)\right|^2 \tag{2.3}$$

With: $X(k)=DTF\left[w(n)\times x(n)\right]$; *A*: Normalization factor; *w*(*n*): weighting window (Hanning window).

The principal interest of weighting windows is to limit the effects of signal truncation. This truncation of signal data corresponds to a temporal filtering of the signal *x* (*t*) by the (rectangular) window *w* (*t*). This consists of realizing the product *x* (*t*) .*w* (*t*). The frequency spectrum *X* (*f*) of a signal *x* (*t*) will be affected by such a filtering. According to Plancherel's theorem:

$$x(t)\times w(t) \;\leftrightarrow\; X(f)*W(f) \tag{2.4}$$

* is a convolution operator

In fact, the principle of calculating the FFT (fast Fourier transform) is to complete with zeros the values of the measurement to go up to the power of two closest to the chosen interval. This brutal zeroing causes the appearance of secondary lobes which, in some cases, can make indistinguishable two neighboring rays. There are windows, called weighting windows, which will allow the signal to be processed by weighting it instead of truncating it.

The windows of analysis or weighting play a very important role in the spectral observation. Let us consider the filter w (t) gate filter (or rectangular filter) which allows the signal to pass only in the interval [0, *T*] (*T* represents the acquisition time of the temporal signal). This is a natural window. Its spectrum (Figure 2.1) is given by the following equation:

$$W(f) = 2T\frac{\sin(2\pi ft)}{\pi ft} \tag{2.5}$$

According to Plancherel's theorem, each spectral line of $x\ (t)$ takes on the appearance of the spectrum of the gate filter.

- The width of the central lobe determines the spectral resolution of the window, that is to say its ability to discriminate two very close frequencies.
- The amplitude of the side lobes determines the spectral spread of the window. Too large a spectral spread will interfere with the detection of a low amplitude signal in the presence of a high amplitude signal.

If T is large enough, $W\ (f)$ remains assimilable to a Dirac pulse and the filtering effect will not be sensitive but it nevertheless exists. Thus, to improve the spectral representation, we introduce weighting windows. The influence of the different windows that can be used is reflected in two ways:

They reduce the rate of ripples and this especially as their spectrum has attenuated side lobes.

Mitigation of the secondary lobes widens the principal lobe and, for a given order N, results in an enlargement of the transition band.

The Hanning window is the most used weighting window in 99% of standard applications. The main lobe of the Hanning window is wider than that of the rectangular window. On the other hand, the amplitude of the side lobes is much lower. This window has a better resistance to "masking" low-level spectral components.

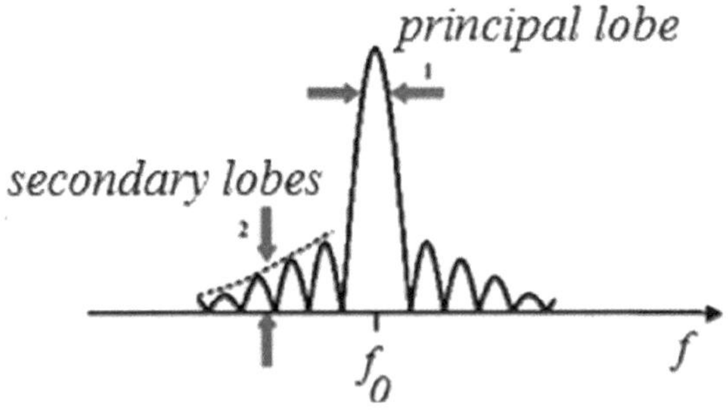

FIGURE 2.1 Spectrum of the natural window (rectangular window).

The Hamming window is derived from the Hanning window (same use) but the focus has been on the frequency resolution (thus at the expense of the amplitude resolution). It is recommended when searching for the "exact" frequency.

The Kaiser-Bessel window is Hamming inverse optimization in the sense that the amplitude resolution is optimized (to the detriment of the frequency resolution). It is recommended when searching for the "exact" level.

The flat top window allows us to have the exact amplitude of a measurement line spectral. On the other hand, its frequency resolution is very bad, which prohibits any use when the spectrum is composed of several characteristic very close frequencies.

2.3 ADVANCED METHODS OF TREATMENT OF VIBRATORY SIGNALS TRANSMITTED IN THE HUMAN BODY

2.3.1 Empirical Mode Decomposition (EMD)

The classical methods decompose a signal on a basis of functions specific to the method, which are therefore independent of the signal studied. Thus, for example, the Fourier transform decompose the signal into a linear combination of sinusoidal functions. The wavelet method, for its part, uses as basic functions the "wavelets", which are localized functions: the physical sense of the different modes is therefore already better. Different from wavelet and Fourier transform, the empirical mode decomposition (EMD) does not use any filter or predetermined function, it is an adaptive decomposition method to better understand intuitively the frequency content of the signal.

The empirical mode decomposition (EMD) is a method of adaptive decomposition, nonparametric and local of nonstationary signals (Equation 2.6). This method was developed in 1998 by N. E. Huang, an engineer in NASA, for the oceanographical study of data (Huang et al., 1998). It aims to break up any signal into a sum of oscillating components extracted directly from this one in an adaptive way. These components (or "intrinsic functions mode"[IMF]) are interpreted like nonstationary forms of waves (i.e. modulated in amplitude and frequency) being able to be possibly associated nonlinear oscillations. The purpose of the EMD method is therefore to decompose signals into different modes having a physical sense. Thus, knowledge of these modes makes it possible to intuitively understand the frequencies present in the signal.

This decomposition method has been the subject of numerous publications (Flandrin et al., 2004; Beya et al., 2010), it has also been used to detect abnormalities in phonocardiogram (PCG) signals and electrocardiograms (ECG). Indeed, the electrocardiogram (ECG) and the phonocardiogram (PCG) reflect the electromechanical behavior of the heart. In patients, the medical interpretation of these signals makes it possible to determine certain pathologies related to the morphology of the heart, to prevent risks of seizures or cardiac insufficiencies. Since the anomalies are localized over time, the treatments to be considered for the signal must keep the location of the events. In contrast to the Fourier transform (TF) and the wavelet transform (TO) methods, the EMD allows precise characterization of these signals by virtue of its very interesting properties on pseudoperiodic signals.

Any signal can be considered as the superposition of a slow component $a(t)$ (low frequency) and an approximation called fast component $d(t)$ (high frequency) called detail. These components are IMFs interpreted as being non-stationary waves. EMD relies on the adaptive decomposition of the signal into a series of IMFs through the sifting process. Each of the IMFs can be considered as a level of scale distinct from the decomposition. This notion of scale is local and the decomposition is non-linear. Decomposition locally describes the signal as a superposition of harmonic components ranging from high to low frequencies and a trend. Each of the IMFs checks the following criteria:

1. Has a mean value of zero.
2. The number of extrema and the number of zero crossings must differ at most by one (that is, has only one extreme between zero crossings).

The EMD method has an adaptive approach: for each signal studied, a new base of functions is constructed. So these modes will better describe the signal. The following four steps illustrate the operating principle of the EMD (at scale 1) (Figure 2.2):

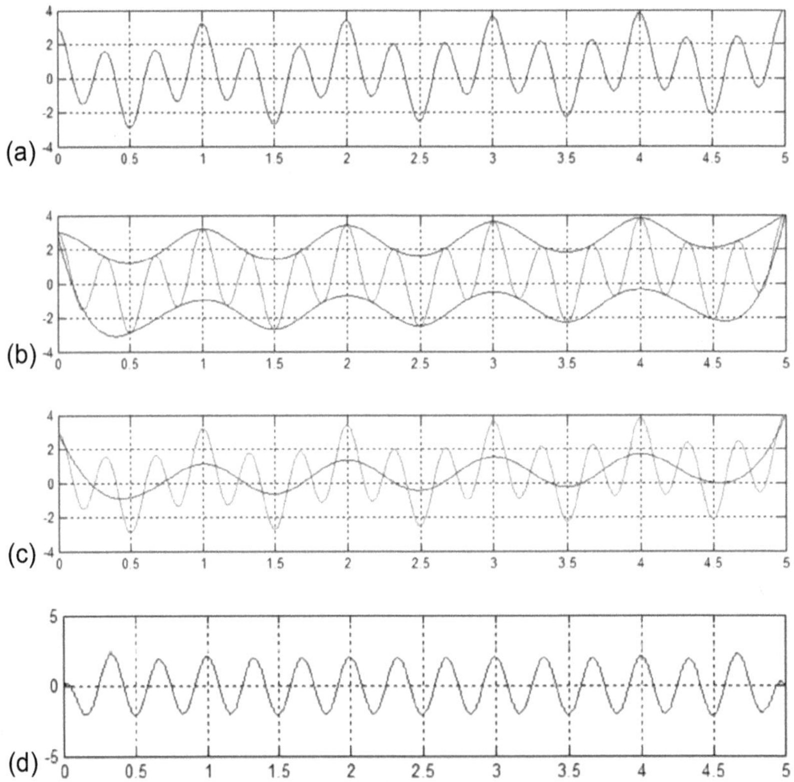

FIGURE 2.2 Graphic illustration of the EMD method (a) raw signal, (b) extrema (minima and maxima envelope), (c) mean envelope ($r_1(t)$), (d) first sifting process (IMF 1: $d_1(t)$).

1. Extract the extrema of $s(t)=r_1(t)$ (the local maxima and minima of the signal) that are interpolated by spline or cubic function to find a higher envelope and a lower envelope.
2. Compute envelope's mean $moy_1\ (r_1(t))$.
3. Extract the mean $moy_1(t)$ from the input signal $r_1(t)$ to obtain the IMF1 ($d_1(t)$). Update $r_2(t)=r_1(t)$ - $moy_1(t)$.
4. Repeat step 1 until the number of extrema is lower than 2.

$$s(t)=d_1(t)+r_1(t)=d_1(t)+d_2(t)+r_2(t)=\cdots=\sum_{k=1}^{N}d_k(t)+r_k(t) \qquad (2.6)$$

The EMD is entirely driven by the data and, unlike the Fourier transform or the wavelets, this decomposition does not rest on any family of functions defined *a priori*. EMD method can be seen as the equivalent of principal component analysis but dedicated to non-stationary signals (Boudraa and Cexus, 2007). EMD has no theoretical basis but only the decomposition principle of EMD is ensured by the so-called "sifting process". This process leads to a decomposition of the signal into different modes or IMFs.

2.3.2 Cyclostationarity Approach

2.3.2.1 Definition of Cyclostationarity

The first studies on cyclostationarity began in 1950 with Benett and Gladyshev. Cyclostationarity began to be attractive (according to publications) in 1980 (William et al., 2006).

In order to define cyclostationarity let us imagine a runner and consider this runner as a random process $X(t)$ that can generate signals. Let us also imagine that it is possible for the runner to perform several times the same race with the same initial condition and position. Each race is a realization of the process. During a race, the runner produces a signal $x(t,r)$ where t is the time and r the realization.

Figure 2.3 illustrates what can be ten realizations of this runner (one per line). It should be noted that each realization is vertically synchronized (same condition). As it will be explained later, we have to cheat a little bit to create this figure (in practice, it is impossible to make ten runs with the same conditions).

Since many realizations are available it is possible to make statistics on these realizations. According to these statistics the process that creates these realizations could be considered as stationary, cyclostationary or non-stationary. Here, the first (expectation) and second order (variance) statistics will be examined.

2.3.2.2 Cyclostationarity at Order 1

The order 1 is linked with the properties of the expectation $\mu_{X(t)}$ of the process $X(t)$ (Equation 2.7). An estimation $\hat{\mu}_{X(t)}$ of this expectation could be done by making a vertical average against the N realizations for each time instant (Equation 2.8).

$$\mu_{X(t)}(t)=\mathbb{E}\{X(t)\} \qquad (2.7)$$

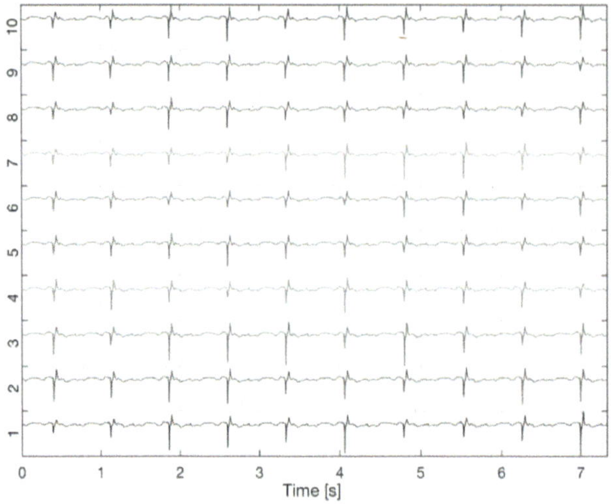

FIGURE 2.3 10 Realization of a run (1 line per realization).

$$\hat{\mu}_{X(t)}(t) = \frac{1}{N}\sum_{r=1}^{N} x(t,r) \tag{2.8}$$

Figure 2.4 shows the result of this vertical average (i.e. an estimation of $\mu_{X(t)}$). If the signal is stationary, the quantity $\mu_{X(t)}$ is constant. If the signal is cyclostationary at the order 1 with the cycle T, the quantity $\mu_{X(t)}$ is periodic at period T:

$$\mu_{X(t)}(t) = \mu_{X(t)}(t+T) \tag{2.9}$$

If the signal is non-stationary, the quantity $\mu_{X(t)}(t)$ is not constant: cyclostationarity is a subset of non-stationarity.

2.3.2.3 Cycloergodicity (or How to Deal with Realizations in Practice)

Generally, only one signal (or realization) is available. It is not possible to repeat the experimentation with the same condition (e.g. runner fatigue). Therefore, it is

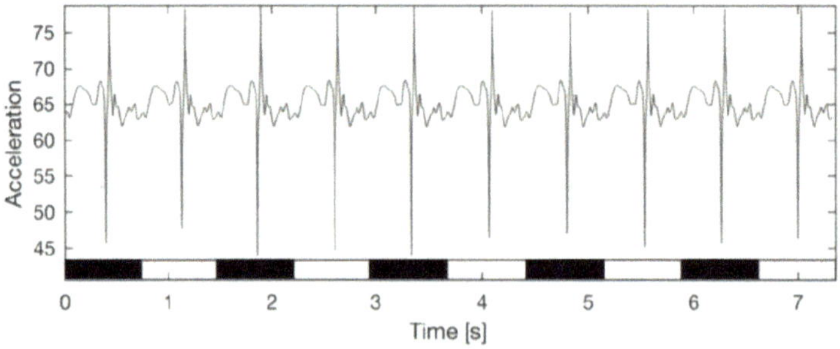

FIGURE 2.4 Average $\hat{\mu}_{X(t)}$ against realization.

common to cut the signal into slices of one period of length T and to consider that each slice corresponds to one realization. It was done on Figure 2.2 where the original signal was split into slices of ten periods.

This procedure is valid only if the signal is cycloergodic (i.e. it is possible to substitute ensemble averaging to block averaging). In practice, the signal will be supposed to be cycloergodic since it is not possible to check it without having realization of the same process.

In this case, Equation 2.8 could be written for a signal of N cycles of period T:

$$\hat{\mu}_{X(t)}(t) = \frac{1}{N}\sum_{n=0}^{N-1} x(t + n.T) \tag{2.9}$$

The obtained quantity is the synchronous average of period T and can be observed in Figure 2.4.

It is also possible to define the residual signal $r(t)$ by subtracting the synchronous average to the signal $x(t)$:

$$\mathrm{r}(\mathrm{t}) = \mathrm{x}(\mathrm{t}) - \hat{\mu}_{X(t)}(t) \tag{2.10}$$

The synchronous average represents the periodic pattern link to the body. The residual signal corresponds to the variation against this periodic part. These variations can be amplified by disease, fatigue, etc. Although these variations are not periodic, they can have a periodic power (i.e. the squared residual can have a periodic part as seen in Figure 2.5). This study of the squared residual is a simplified study at order 2.

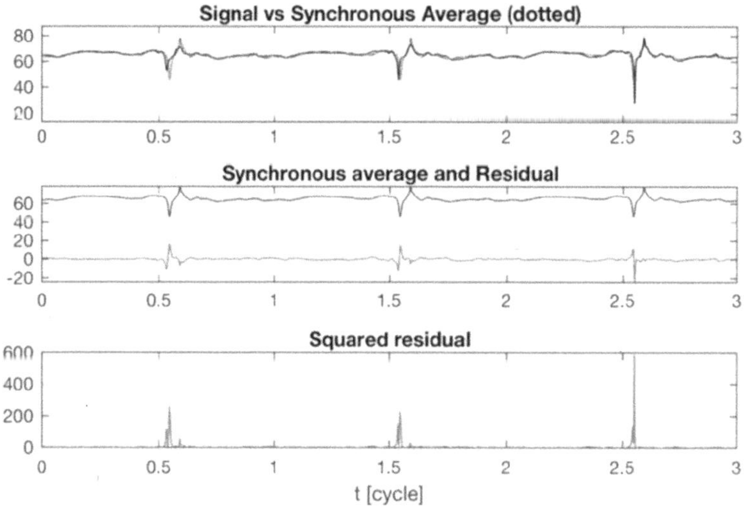

FIGURE 2.5 Decomposition into a periodic part and a residual.

2.3.2.4 Cyclostationarity at Order 2

The order 2 cyclostationarity is linked with the property of the (auto)correlation of the process $X(t)$:

$$R_{X(t)}(t_1,t_2) = \mathbb{E}\{X(t_1).X^*(t_2)\} \tag{2.11}$$

$$\hat{R}_{X(t)}(t_1,t_2) = \frac{1}{N}\sum_{r=1}^{N} x(t_1,r).x^*(t_2,r) \tag{2.12}$$

Since the periodic part is already extracted by the synchronous average, it could be interesting to compute the correlation of the residual signal instead of the signal $x(t)$. In this case, the correlation become a variance, and is only focused on the variation around the average.

The correlation denotes the link between two shifted versions of a signal. If the process is stationary, the autocorrelation just depends on the delay between the two time instants $\tau = t_2 - t_1$.

$$R_{X(t)}(t_1,t_2) = R_{X(t)}(t_2 - t_1) \tag{2.13}$$

By definition, a process is cyclostationary at order 2, the autocorrelation is periodic:

$$R_{X(t)}(t_1,t_2) = R_{X(t)}(t_1 + T, t_2 + T) \tag{2.14}$$

Generally, correlation is expressed according to the time $t = (t_2 - t_1)/2$ and the lag $\tau = t_2 - t_1$. Cyclostationarity at order 2 means that the correlation between the signal and itself does no longer depend on the distance τ but also varies periodically against the time t. Figure 2.6 shows on the left top corner the correlation of the residual signal, it is clearly periodic. For the lag $\tau = 0$, the correlation of the residual signal corresponds to the synchronous average of the squared signal shown in Figure 2.4.

Due to the periodicity, the autocorrelation could be represented by a Fourier series (i.e. a collection of frequency at fundamental and harmonics of the period $1/T$). The time is transformed into the cyclic frequency α. This Fourier series is called the cyclic autocorrelation function $R^{\alpha}_{X(t)}$:

$$R_{X(t)}(t,\tau) = \sum_{\alpha} R^{\alpha}_{X(t)}(\tau).e^{2\pi i\alpha t} \tag{2.15}$$

The cyclic correlation is shown on the right top corner of Figure 2.6. It can be seen that it is 0 for cyclic frequencies $\alpha \neq k/T$ (in the graph frequency have been normalized so that $1/T = 1$).

It is also possible to transform the delay τ into a frequency f. The result is the spectral correlation function:

$$S^{\alpha}_{X(t)}(f) = \int_{-\infty}^{+\infty} R^{\alpha}_{X(t)}(\tau).e^{-2\pi i f\tau} d\tau \tag{2.16}$$

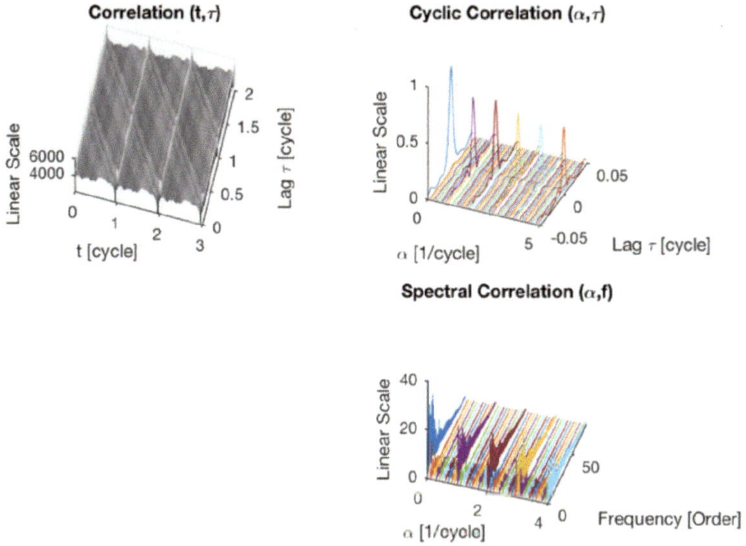

FIGURE 2.6 Representation at order 2 of cyclostationary process.

It can be checked on Figure 2.6 that the spectral correlation is zero for $\alpha \neq k/T$. For $\alpha = 0$, the spectral correlation corresponds to the residual signal spectrum.

More details on cyclic spectral analysis can be found in (Antoni, 2007).

2.3.3 Huberian Robust ARMA Model

2.3.3.1 Huberian Robust Estimator

Let Θ be a parametric subset of $\mathbb{R}^d$ and Γ a compact subset of $\mathbb{R}$. Let $\rho^H : \mathcal{S}\times\Theta\times\Gamma \to \mathbb{R}$ be a symmetric function such that $\rho^H\left(\varepsilon_k(\theta,\gamma)\right)$ is measurable for each $\theta \in \Theta$, $\gamma \in \Gamma$, with $\mathcal{S}$ a probability space and defined by

$$\rho^H\left(\varepsilon_k(\theta,\gamma)\right) = \frac{\varepsilon_k^2(\theta,\gamma)}{2} 1_{|\varepsilon_k| \le \gamma} + \left(\gamma\left|\varepsilon_k(\theta,\gamma)\right| - \frac{\gamma^2}{2}\right) 1_{|\varepsilon_k| > \gamma} \tag{2.17}$$

where γ is a threshold, $\varepsilon_k(\theta,\gamma) = \Delta t_k - \Delta\hat{t}_k(\theta,\gamma)$ the prediction error signal (estimated residuals) and 1_Y the unit function for a condition Y. The measures are Δt_k and the prediction model is $\Delta\hat{t}_k(\theta,\gamma)$. A Huberian robust estimator $\hat{\theta}_N^{H,\hat{\gamma}}$ is given from a minimization of the matrix form

$$\begin{aligned}\hat{\theta}_N^{H,\hat{\gamma}} &= inf_{\theta\in\Theta,\gamma\in\Gamma} W_N(\theta,\gamma)\\ &= inf_{\theta\in\Theta,\gamma\in\Gamma}\left[\frac{1}{2N}\left\|Y_{\mathrm{v}_2} - \Phi_{\mathrm{v}_2}(\theta,\gamma)\theta\right\|^2 + \frac{\gamma}{N}\left[\left|Y_{\mathrm{v}_1} - \Phi_{\mathrm{v}_1}(\theta,\gamma)\theta\right| + \frac{\gamma}{2}\left\|S_{\mathrm{v}_1}\right\|^2\right]\right]\end{aligned} \tag{2.18}$$

where $\nu_2(\theta,\gamma)=\{k:|\varepsilon_k(\theta,\gamma)|\leq\gamma\}$ and $\nu_1(\theta,\gamma)=\{k:|\varepsilon_k(\theta,\gamma)|>\gamma\}$ are two compact subsets. Here

$$Y_{\nu_i}=\left[\Delta t_{1,\nu_i}\ldots\Delta t_{N,\nu_i}\right]^T,\ \Phi_{\nu_i}(\theta,\gamma)=\begin{bmatrix}\varphi_{1,\nu_i}^T(\theta,\gamma)\\ \vdots\\ \varphi_{N,\nu_i}^T(\theta,\gamma)\end{bmatrix},\ \varphi_{k,\nu_i}^T(\theta,\gamma)$$

is the regressor such that $\Delta\hat{t}_{k,\nu_i}(\theta,\gamma)=\varphi_{k,\nu_i}^T(\theta,\gamma).\theta$, $S_{\nu_i}=\left[s_{1,\nu_i}\ldots s_{N,\nu_i}\right]^T$ is the sign vector, $\|X\|^2=\sum_i x_i^2$, $|X|=\sum_i|x_i|$ and for a general case $X_{k,\nu_i}=\begin{cases}X_k & \text{for } k\in\nu_i\\ 0 & \text{otherwise}\end{cases}$.

2.3.3.2 Choice of γ

In the prediction error procedure, there exists an inner feedback loop to compute the pseudolinear prediction model $\Delta\hat{t}_{k,\nu_i}(\theta,\gamma)$. The estimated residuals $\varepsilon_k(\theta,\gamma)$ are treated by a parametric adaptive algorithm which includes $W_N(\theta,\gamma)$ to be minimized. The presence of NO in the measurements Δt_k induces large values in $\varepsilon_k(\theta,\gamma)$. A convenient choice of γ improves the robustness by reducing the effects of these large deviations. In the literature, γ is chosen in the interval range [1; 2] for linear models. However, this choice does not ensure convergence, consistency or stability of $\hat{\theta}_N^{H,\hat{\gamma}}$. Accordingly, the probability density function (pdf) of $\varepsilon_k(\theta,\gamma)$ is strongly disturbed and presents heavy tails. It is shown that Huberian robust estimators are not always robust and efficient when [1; 2]. In a recent paper (Corbier, 2012) on piezoelectric systems, the use of small values of γ in [0.05; 0.5] led to derive relevant output error models. In this work, even though the prediction errors were disturbed by numerous NO, the choice of the small values of γ around 0.05 allowed to obtain interesting results in the frequency interval range for vibration drilling control.

In the sequel, we introduce a new curve ensuring a reduction of the bias and we show the choice of γ in low values. In (Corbier, 2012), we studied the quality of the robustness through influence function of the robust estimator. We showed that the upper bound of the bias is proportional to the high NO, denoted $\mathcal{L}^p$ and a new function named *tuning function*, denoted $f^{\omega}(\gamma)$. Figure 2.7 shows this curve. It appears the *classical interval*, denoted C_γ where $\gamma\in[1;1.5]$ and a new interval, named *extended interval*, denoted E_γ where $\gamma\in[0.001;0.2]$. From a linearization of $f^{\omega}(\gamma)$ in C_γ and E_γ, in absolute value, the slope in E_γ is six times as important as that of the slope in C_γ. Accordingly, the sensitivity to reduce the influence of high NO in E_γ is six times as important. Therefore, this new curve allows locating a new investigation interval of γ in low values in order to get low values of $f^{\omega}(\gamma)$ to decrease the effects of NO.

2.3.3.3 Convergence Properties of the Huberian Robust Estimator

Now, consider an ARMA process $\mathcal{P}$ with output signal Δt_k for $k=1\ldots N$. Let $Z^N=\{\Delta t_1\ldots\Delta t_N\}$ be the dataset. The main purpose is to decide upon how to use information contained in Z^N to select a proper value of the Huberian

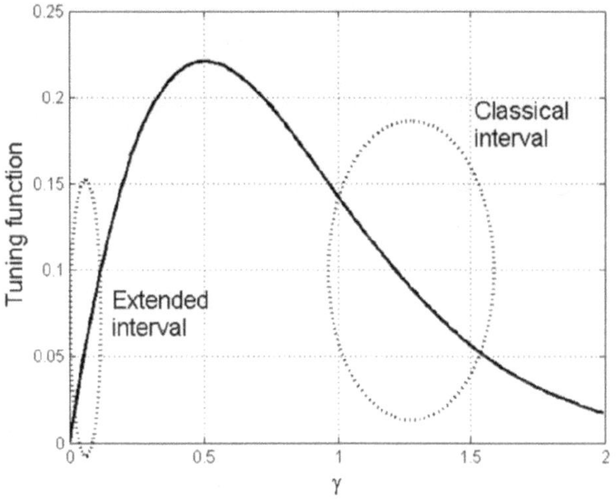

FIGURE 2.7 Tuning function with two intervals. The classical with $\gamma \in [1;1.5]$ and the extended with $\gamma \in [0.001;0.2]$.

robust estimator $\hat{\theta}_N^{H,\hat{\gamma}}$. We have to determine a mapping from Z^N to the compact $D_{\mathbf{M}} \times D_\gamma : Z^N \rightarrow \hat{\theta}_N^{H,\hat{\gamma}} \in D_{\mathbf{M}} \times D_\gamma$. Moreover, we shall assume that P belongs to the model's structure **M**] (Ljung, 1999). Let θ_0 be the true parameter vector describing exactly the process, then, from the central limit theorem, $\sqrt{N}\left(\hat{\theta}_N^{H,\hat{\gamma}} - \theta_0\right) \in \mathcal{N}\left(0, P^H_{\theta_0,\hat{\gamma}}\right)$, where $P^H_{\theta_0,\hat{\gamma}}$ is the asymptotic covariance matrix given by $P^H_{\theta_0,\hat{\gamma}} = [\bar{W}''(\theta_0,\hat{\gamma})]^{-1} Q^H(\theta_0,\hat{\gamma}) [\bar{W}''(\theta_0,\hat{\gamma})]^{-1}$, with $\bar{W}''(\theta_0,\hat{\gamma}) = \lim_{N\rightarrow\infty} \mathbb{E} W_N''(\theta_0,\hat{\gamma})$ the limit Hessian and $Q^H(\theta_0,\hat{\gamma}) = \lim_{N\rightarrow\infty} \frac{1}{N} \sum_{t=1}^{N} \sum_{u=1}^{N} \mathbb{E}\Psi_t(\theta_0,\hat{\gamma}) \Psi_u(\theta_0,\hat{\gamma})^T$ the Q^H-matrix with. Hence, the asymptotic covariance matrix of Huberian robust estimator is $\text{cov}\hat{\theta}_N^{H,\hat{\gamma}} \approx \frac{P^H_{\theta_0}}{N}$ with $\psi_t(\theta_0,\hat{\gamma}) = -\varphi_{t,\nu_2}(\theta_0,\hat{\gamma})\, \varepsilon_{t,\nu_2}(0_0,\hat{\gamma}) - \hat{\gamma}\, \varphi_{t,\nu_1}(0_0,\hat{\gamma})\, s_{t,\nu_1}$ and $W_N''(\theta_0,\hat{\gamma}) \approx \frac{1}{N} \sum_{t=1}^{N} (\varphi_{t,\nu_2}(\theta_0,\hat{\gamma})\, \varphi_{t,\nu_2}^T(\theta_0,\hat{\gamma}))$.

In the literature, the standard asymptotic normality for MLE requires that $W_N(\theta,\gamma)$ be twice continuously differentiable, which is not the case here by the presence of the sign function. There exists, however, asymptotic normality results for non-smooth functions and we will hereafter use the one proposed by Newey and McFadden (1994; Andrews, 1994). The basic insight of their approaches is that the smoothness condition of $W_N(\theta,\gamma)$ can be replaced by a smoothness of its limit, which in the standard maximum likelihood case corresponds to the expectation $-\bar{\mathbb{E}} ln f_H(\varepsilon_k(\theta,\gamma)) = \bar{W}(\theta,\gamma)$, where $f_H(\varepsilon_k(\theta,\gamma))$ is the pdf of the prediction error, with the requirement that certain remainder terms are small. Hence, the standard differentiability assumption is replaced by a *stochastic differentiability* condition, which can then be used to show that the MLE $\hat{\theta}_N^{H,\hat{\gamma}}$ is asymptotically normal. Recall that the derivative w.r.t θ of ρ^H is the ψ-function. If this function is differentiable in θ, one can establish the asymptotic normality of $\hat{\theta}_N^{H,\hat{\gamma}}$ by expanding $\sqrt{N}\left(\hat{\theta}_N^{H,\hat{\gamma}} - \theta_0\right)$ about θ_0 using element by element mean value expansions. This is the standard way of establishing asymptotic normality of the estimator. In a variety of applications, however, $\psi_t(\theta_0, \hat{\gamma})$ is not differentiable in θ, or not even continuous, due to the appearance of a sign function. In such a case, one can still establish asymptotic normality of the estimator provided $\bar{\mathbb{E}}\psi_t(\theta_0, \hat{\gamma})$ is differentiable in θ. Since the expectation operator is a smoothing operator, $\bar{\mathbb{E}}\psi_t(\theta_0, \hat{\gamma})$ is often differentiable in θ, even though $\psi_t(\theta_0, \hat{\gamma})$ is not.

2.4 CASE STUDY

2.4.1 Application of the Cyclostationary Approach to Walking

2.4.1.1 Problem with Walking Speed Fluctuation

The cyclostationary approach needs to know the cycle precisely. Unfortunately, there can be some walking speed fluctuation against the time, especially for a long walk or a run. Figure 2.8 shows on the left the variation of the stride duration at the end and the beginning of a 30 min run. The stride duration is estimated by detecting the local maximums on the acceleration signal (on the right) and measuring the time between each maximum.

Figure 2.8 shows the effect of fluctuations on synchronous averaging of a 4 min 20 sec run acceleration. On the left, the signal is split in blocks of a length that correspond to the average period. Next, each block is superposed and shown in the left top

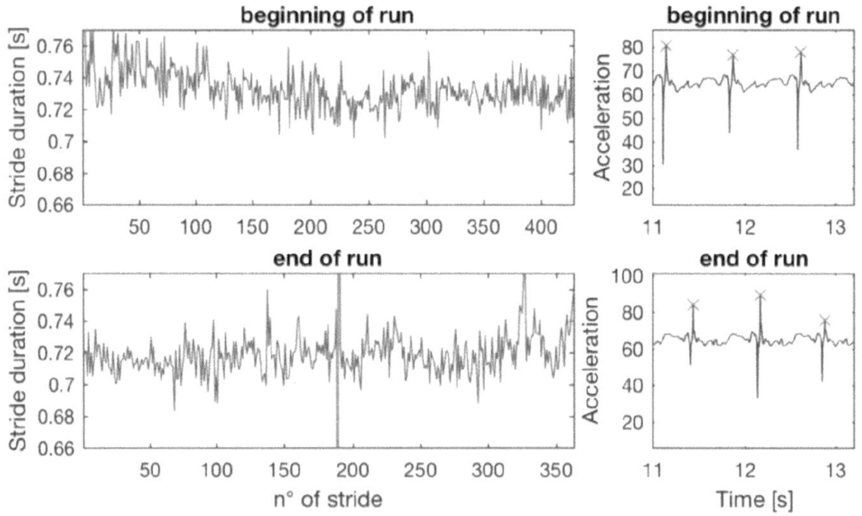

FIGURE 2.8 Stride duration and its measurement.

corner of the figure. The fluctuation of the peaks around average period can easily be observed. As a result, when a "vertical averaging" is performed to compute the synchronous average, the cyclic pattern is totally destroyed (bottom left). Therefore, cyclostationarity cannot be exploited.

On the right part of the figure, the fluctuation is compensated before block superposition. All patterns are in phase and the synchronous average on the bottom clearly estimates the periodic pattern.

Therefore, if synchronous average is used, period fluctuation compensation is essential to exploit cyclostationarity. This pre-processing has been made on all signals shown in the previous part. Other methods exist that are less sensitive to speed fluctuations.

2.4.1.2 Speed Fluctuation Removal

Speed fluctuation compensation is made in two steps. The first step is to detect each stride in order to estimate the stride number against the time. Figure 2.9 illustrates this detection by looking at the local maximum of the acceleration (top right) and the result: the stride number against the time (bottom right). Other methods like correlation, cepstrum, etc. can be used to detect the stride.

By using phase estimation as describe in (Bonnardot et al., 2005) for mechanical signals and applied in (Maiz, 2014) in biomechanics signals, it is possible to estimate

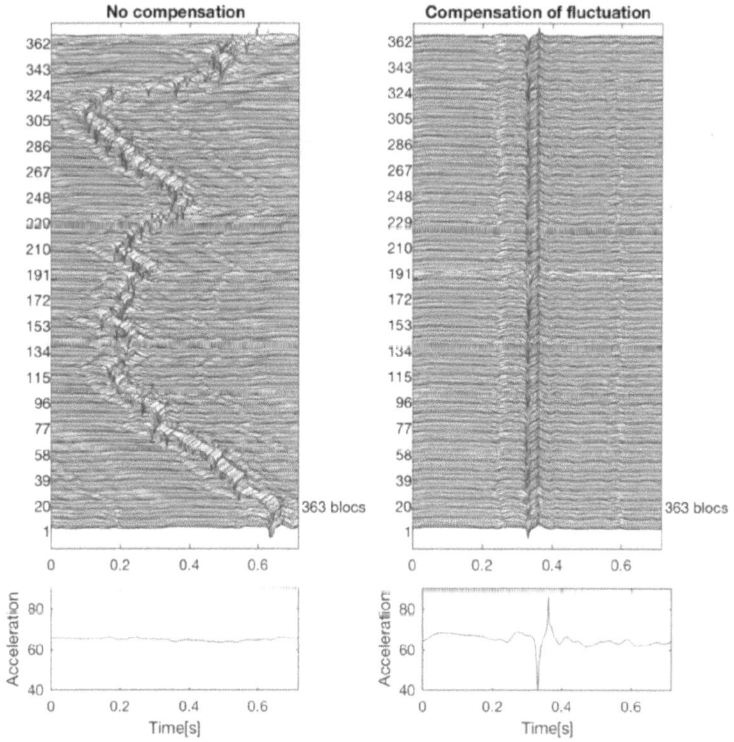

FIGURE 2.9 Effect of stride duration compensation: synchronous average computation.

more elements inside the stride (the progression inside the walk cycle). Unfortunately, it could be hard to physically interpret this progression.

Removing speed fluctuation is equivalent to looking at the signal against the stride instead of the time. The stride scale is not sensitive to speed fluctuations. Therefore, in a second step a vector containing uniformly spaced stride position is generated. The step between stride position is lower than one and can be chosen so that the size of this vector is similar to the temporal signal size. By using interpolation, it is possible to compute the signal against the stride as shown in Figure 2.10: given a stride position, it could be possible to retrieve the associated time and signal value.

2.4.1.3 Application for Fatigue Detection

In Sabri (2010), Maiz et al., (2014), and Zakaria (2015), it was shown that the cyclostationarity at order 2 gives information about the fatigue of a runner. This indicator was also used in order to detect diseases in (Zakaria, 2015).

Figure 2.11 shows the evolution of spectral correlation during a run of half an hour. The first three figures are similar. The last figures that correspond to the end of the race have more power on all cyclic frequencies. It means that there is more cyclic variation in the signal than the previous ones. This variation could be explained by more fatigue and less regularity (Figure 2.12).

2.4.2 Application on Huberian Robust ARMA Modeling to Neurodegenerative Disorders

The process output data are denoted Δt_k corresponding to the stride signal. Figure 2.3 shows an example of the left gait signal from heel–toe force sensors underneath the left foot where three phases appear.

Now assume that Δt_k is generated according to $\Delta t_k = H_0(q)e_k$ where $H_0(q)$ is the noise filter and e_k a random variable sequence with zero mean and variances λ. The ARMA model set is parameterized by a d-dimensional real-valued parameter vector θ, i.e.

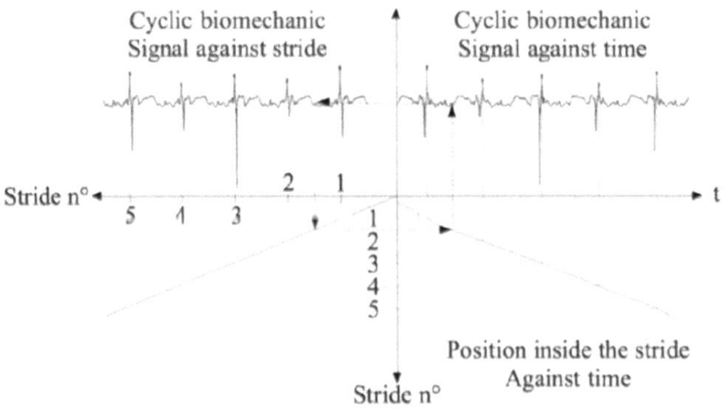

FIGURE 2.10 Speed fluctuation compensation by interpolation.

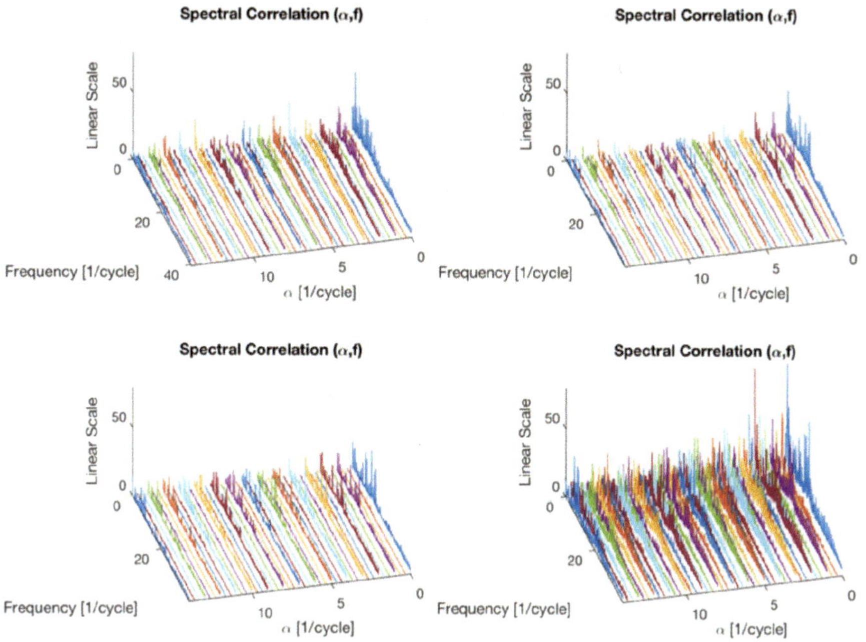

FIGURE 2.11 Evolution of spectral correlation during a run.

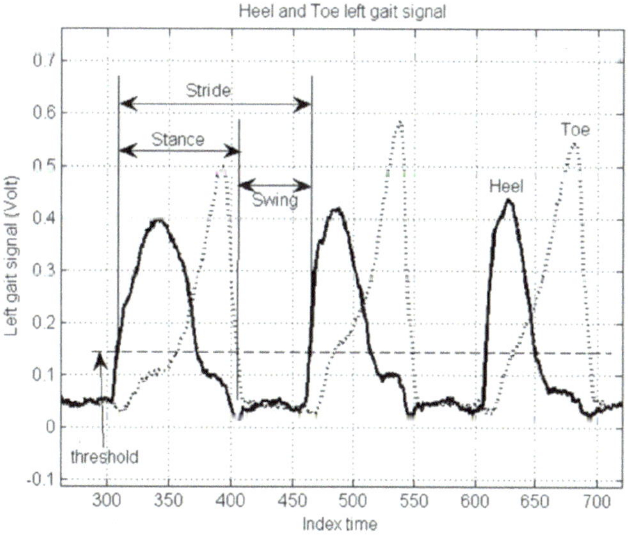

FIGURE 2.12 Example of gait signals from heel and toe sensors underneath the left foot.

$$\Delta t_k = H\left(q^{-1},\theta\right)e_k = \frac{\mathcal{C}\left(q^{-1},\theta\right)}{\mathcal{A}\left(q^{-1},\theta\right)}e_k \tag{2.19}$$

with $\mathcal{C}\left(q^{-1},\theta\right)$ and $\mathcal{A}\left(q^{-1},\theta\right)$ two monic polynomials such that $\mathcal{C}\left(q^{-1},\theta\right)=1+c_1q^{-1}+\cdots+c_{nc}q^{-nc}$ and $\mathcal{A}\left(q^{-1},\theta\right)=1+a_1q^{-1}+\cdots+a_{n_A}q^{-n_A}$, where q^{-1} is the lag operator such that $X_{t-1}=q^{-1}X_t$, and $\theta=\left[a_1,\ldots,a_{n_A},c_1,\ldots,c_{nc}\right]^T$ the parameter vector. The Huberian robust prediction model is then given from (3) by $\Delta\hat{t}_k(\theta,\gamma)=\varphi_k^T(\theta,\gamma).\theta$ with its regressor $\varphi_k^T(\theta,\gamma)=\left[-\Delta t_{k-1} \quad \ldots \quad -\Delta t_{k-n_A}\ \varepsilon_{k-1}(\theta,\gamma) \quad \ldots \quad \varepsilon_{k-nc}(\theta,\gamma)\right]$.

Experimental results are presented over 16 CO, 15 PD, and 19 HD, left and right feet for different estimation norms. The L_2 norm corresponds to the *least-square estimation* (LSE), L_1 norm to the least sum absolute deviation (LSAD) and L_∞ norm to the supremum norm. In the Huberian context, a campaign of estimations is carried out in C_γ with $\hat{\gamma}=1.5$ (Huber et al., 2009) and E_γ with $\gamma\in\left[0.001;0.2\right]$.

For each estimator, comparisons between CO vs PD and HD for left and right feet are given. Table 2.1 shows the means of $\hat{\gamma}$, RMSE, FIT, L_2C, L_1C and the total number of parameters $d=n_A+n_C$. The RMSE is the root mean square error between process output and Huberian robust prediction model output.

The FIT is given by $100\left(1-\frac{y-\hat{y}}{y-\bar{y}}\right)$ where y, $\hat{y}$ and $\bar{y}$ are the process outputs, the Huberian robust prediction model output and the mean of the process output, respectively. L_2C and L_1C are the L_2 and L_1 contributions, respectively, given by $L_iC=\frac{card\left[\nu_i\left(\hat{\theta}_N^{H,\hat{\gamma}},\hat{\gamma}\right)\right]}{N}$. These are indicators of the density of NO in the prediction errors. For instance, $L_2C=40\%$ means that 40% of prediction errors belong to $\left[-\hat{\gamma},\hat{\gamma}\right]$ and deal with the L_2 norm in the Huberian robust function. Here, the threshold γ in E_γ was varied among the range [0.001;0.2] with an incremental step of 0.001 for CO, PD and HD. We focus on the main results in Table 2.2. First, L_2, L_1 and L_∞ norms give non-relevant results with large RMSE, low FIT and large d between 40 and 70. The lack of robustness and degree of freedom (DOF) in these norms lead to an over-estimation of d. On the other hand, each FIT presents a low value. In C_γ for $\hat{\gamma}=1.5$, the number of parameters is reduced with $25\le d\le 32$ but not sufficient for a reduced order ARMA modeling. We can notice a high L_2C, meaning a too large contribution of the L_2 norm, which is very sensitive to the large NO in the prediction errors. The Huberian approach in E_γ leads to relevant results. Indeed, this remains in agreement with the formal point of view related to the bias and the new curve in Figure 2.2: low values of γ involve reduced bias and improve the FIT of the reduced order model. In Corbier and Carmona (2015), we showed that the Huberian model order denoted d_{M}^H is such that $d_{\mathrm{M}}^H<d_{\mathrm{M}}^{L_1}<d_{\mathrm{M}}^{L_2}$ since the Huberian function has one DOF and can be tuned from γ by improving the estimation and reducing the number of parameters for pseudolinear models. First, we notice that $\left\langle\hat{\gamma}_{\text{control}}\right\rangle\approx2\left\langle\hat{\gamma}_{\text{disease}}\right\rangle$, meaning that there are twice more NO in STS-PD and STS-HD than STS-CO. Indeed, for PD and HD,

TABLE 2.1
Means of $\hat{\gamma}$, RMSE, FIT, L_2C, L_1C and the Total Number of Parameters $d = n_A + n_C$ for CO

	$\hat{\gamma}$	*RMSE*	*FIT*	L_2C	L_1C	*d*
			CO Left			
L_2	–	11.2	10	100	0	70
L_1	–	4.3	42	0	100	41
L_∞	–	4.2	25	–	–	45
In C_γ	1.5	2.4	42	95	5	25
In E_γ	0.17	0.09	92	41	59	9
			CO Right			
L_2	–	10.2	9	100	0	70
L_1	–	5.3	44	0	100	39
L_∞	–	3.2	26	–	–	46
In C_γ	1.5	2.3	44	96	4	27
In E_γ	0.18	0.08	92	43	57	9
			PD Left			
L_2	–	13	9	100	0	70
L_1	–	5.2	38	0	100	46
L_∞	–	5.3	26	–	–	56
In C_γ	1.5	3.1	31	96	4	28
In E_γ	0.09	0.34	78	30	70	9
			PD Right			
L_2	–	13	9	100	0	70
L_1	–	6.2	35	0	100	46
L_∞	–	5.5	28	–	–	54
In C_γ	1.5	3.3	31	96	4	30
In E_γ	0.09	0.29	78	32	68	9
			HD Left			
L_2	–	8	17	100	0	70
L_1	–	4.1	36	0	100	44
L_∞		6.3	24	–	–	54
In C_γ	1.5	3.2	32	96	4	31
In E_γ	0.08	0.28	78	29	71	9

(Continued)

TABLE 2.1 (CONTINUED)
Means of $\hat{\gamma}$, RMSE, FIT, L_2C, L_1C and the Total Number of Parameters $d = n_A + n_C$ for CO

	$\hat{\gamma}$	*RMSE*	*FIT*	L_2C	L_1C	*d*
			HD Right			
L_2	–	13	9	100	0	70
L_1	–	6.2	35	0	100	46
L_∞	–	5.1	32	–	–	56
In C_γ	1.5	3.5	29	95	5	32
In E_γ	0.07	0.16	87	27	73	9

TABLE 2.2
Some Temporal Indicators

Feature	**Equation**
Root mean square	$x_{\text{rms}} = \sqrt{\frac{1}{N}\sum_{n=1}^{N}\left(x(n)\right)^2}$
Kurtosis	$ku = \frac{\sum_{n=1}^{N}\left(x(n) - x_{\text{avg}}\right)^4}{(N-1)\sigma^4}$
	$x_{\text{avg}} = \frac{1}{N}\sum_{n=1}^{N} x(n)$
Skewness	$skew = \frac{\sum_{n=1}^{N}\left(x(n) - x_{\text{avg}}\right)^3}{(N-1)\sigma^3}$
Crest factor	$CF = \frac{\max\left(x(n)\right)}{x_{\text{rms}}}$
Impulse factor	$IF = \frac{\max\left(x(n)\right)}{\frac{1}{N}\sum_{n=1}^{N}\lvert x(n)\rvert}$

the estimation requires a low value of $\hat{\gamma}$ involving a large value of $L_1C \approx 70\%$. For CO, $\hat{\gamma} = 0.19$ and $L_1C \approx 58\%$. Table 2.3 shows a list of parameters and the corresponding variance of each parameter for CO and PD left with $\hat{\gamma} = 0.05$ and $\hat{\gamma} = 0.003$ respectively. Figure 2.13 show three $L_2 - L_1$ ARMA models for left PD ($\hat{\gamma} = 0.003$),

TABLE 2.3
Parameters of CO Left ($\hat{\gamma} = 0.05$) and PD Left ($\hat{\gamma} = 0.003$) ARMA Models and Variance λ^H of Each Parameter

I	1	2	3	4	5
			CO Left		
a_i	−0.877	−0.152	0.173	−0.215	0.073
b_i	−0.236	−0.065	0.141	−0.098	–
$\lambda_{a_i}^H$	0.0021	0.0032	0.0015	0.0035	0.0026
$\lambda_{b_i}^H$	0.0012	0.0075	0.0056	0.0074	–
			PD Left		
a_i	−0.712	−0.022	−0.018	−0.181	−0.060
b_i	−0.166	0.119	0.160	0.133	–
$\lambda_{a_i}^H$	0.0031	0.0022	0.0095	0.0015	0.0086
$\lambda_{b_i}^H$	0.0002	0.0005	0.0066	0.0024	–

left HD ($\hat{\gamma} = 0.005$) and left ALS ($\hat{\gamma} = 0.003$) with FIT≈83%. In Figure 2.4 (up) four NO clearly appear in index-times $k = 52$, $k = 113$, $k = 190$ and $k = 247$ with high levels corresponding to the turnaround during the walking period. In this phase, the classical estimators are highly disturbed and sometimes achieve the leverage point. We can notice the good behavior of the Huberian robust reduced order ARMA model during this phase. Equation 2.20 shows the reduced order ARMA model of left PD for $\hat{\gamma} = 0.003$

$$\begin{aligned}\Delta t_k &= 0.712\Delta t_{k-1} + 0.022\Delta t_{k-2} + 0.018\Delta t_{k-3} + 0.181\Delta t_{k-4} \\ &\quad + 0.060\Delta t_{k-5} + e_k - 0.236 e_{k-1} - 0.065 e_{k-2} + 0.141 e_{k-3} - 0.098 e_{k-4}\end{aligned} \tag{2.20}$$

The limited number of ARMA parameters contradicts conclusions in (Hausdorff et al., 1995) and recently in (Ahn and Hogan, 2013). These studies showed a stride interval of normal human walking which exhibit long-range temporal correlations. They presented a highly simplified walking model by reproducing the long-range correlations observed in stride intervals without complex peripheral dynamics. Based on a fractal approach they showed an important point of view related to the *long-range memory effect* of human walking. Our new approach shows a *short-range memory effect* for normal and diseased human walking. It remains to investigate this *memory effect* and to try to interpret in physiological and cognitive terms the correlations with the central nervous system.

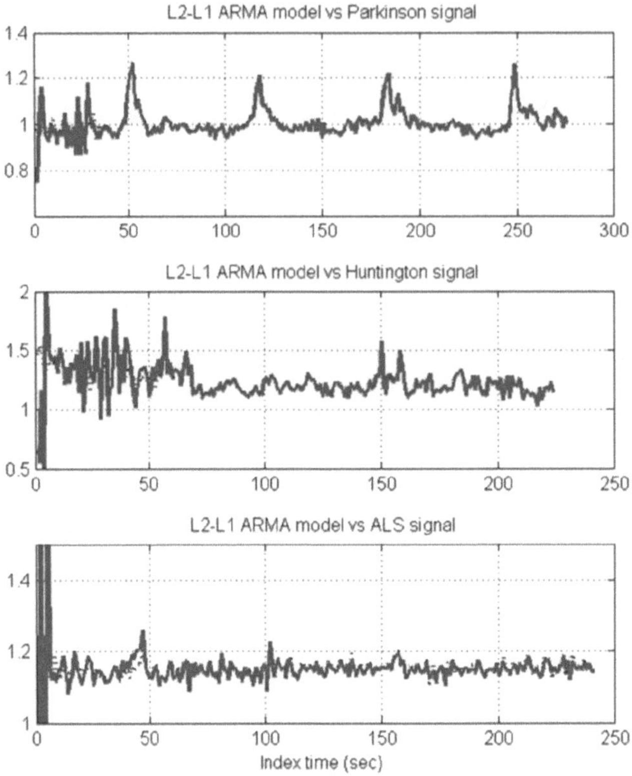

FIGURE 2.13 $L_2 - L_1$ ARMA models (thick line) for left PD ($\hat{\gamma} = 0.003$), left HD ($\hat{\gamma} = 0.005$) and left ALS ($\hat{\gamma} = 0.003$) vs gait output signal (dashed line).

2.4.3 Application to Classification of Neurodegenerative Diseases

A second application is focused on a classification of neurodegenerative diseases based on a Huberian robust mixed kernel function to estimate the nonparametric probability density function (NPDF). From Equation 2.17 the Huberian robust mixed kernel function is yielded as

$$K^{\mathcal{H}}(X,\gamma) = C_f^{\mathcal{H}}(\gamma,\phi)e^{\frac{-X^2}{2}}1_{\{|X|\leq\gamma\}} + C_f^{\mathcal{H}}(\gamma,\phi)e^{-\gamma|X|+\frac{\gamma^2}{2}}1_{\{|X|>\gamma\}} \tag{2.21}$$

where:

$C_f^{\mathcal{H}}(\gamma,\phi)$ is a normalization constant defined by $C_f^{\mathcal{H}}(\gamma,\phi) = \dfrac{1}{\phi\sqrt{2\pi}\left[1+2\left(\dfrac{\varphi(\gamma)}{\gamma} - \Phi(-\gamma)\right)\right]}$.

Here ϕ is a constant carried out from a minimization between mean square error and histogram for a given threshold γ, $\varphi(\gamma)=\frac{1}{\sqrt{2\pi}}e^{\frac{-\gamma^2}{2}}$ and $\Phi(-\gamma)=\int_{-\infty}^{-\gamma}\varphi(X)dX$. From Equation 2.21, a NPDF can be expressed as

$$\hat{f}_N^{\mathcal{H}}(\Delta t,\gamma)=\frac{1}{Nh_N^{\mathcal{H}}(\gamma)}\sum_{i=1}^{N}K^{\mathcal{H}}\left(\frac{\Delta t-\Delta t_i}{h_N^{\mathcal{H}}(\gamma)},\gamma\right) \tag{2.22}$$

where $h_N^{\mathcal{H}}(\gamma)$ is the Huberian bandwidth given by

$$h_N^{\mathcal{H}}(\gamma)=\left[\frac{R\left(K^{\mathcal{H}}(\gamma)\right)}{\left(\sigma_K^2(\gamma)\right)^2 R\left(f_{\mathcal{H}}''(\gamma)\right)}\right]^{1/5} N^{-1/5} \tag{2.23}$$

with $R\left(G(\gamma)\right)=\int_{\mathbb{R}}G(w,\gamma)dw$, $\sigma_K^2(\gamma)=\int_{\mathbb{R}}w^2K^H(w,\gamma)dw$.

In order to classify diseases, let us consider the first statistical indicator named DGD defined as follows. Let $L_2^S(\gamma)=\left\{k,|\varepsilon_k|\le h_N^{\mathcal{H}}(\gamma)\gamma\right\}$ be a compact subset and let $\varepsilon_k=\Delta t-\Delta t_k$ be estimation errors. From $L_2^S(\gamma)$, define $C_N^{L_2}(\gamma,\Delta t)=\frac{1}{N}\sum_{k\in L_2^S(\gamma)}|s_k(\Delta t)|$ as a L_2-contribution in the NGKF framework where $s_k(\Delta t)=s(\Delta t-\Delta t_k)$ with $s(\blacksquare)$ the sign function. Let $\mathbb{D}_\Delta$ and $\mathbb{D}_\gamma$ two compact subsets containing all values of Δt and γ. The DGD is then defined as the gap between control L_2-contribution and disease L_2-contribution by

$$\mathrm{DGD}_N^{L_2}(\gamma)=\left|\sup_{\Delta t\in\mathbb{D}_\Delta^h}C_N^{L_2,co}(\gamma,\Delta t)-\sup_{\Delta t\in\mathbb{D}_\Delta^d}C_N^{L_2,di}(\gamma,\Delta t)\right| \tag{2.24}$$

In practice, $\mathrm{DGD}_N^{L_2}(\gamma)$ can be calculated for $\gamma_{\text{opt}}\in[0.9,1.5]$ with $-h_N^{\mathcal{H}}(\gamma_{\text{opt}})\le\varepsilon_k\le h_N^{\mathcal{H}}(\gamma_{\text{opt}})$.

The second new statistical indicator focuses on the variability of two consecutive time-signals Δt_k compared to a threshold k such that $0<\kappa<2$. Indeed, this interval allows to increase the accuracy of the variability, since sometimes the gap between Δt_k and Δt_{k+1} can be either very tiny for controls or large for diseases. Let us define a general subset $\nu_\kappa^X=\left\{k,\delta_{k-1}^X\delta_k^X<0,|\delta_k^X|\ge\kappa\right\}$ where $X=\{co,di\}$ and $\delta_k^X=\Delta t_{k+1}-\Delta t_k$ for a corresponding X. The VGD is then defined as

$$\mathcal{V}_{GD}(\kappa)=\left|\sum_{k\in\nu_\kappa^{co}}|s_k(\Delta t)|-\sum_{k\in\nu_\kappa^{di}}|s_k(\Delta t)|\right| \tag{2.25}$$

For small values of $\kappa=\kappa_0\left(\kappa_0\in\mathbb{D}_\kappa\right)$ and large values of $\kappa=\kappa_1\left(\kappa_1\in\mathbb{D}_\kappa\right)$ $\mathcal{V}_{gd}(\kappa_0)$ tends to 0. Then it is shown that for $\kappa_0<\kappa_m<\kappa_1$, $\mathcal{V}_{GD}(\kappa)$ is the maximum. This means that the variability of two consecutive time-signals is maximum for a considered threshold κ_m.

Below some results of DGD and VGD of gait rhythm disorders from a database containing 16 CO, 15 PD, 19 HD and 13 ALS are presented.

Figure 2.14 shows DGD of the left strides with respect to γ. Optimally tuned threshold is given in the interval range $\gamma_{opt} \in \left[0.9,1.5\right]$. In order to have both the optimal and a sufficient number of points, we decided to extend this interval to $\gamma_{opt} \in \gamma_{exp} \in \left[0.9,1.5\right]$. The minimum of MSE between estimated NPDF and histogram for $\gamma \in \gamma_{exp}$ with an incremental step of 0.01 involving $\phi \approx 0.12$. This new statistical tool clearly differentiates and discriminates each disease and we noticed upper values for HD, meaning a maximum topological distance between CO and HD. Our result correlated with a study (Paulsen et al., 2008) in which HD is the neurological disease having the greatest tremors.

For VGD, k was varied over the range [0.01 s, 05 s] with an incremental step of 0.01 s. The variable k corresponds to a time-window beyond which one evaluates the variability of two consecutive gaits. Recall that VGD means the variability of two consecutive gait rhythms between CO and patient. For the latter, this is crucial information of the time-evolution of each disease and is related to non-stationary behavior of walking. Figure 2.6 shows an example of VGD of each disease for the left stance signal. It appeared for each patient a maximum corresponding to a threshold $\kappa \approx 0.04$ s. HD was then the largest disorder in gait rhythm and VGD had a different maximum for each disease, equal to 105%, 90% and 46% for HD, ALS and PD, respectively. The classification of each disease was well carried out from the maximum of VGD. We see that DGD and VGD are complementary tools and converge toward the same result: HD presents the greatest tremors in VGD (Figure 2.15).

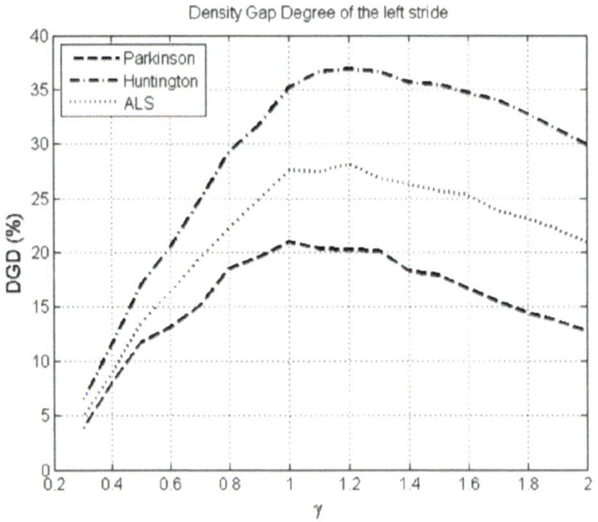

FIGURE 2.14 DGD of the left stride where the maximums and diseases are distinguished with 21%, 37% and 28% for PD, HD and ALS, respectively, at $\gamma_{opt} \approx 1.1$.

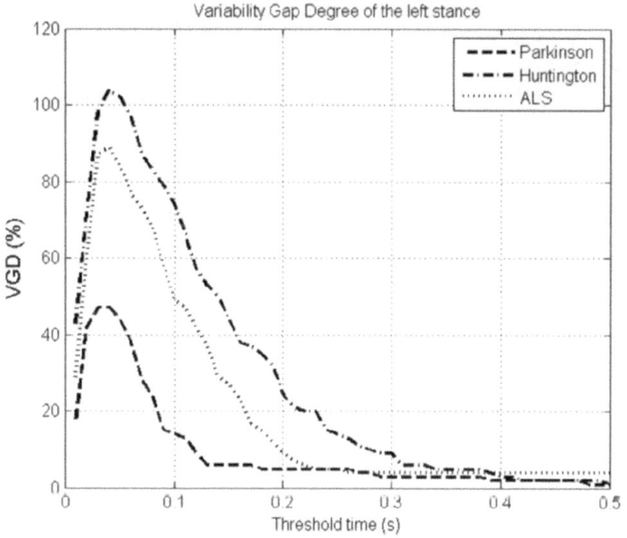

FIGURE 2.15 VGD of two consecutive gait rhythms between control and patient.

2.5 CONCLUSION

Exposure to whole-body vibration causes movement and effects in the body, which can cause discomfort, negatively affect performance and pose a health and safety risk. The study and analysis of the signals from the various sensors taking into account the environmental constraints make it possible to prevent the risks related to exposure to vibrations and/or shocks by using the techniques of signal processing.

REFERENCES

Ahn, J. and Hogan, N. (2013). Long-range correlations in stride intervals may emerge from non-chaotic walking dynamics. *PLOS ONE*, 8(9):2–10.

Andrews, D. (1994). Large sample estimation and hypothesis testing. *Handbook of Econometric*, volume 4. Elsevier Science.

Antoni, J. (2007). Cyclic spectral analysis in practice. *Mechanical Systems and Signal Processing*, 21(2):597–630. ISSN 0888-3270.

Beya, O., et al. (2010). Empirical modal decomposition applied to cardiac signals analysis. *SPIE 2010 San José – USA*.

Bonnardot, F., et al. (2005). Use of the acceleration signal of a gearbox in order to perform angular resampling (with limited speed fluctuation). *Mechanical Systems and Signal Processing*, 19(4):766–785. ISSN 0888-3270.

Boudraa, A. and Cexus, J.-C. (2007). EMD-based signal filtering. *IEEE Transactions on Instrumentation and Measurement*, 56(6):2196–2202.

Corbier, C. (2012). Contribution à l'estimation robuste de modèles dynamiques: application à la commande de systèmes dynamiques complexes. PhD Thesis, Arts et Métiers ParisTech.

Corbier, C. and Carmona, J.-C. (2015). Extension of the tuning constant in the Huber's function for robust modeling of piezoelectric systems. *International Journal of Adaptive Control and Signal Processing*, 28(8):1–16.

Debbal, S. and Bereksi-Reguig, F. (2007). Spectral analysis of PCG signals. *The Internet Journal of Medical Technology*, 4(1).

Flandrin, P., Rilling, G., and Goncalves, G. (2004). Empirical mode decomposition as a filter bank. *IEEE Signal Processing Letters*, 11(2): 112–114.

Hausdorff, J., et al. (1995). Is walking a random walk? Evidence for long-range correlations in stride interval of human gait. *Journal of Applied Physiology*, 78(1):349–358.

Huang, N. (1998). The empirical mode decomposition and Hilbert spectrum for nonlinear and non-stationary time series analysis. *Proceedings of the Royal Society of London A*, 454:903–995.

Huber, P. and Ronchetti, E. (2009). *Robust Statistics*. John Wiley and Sons, New York, NY, 2nd edition.

Ljung, L. (1999). *System Identification: Theory for the User*. Prentice Hall PTR, New York.

Maiz, S. (2014). Estimation et detection de signaux cyclostationnaires par les methodes de ré-échantillonnage statistique: Applications à l'analyse des signaux biomécaniques. PhD Thesis, Université Jean Monnet.

Maiz, S., et al. (2014). New second order cyclostationary analysis and application to the detection and characterization of a runner's fatigue. *Signal Processing*, 102:188–200. ISSN 0165-1684.

Newey, W. and McFadden, D. (1994). *Large sample estimation and hypothesis testing*. Handbook of Econometric, vol. 4:2113–2247, Elsevier Science.

Pachaud, C. (1997). Crest factor and kurtosis contributions to identify defects inducing periodical impulsive forces. *Mechanical Systems and Signal Processing*, 11(6):903–916.

Paulsen, J., et al. (2008). Detection of Huntington's disease decades before diagnosis: The predict-HD study. *Journal of Neurology, Neurosurgery, and Psychiatry*, 79:874–880.

Sabri, K., et al. (2010). Cyclostationary modeling of ground reaction force signals. *Signal Processing*, 90(4):1146–1152. ISSN 0165-1684.

William, A., Gardner, Napolitano, A., and Paura, L. (2006). Cyclostationarity: Half a century of research. *Signal Processing*, 86(4):639–697. ISSN 0165-1684.

Zakaria, F. (2015). Human locomotion analysis: Exploitation of cyclostationarity properties of signals. PhD Thesis, Université Jean Monnet.

3 Numerical and Experimental Assessment of Mechanical Vibrations

Georges Kouroussis, Roger Serra, and Sébastien Murer

CONTENTS

3.1 INTRODUCTION

In Chapter 1, the instrumentation for mechanical vibration analysis established the importance of monitoring the human body, i.e. testing, acquisition and analysis of the human body. It appeared that the human body can be considered as a mechanical system and all the vibration interaction issues can be covered under this assumption.

In addition, humans generally interact with equipment. It is therefore natural to consider a coupled man-machine system where both subsystems interact.

The objective of this chapter is not to define the mechanical vibration but to cross the bridge between the human body behavior and structural dynamics, including the modal analysis. To do this, the experimental approach offers a systematic way of analyzing this mechanical behavior. This step is not trivial and many parameters could introduce a bias or uncertainties in the measured data without rigorous testing procedures, particularly in the human context. It should lead to the "real" representation of the dynamics of the structure, either in the time, modal or frequency domain and in the case of the interaction of human with equipment.

The origin of modal analysis is based on the decoupling between space and time in steady-state motion. Any motion y is therefore considered as two separated variable functions

$$y(x,t) = Y(x)\Phi(t) \tag{3.1}$$

where the spatial-dependence term $Y(x)$ is explicitly separated to the time-dependence term $\Phi(t)$. This allows a direct analysis by considering the vibration mode superposition. With the development of the signal processing tools, any motion—transient or steady-state motion—can be treated without loss of accuracy and efficiency. The main hypothesis of this approach is the linearity between forces and mechanical responses, assuming a linear mechanical behavior of man-machine systems studied.

This chapter presents therefore the concept of structural dynamics, modal analysis theory and finite element or multibody model updating using measured data. Some examples are provided with a freeware in-house toolbox *EasyMod*.

3.2 DYNAMIC MODELS AND MODAL CONCEPT

Both numerical (theoretical) and experimental vibration analyses are based on one of the most prolific and powerful theory in structural dynamics: the modal model of a dynamic system. It provides an efficient means of determining the dynamic characteristics of a system, independently of the nature of the dynamic excitation acting on the system. Different paths from which the modal model can be derived are shown in Figure 3.1. Algebraic and numerical tools are used to link the spatial, the modal and the frequency models together. Theoretical and numerical modal analyses mainly rely on the description of the physical system through the calculation of the mass, stiffness and damping matrices of the system. The frequency analysis (with the Fourier transform or, in practice, the discrete Fourier transform) establishes a direct use of the frequency response function (FRF), defined as a function of the geometric and dynamic properties of the physical system only and often considered as its vibratory signature. Finally, to establish the link between the modal model and the experiments, the experimental modal analysis (and the operational modal analysis to a lesser extent) closes the relation between modes and frequency analysis from measurements.

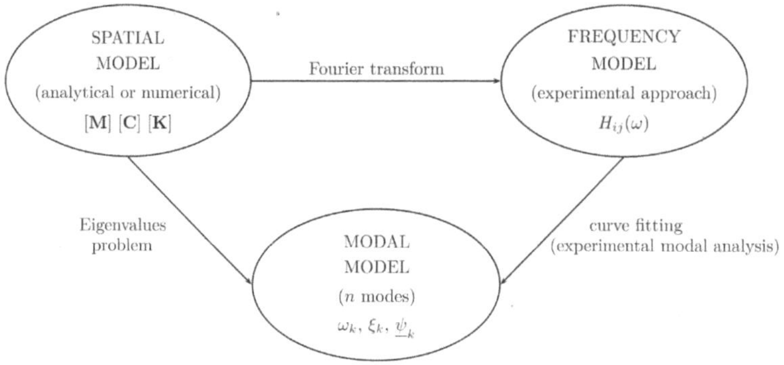

FIGURE 3.1 From a mathematical model to another model: link between the spatial model, the frequency model and the modal model (annotations, symbols and parameters are defined in the next sections).

In the following sections, some theoretical reminders are presented in order to further explain the present possibilities concerning numerical modeling and experimental assessment of vibration data. Detailed theoretical analysis of these applications and more will be found in the literature.

3.2.1 Representation of Damping in Vibration Analysis

In characterizing dynamic motion in a mechanical system, the most important knowledge is probably the definition of damping, which influences the amplitude of the response and involves the major mechanisms of mechanical energy dissipation in the system. However, contrary to other mechanical parameters, it is difficult to quantify its value, mainly due to the different types of damping encountered in practice.

By definition, the damping is related to deformation energy during motion. It is usually mapped using the stress-strain relationship through the area covered by its hysteresis loop (Figure 3.2), concluding in some ways that damping effect has to depend on a relative velocity. In practice, damping occurs when:

- internal energy dissipation associated with micro-structure defects appears inside the material constituting the system (or some components of the system); two general types of internal damping are usually identified in practice: viscoelastic and hysteretic behavior,
- mechanical energy dissipation is caused by friction due to the relative motion between components of the system; among all the means to develop mathematical models to describe structural damping, the Coulomb friction model is often adopted in engineering,
- fluid resistance causing mechanical energy dissipation generally intervenes at the fluid–structure interface.

Usually, viscous damping is considered in practice, showing a linear relationship between damping force and velocity. The other damping types are generally

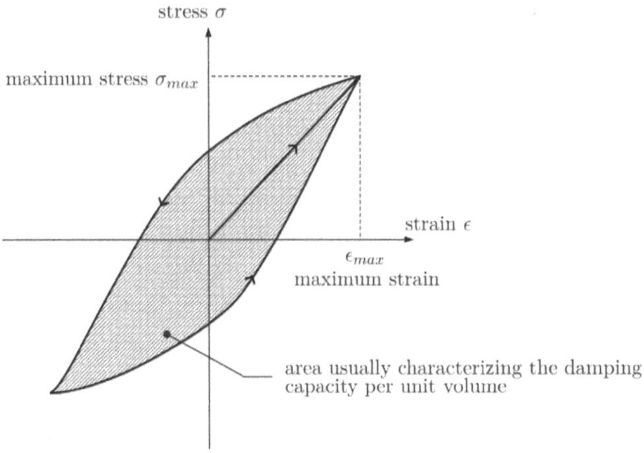

FIGURE 3.2 A typical hysteresis loop for mechanical damping.

reformulated in order to derive an equivalent viscous damping and the associated damping ratio ξ, theoretically defined as the ratio of the actual damping to a critical damping introduced for damping stability concepts. Hysteretic damping is characterized in the same way by its loss factor η. Both damping ratio and loss factor are usually retained in modal analysis, hence for low damping, the simple mathematical relationship is approximately given by

$$\eta = 2\xi \tag{3.2}$$

The loss factor is often used as the most comprehensive damping parameter (Carfagni et al., 1998) and approximate values of some common materials are given in Table 3.1.

TABLE 3.1
Approximate Loss Factors of some Useful Materials (de Silva, 2006)

Material	Loss Factor $\eta \cong 2\xi$
Aluminum	2×10^{-5} to 2×10^{-3}
Concrete	0.02 to 0.06
Glass	0.001 to 0.002
Rubber	0.1 to 1.0
Steel	0.002 to 0.01
Wood	0.005 to 0.01

3.2.2 Frequency Response Functions of a Single-Degree-of-Freedom System

A single-degree-of-freedom (SDOF) system is usually used to explain the basic theory of linear vibration. SDOF systems are also used to idealize mechanical and structural systems. Moreover, as explained in the next section, using the modal basis provides a convenient way to consider a multiple-degree-of-freedom system (MDOF) as a series of SDOF systems, each of them characterizing the studied structure for a specific vibration mode. It is convenient to consider a SDOF system as a lumped mass m connected to a fixed frame by means of a spring (of stiffness k) and a viscous damper (with damping coefficient c), as illustrated in Figure 3.3. The effect of static equilibrium is taken into account in the relative motion x of the mass, so as

$$m\ddot{x} + c\dot{x} + kx = f(t) \tag{3.3}$$

represents the equation of motion of the SDOF system subjected to a force $f(t)$ that varies with time. Sometimes, the canonical form is preferred, which is obtained by dividing Equation 3.3 by the coefficient of $\ddot{x}$ (m in this case):

$$\ddot{x} + 2\xi\omega_0\dot{x} + \omega_0^2 x = \frac{1}{m} f(t) \tag{3.4}$$

involving the natural circular frequency (f_0 is the corresponding natural frequency) of the system

$$\omega_0 = 2\pi f_0 = \sqrt{\frac{k}{m}} \tag{3.5}$$

and its damping ratio

$$\xi = \frac{c}{2\sqrt{km}} = \frac{c}{c_c} \tag{3.6}$$

where c_c is the aforementioned critical damping. Damping ratio ξ is always considered to be small, and, in vibration analysis, $\xi < 1$.

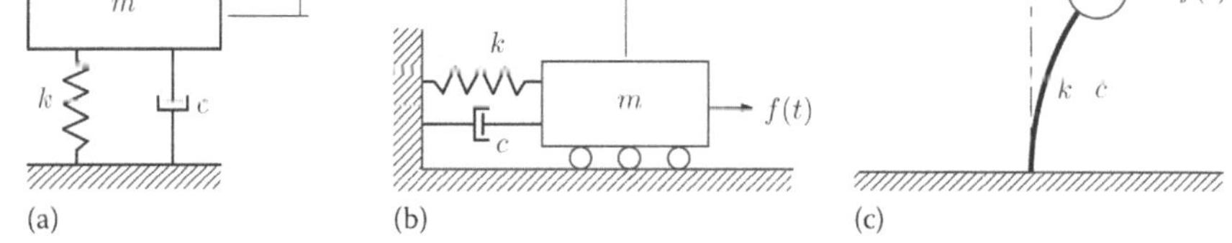

FIGURE 3.3 Different representations of a SDOF system with a time-varying excitation: (a) vertically connected, (b) horizontally connected, (c) massless beam with a lumped mass.

For a harmonic force $f(t) = F\sin(\omega t)$, the response of the system defined by Equation 3.3 is also a harmonic function, in the form $x(t) = X\sin(\omega t - \phi)$ where X is the amplitude and ϕ the phase (the minus sign indicates that a certain—positive—delay exists between the force and the response). Both parameters depend on the circular frequency ω (or the frequency f since $\omega = 2\pi f$). Various methods are available to calculate the unknown amplitude X and phase ϕ:

- The complex substitute, assuming that the harmonic response can be rewritten as $x = X\ \text{Im}\left(e^{j}\left(\omega t - \phi\right)\right) = \underline{X}\ \text{Im}\left[e^{j\omega t}\right]$ where $\underline{X} = X\ Im\left[e^{-j\phi}\right]$.
- The Laplace transform, based on the unilateral transform defined by $F(p) = \int_0^{\infty} e^{-pt} f(t)\,dt$ for the force and by $X(p) = \int_0^{\infty} e^{-pt} x(t)\,dt$ for the response.
- The Fourier transform, using the definitions $F(\omega) = \frac{1}{2\pi}\int_{-\infty}^{+\infty} f(t)e^{-j\omega t}dt$ and $X(\omega) = \frac{1}{2\pi}\int_{-\infty}^{+\infty} x(t)e^{-j\omega t}dt$, for the force and the response, respectively. Note that the Fourier transform is intimately related to the Laplace transform, using the change of variable $p = j\omega$, for causal functions.

These methods offer three different ways to calculate the same response, even though the complex substitute represents a simple and efficient calculation. For instance, the ratio of the response to force input can be derived as

$$\frac{X(\omega)}{F(\omega)} = \frac{1}{k - \omega^2 m + j\omega c} \tag{3.7}$$

$$= \frac{1/k}{1 - \omega^2/\omega_0^2 + j2\xi\omega/\omega_0} \tag{3.8}$$

$$= \frac{1/m}{\omega_0^2 - \omega^2 + j2\xi\omega\omega_0} \tag{3.9}$$

for a viscous damping. In the case of a hysteretic damping, the ratio becomes

$$\frac{X(\omega)}{F(\omega)} = \frac{1}{k - \omega^2 m + jd} \tag{3.10}$$

$$= \frac{1/k}{1 - \omega^2/\omega_0^2 + j\eta} \tag{3.11}$$

$$= \frac{1/m}{\omega_0^2 - \omega^2 + j\eta\omega_0^2} \tag{3.12}$$

reminding the link between the viscous and the hysteretic coefficients (Equation 3.2). Notice that no real-time equivalent of Equation 3.3 exists for a hysteretic damping, that can be somewhat written

$$m\ddot{x} + jdx + kx = f(t) \tag{3.13}$$

where d is the hysteretic damping coefficient. The ratio defined by Equations 3.7 or 3.10 is obviously complex and is typically denoted by $H(\omega)$. It uniquely defines the frequency response function (FRF) of the SDOF system and presents interesting properties:

- the FRF is independent on the force and the response, if the system behavior remains linear;
- it can be seen as the response of the system to a unit force;
- it provides a rapid means to calculate the amplitude (and the phase) of the system studied, especially at the resonance $\omega = \omega_0$;
- a simple multiplication by $j\omega$ (by $-\omega^2$) allows the transposition from displacement to velocity (to acceleration) and vice versa. To avoid misinterpretation, the ratio of displacement to force is called receptance, often denoted $R(\omega)$, mobility $M(\omega)$ in the case of velocity-based FRF or accelerance $A(\omega)$ for acceleration-based FRF;
- the corresponding graphical representations provide useful information and knowledge.

The latter are illustrated in Figures 3.4 through 3.7. These show that $R(\omega)$, $M(\omega)$ and $A(\omega)$ are easily interchangeable. In addition, these different graphics are noteworthy of interpretation and observation. Bode and imaginary curves directly show an extremum value at the resonance frequency (always a maximum for the Bode curves—Figure 3.4). The real part of the FRF (Figure 3.6) cancels at the resonance and the phase (Figure 3.5) shows a 180° variation at the resonance area (exactly for a viscous damping, almost for a hysteretic damping), for any kind of form (receptance, mobility or accelerance). Last but not least, Nyquist curves illustrate the circularity of an FRF: a perfect circle is obtained for a receptance FRF with viscous damping although the mobility FRF with hysteretic damping confirms the true circularity.

Finally, the FRF of a SDOF system can be presented in other different forms. In the case of viscous damping, the factorized FRF is often used

$$R(\omega) = \frac{r}{j\omega - \lambda} + \frac{r^*}{j\omega - \lambda^*} \tag{3.14}$$

introducing the residual

$$r = \frac{1}{2jm\omega_0\sqrt{1-\xi^2}} \tag{3.15}$$

the complex pole

$$\lambda = -\xi\omega_0 + j\omega_0\sqrt{1-\xi^2} \tag{3.16}$$

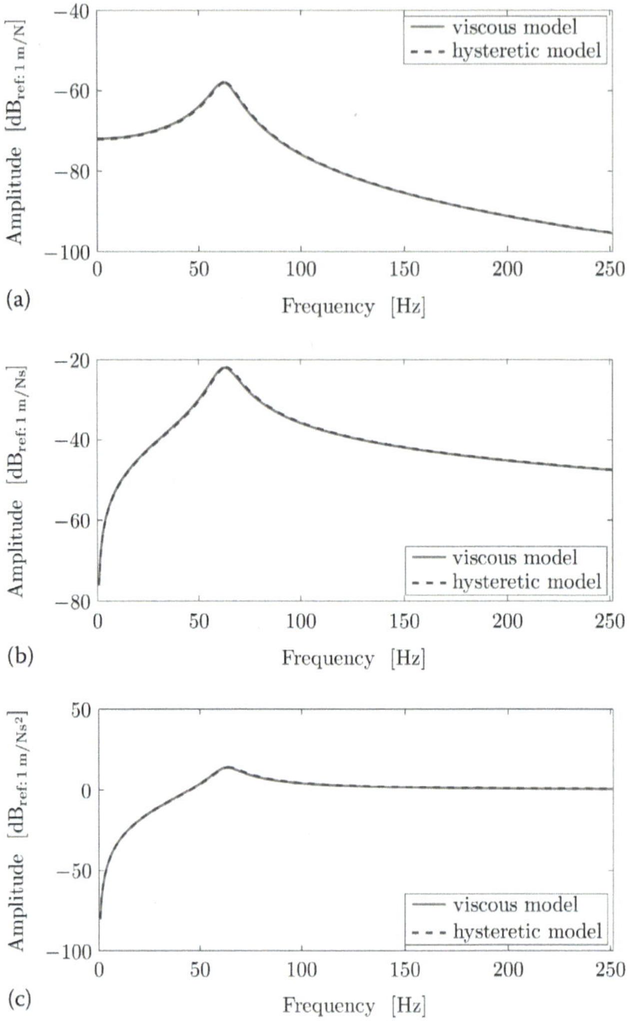

FIGURE 3.4 Bode representation of an FRF (linear-log plot): (a) receptance, (b) mobility, (c) accelerance.

and their conjugates. For time domain representation, the impulse response of the system, obtained using the Fourier transform (or the discrete Fourier transform)

$$h(t) = \mathcal{F}^{-1}\left(R(\omega)\right) = \int_{-\infty}^{+\infty} R(\omega) e^{j\omega t} dt \tag{3.17}$$

is also of great importance.

Each of these graphical properties is used by the modal analysis methods to extract precisely the natural frequency, and, therefore, the other modal parameters.

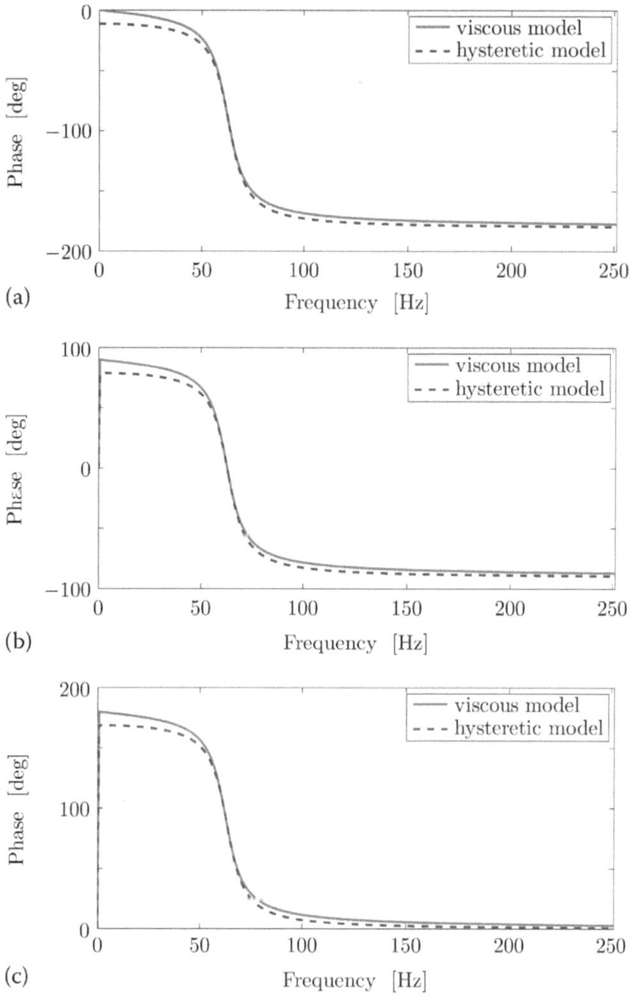

FIGURE 3.5 Phase representation of an FRF: (a) receptance, (b) mobility, (c) accelerance.

3.2.3 Frequency Response Functions of a Multiple-Degree-of-Freedom System

The purpose of this section is not to develop in a detailed manner the theory of modal analysis for MDOF systems but to put emphasis on the important particularities that make the link with the SDOF system theory. Several books (e.g. de Silva, 2006; Maia, 1997; Fu and He, 2001) have addressed more comprehensive reviews of the modal analysis theory. This section summarizes the main parts of these contributing references.

The matrix form of the equations of motion of an MDOF system with n degrees of freedom is

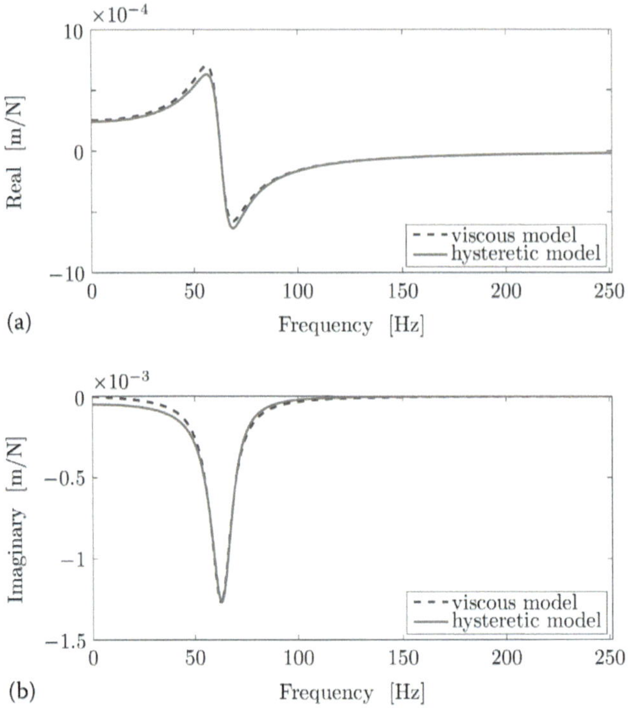

FIGURE 3.6 Real and imaginary parts of an FRF.

$$[\mathrm{M}]\{\ddot{\underline{x}}\}+[\mathrm{C}]\{\dot{\underline{x}}\}+[\mathrm{K}]\{\underline{x}\}=\{\underline{\mathrm{f}}(\mathrm{t})\} \tag{3.18}$$

considering the case of a viscous damping. Equation 3.18 involves the mass matrix M, the stiffness matrix K and the damping matrix C.

The knowledge of the mass and stiffness matrices straightforwardly leads to the following eigenvalue problem:

$$([\mathrm{K}]-\lambda[\mathrm{M}])\{\underline{\psi}\}=0 \tag{3.19}$$

providing, as solutions, n eigenvalues λ_r and n eigenvectors $\underline{\psi}_r$. The undamped natural frequencies ω_r of the system are obtained from the square roots of these eigenvalues. The undamped eigenvectors give the form of the mode shapes to within a constant. In addition to these results, another important property arises, based on the orthogonality of the eigenvectors:

$$\{\underline{\psi}_r\}^T[\mathrm{M}]\{\underline{\psi}_r\}=m_r \tag{3.20}$$

and

$$\{\underline{\psi}_r\}^T[\mathrm{K}]\{\underline{\psi}_r\}=k_r \tag{3.21}$$

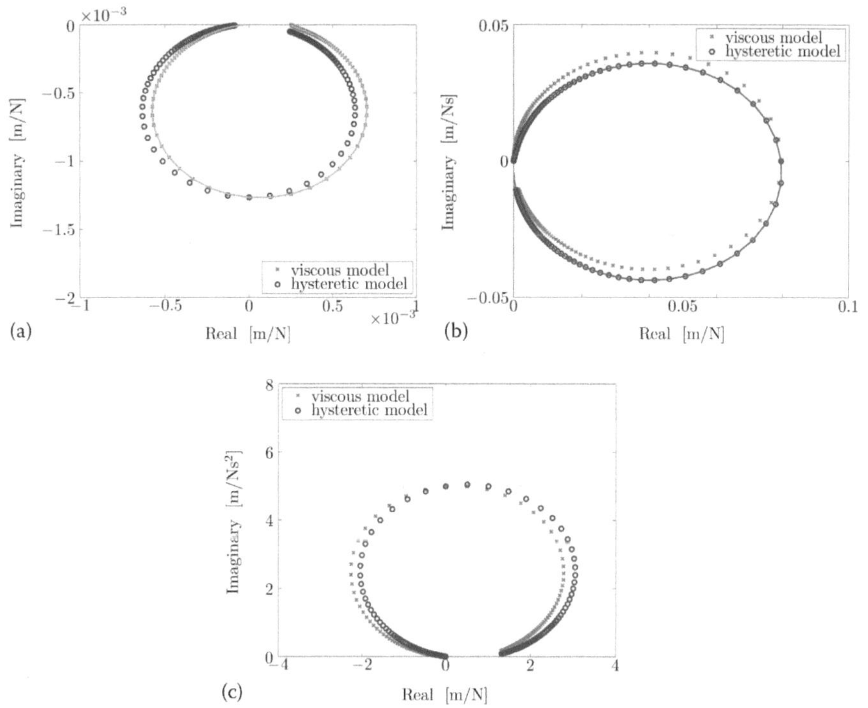

FIGURE 3.7 Nyquist representation of an FRF: (a) receptance, (b) mobility, (c) accelerance.

This gives another basis for calculating the response of the MDOF system: the modal basis manifests a new space where each degree of freedom defining the system is replaced by a combination of modal coordinates that creates a new system of decoupled equations, each having an equivalent mass m_r and an equivalent stiffness k_r. This is the fundamental property in vibration where the vibration of a system consists in a linear combination of all modes. This property is illustrated in Figure 3.8 showing a Bode curve of a MDOF system with multiple resonances defining the contribution of n SDOF resonances.

It is evident that mode shapes are not unique since multiples of $\underline{\psi}_r$ also satisfy Equation 3.19. Mode shapes $\underline{\psi}_r$ are referred to as principal modes, or normal modes of the system. They are also called the undamped modes for obvious reasons. Due to the non-uniqueness of mode shapes $\underline{\psi}_r$, a normalization is preferred, and often the mass–normalized mode shapes are retained, defined by

$$\{\underline{\Phi}_r\} = \frac{1}{\sqrt{m_r}}\{\underline{\psi}_r\} \quad (r = 1,2,\ldots,n) \tag{3.22}$$

and involving the new orthogonality relationships

$$[\Phi]^{\mathrm{T}}[\mathrm{M}][\Phi] = [\mathrm{I}] \tag{3.23}$$

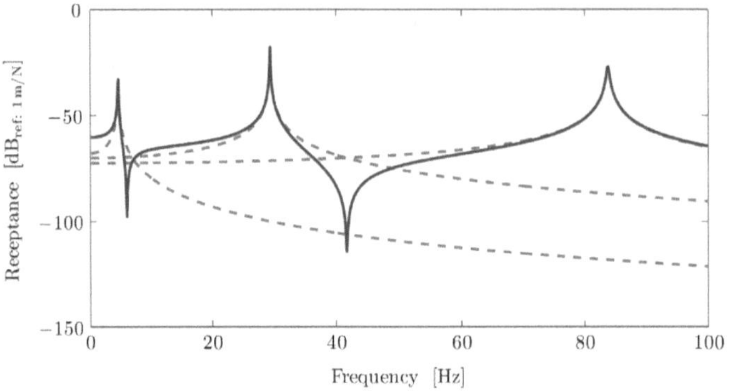

FIGURE 3.8 A MDOF system: superposition of *n SDOF* systems.

$$[\Phi]^T[K][\Phi] = \begin{bmatrix} \ddots & \square & \square \\ \square & \omega_r^2 & \square \\ \square & \square & \ddots \end{bmatrix} \tag{3.24}$$

The non-physical senses of m_r and k_r are no longer topical, replaced directly by the natural frequencies ω_r. However, the knowledge of the modal mass m_r provides additional information for efficient comparison (Lang, 2016).

If damping is taken into account in the mechanical system, the representation of FRF is still self-sufficient. However, a distinction between hysteretic and viscous damping must be done, as well as the possibility of proportional damping.

The most simple form (and the easiest to obtain) is related to the viscous proportional damping. Indeed, the damping matrix can be defined by

$$[C] = \alpha[M] + \beta[K] \tag{3.25}$$

Here, both α and β are real positive constants. A simple mathematical reasoning, based on the modal basis transformation, provides the expression of an FRF related to the response at coordinate i due to a single force applied at coordinate j

$$H_{ij}(\omega) = \sum_{k=1}^{n} \frac{\phi_{ik}\phi_{jk}}{\omega_k^2 - \omega^2 + j2\xi_k\omega\omega_k} \tag{3.26}$$

$$= \sum_{k=1}^{n} \frac{B_{ijk}}{\omega_k^2 - \omega^2 + j2\xi_k\omega\omega_k} \tag{3.27}$$

where B_{ijk} is the real modal constant, obtained by multiplying together the ith and j^{th} elements of the kth mode shape ϕ_k.

When the damping matrix C is not proportional, the changeover to a new dynamic system is necessary, needing a new eigenproblem calculation based on a state-space approach. This induces, in the calculation, doubling the size of the system resulting in n pairs of complex and conjugate poles and n eigenvectors. A factorized FRF is therefore used to define the linear relation between excitation and response

$$H_{ij}(\omega) = \frac{X_i(\omega)}{F_j(\omega)} = \sum_{k=1}^{n} \left(\frac{r_{ijk}}{j\omega - \lambda_k} + \frac{r_{ijk}^*}{j\omega - \lambda_k^*} \right) \tag{3.28}$$

where r_{ijk} is called the residue and has the same property as an eigenvector. Equation 3.28 represents the most general form of frequency response function and can be degenerated into Equation 3.26 if damping is proportional.

The case of hysteretic damping is straightforward if the system is mathematically considered as undamped with a new stiffness matrix

$$\left[\mathrm{K}^*\right] = \left[\mathrm{K}\right] + j\left[\mathrm{D}\right] \tag{3.29}$$

where D is the hysteretic damping matrix. Therefore, the FRF has the following form:

$$H_{ij}(\omega) = \sum_{k=1}^{n} \frac{\phi_{ik}\phi_{jk}}{\omega_k^2 - \omega^2 + j\eta_k\omega_k^2} \tag{3.30}$$

$$= \sum_{k=1}^{n} \frac{B_{ijk}}{\omega_k^2 - \omega^2 + j\eta_k\omega_k^2} \tag{3.31}$$

where B_{ijk} is still the modal constant. This constant is however complex. Analogously, the damping matrix related to the hysteretic proportional behavior relies on the following definition

$$\left[\mathrm{D}\right] = \mu\left[\mathrm{M}\right] + \nu\left[\mathrm{K}\right] \tag{3.32}$$

with, similarly to Equation 3.25, two real positive constants μ and ν. By considering the same mathematical principle, the proportional behavior gives the same expression for a FRF as Equation 3.30 with the interesting particularity that B_{ijk} is real.

It turns out that the most significant outcome of having non-proportional damping is that the vibration modes are complex, slightly different from the physical interpretation of in-phase and out-of-phase motions described in the introduction.

3.2.3.1 Some Important Considerations

The knowledge of modal analysis theory brings valuable information on the nature of the system studied. The main hypothesis of linearity can be verified by generating some FRFs with different force levels and verifying that all of them are identical. Another way to verify the linearity properties is to use the Betti–Maxwell reciprocal work theorem, which states that, for a linear elastic structure subject to two sets of forces $P_i, i = 1,\ldots,m$ and $Q_j, j = 1,\ldots,n$, the work done by set $\underline{P}$ through the displacements produced by set $\underline{Q}$ is equal to the work done by set $\underline{Q}$ through the displacements produced by set $\underline{P}$. This theorem has applications in modal analysis where it is used to define the reciprocity of frequency response functions for each case (undamped and damped structures)

$$H_{ij}(\omega) = H_{ji}(\omega) \tag{3.33}$$

A continuous system (with an infinite number of degrees of freedom) presents, by definition, an infinite number of modes. In practice, the number of degrees of freedom is high but limited to a finite value (due to a maximum frequency to analyze). That means that the number of modes is also finite. In addition, either in numerical or experimental analysis, the frequency range of interest is imposed by the user due to other considerations (dedicated applications, specific outcomes, etc.). Inside this frequency band, the modal parameters need to be consistent with the theoretical representation of FRFs (e.g. Equation 3.28). In the lower- and higher frequency ranges, residual terms can be included to account for modes in these ranges. In this case, Equation 3.28 can be rewritten as:

$$H_{ij}(\omega) = \sum_{k}^{N} \left(\frac{r_{ijk}}{j\omega - \lambda_k} + \frac{r_{ijk}^{*}}{j\omega - \lambda_k^{*}} \right) + R_{ij} \tag{3.34}$$

where R_{ij} is the residue, compensating the effect of lower and upper modes outside the frequency band studied, and N the number of modes considered in the same frequency band $N \leq n$.

3.2.3.2 The Interest of a Modal Basis

By transforming ordinary differential equations governing the dynamic behavior of mechanical system into algebraic (and complex) equations, the modal analysis gives the users new insights in terms of analysis of mechanical properties. The knowledge of the SDOF system theory is sufficient to take better account of any complex mechanical system such as those related to the human body. Modal models are not only useful for analyzing experimental data (as explained in the next section) but also invaluable for simulation and design studies, as well as structural dynamic modifications (Avitabile, 2001).

Combined with the property of a SDOF system and the establishment of FRFs for complex mechanical systems, plenty of information (properties at the resonance) allows the definition of robust procedures to perform tests on structures and to evaluate the performance under extreme dynamic loading (Avitabile, 2017). Along with numerical simulation, the modal analysis completes the expected information by understanding how each of the modes contributes to the total response of the system, in addition to correlation and correction of these models.

3.3 NUMERICAL MODELING

In the case of human mechanical behavior, two kinds of numerical modeling are most often retained for solving problems of engineering and mathematical physics, regarding the nature of the motion. The purpose of this section is to present both of them in a brief way, by explaining how the equations of motion of the mechanical system are obtained. By making the link with the human response to vibration, it is possible, as an initial approach, to represent the human body (and all the required equipment) as a linear model, composed by lumped masses or flexible bodies interconnected by actuators (using spring and damper elements). An example, depicted in Figure 3.9, aims at determining the vertical response to a vibration excitation applied at the foot.

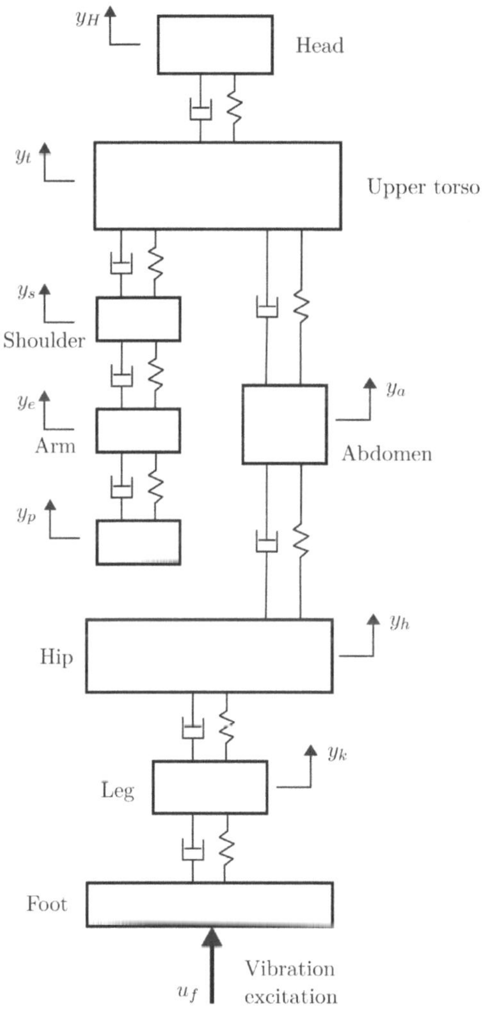

FIGURE 3.9 A linear model for representing the dynamic response of a (half and symmetrical) human body to vibration (u_i: displacement of each body i; H: head, t: upper torso, s: shoulder, e: arm, a: abdomen, p: hand, h: hip, k: leg, f: foot) (de Silva, 2006).

3.3.1 Multibody System Dynamics

Generally speaking, any physical system which can be seen as a set of rigid or flexible bodies, connected to each other by kinematic joints restraining their relative motion (revolute, prismatic, cylindrical, spherical, etc.) or force elements (springs, dampers, actuators, etc.) can be considered as a multibody system (MBS). The human body is one of the typical applications of the domain covered by MBS (Figure 3.10), although the main field of interest is vehicle dynamics (Popp and Schiehlen, 2010).

To create a multibody model, users classically implement generic multibody codes available in commercial software packages, which are nowadays widely used to assess the dynamic performance of mechanical systems. Let us mention also free software tools mainly developed for academic purposes which can be advantageously

FIGURE 3.10 An example of a multibody model related to human walking: bones are modelled as rigid bodies and soft tissues, cartilages, ligaments and muscles are considered as force elements.

used for education (Fisette and Samin, 2007; Lipinski et al., 2012) as well as for research purposes (Verlinden et al., 2013; Masarati and Morandini, 2008).

A multibody system simulation code (Figure 3.11) allows the assembly of simple elements like bodies (considered as rigid or flexible), joints and force elements in order to build the mathematical model of the mechanical system considered. Eventually, special elements can be foreseen to account for the peculiarities of each application: for example, dedicated controlled actuators mimicking the behavior of muscles in human modeling. The distinction between the multibody approach and an approach based on the assembly of lumped masses (mass—spring—damper) is sometimes not clear for the user, but we may slightly rephrase (Popp and Schiehlen, 2010):

> The elements of multibody systems include rigid bodies which may also degenerate to particles, coupling elements like springs, dampers or force-controlled actuators as well as ideal, i.e. inflexible, kinematical connecting elements like joints, bearings, rails and motion control led actuators. The coupling and connection elements, respectively, are generating internal forces and torques between the bodies of the system or external forces with respect to the environment and defined generally as non-linear. Both of them are considered as massless elements. Each body motion may undergo large translational and rotational displacements.

It clearly appears that one main particularity of the multibody approach is that large displacement is preferred to large deformation, without limiting the analysis to linear behaviors. This also makes the difference with the finite element method (see next section). In order to analyze a multibody system, the user must be able to express the spatial configuration of each body of the system. Indeed, once the

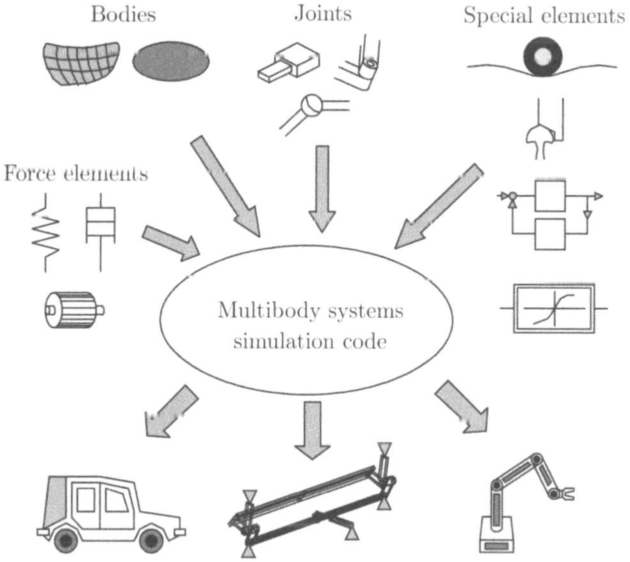

FIGURE 3.11 Principle of multibody simulation software.

position is known, velocities and accelerations can be obtained by derivation and all forces, including inertia forces, can be calculated, so that the equations of dynamic equilibrium can be written. Practically, the first thing to do when building the model of a multibody system is then to choose a set of configuration parameters $\underline{q}$ unambiguously defining the situation of each body. The nature of the configuration parameters essentially depends on the type of coordinates used to express the kinematics of the system. Among all the types of coordinates that can be enumerated, the Cartesian coordinates approach is often used in commercial software packages.

The approach based on Cartesian coordinates first considers independent bodies and assigns to each of them as many configuration parameters as needed to let it move freely in space, i.e. three or six parameters for planar or spatial case, respectively. All the joints are then systematically treated in the form of constraint equations. The configuration of a multibody system is therefore described by n Cartesian coordinates $\underline{q}$, and a set of one algebraic kinematic independent holonomic constraints $\underline{\lambda}$

$$\left[\mathrm{M}\left(\underline{q}\right)\right]\left\{\underline{\ddot{q}}\right\}+\left[\mathrm{B}^{\mathrm{T}}\left(\underline{q}\right)\right]\left\{\underline{\lambda}\right\}=\left\{\underline{\mathrm{f}}\left(\underline{q},\underline{\dot{q}},t\right)\right\}$$

$$\left[\mathrm{B}\left(\underline{q}\right)\right]\left\{\underline{\dot{q}}\right\}=0 \tag{3.35}$$

where:

- M is the mass matrix associated to the Cartesian approach
- B the Jacobian matrix of constraints
- $\underline{f}$ the vector of all the external forces acting on each body (excluding the joint forces). Equation 3.35 is by nature non-linear but a linearization around the static equilibrium configuration offers a way to consider a linear behavior defined by

$$[\mathrm{M}]\left\{\underline{\ddot{q}}\right\}+[\mathrm{CT}]\left\{\underline{\dot{q}}\right\}+[\mathrm{KT}]\left\{\underline{q}\right\}=\left\{\underline{f}\right\} \tag{3.36}$$

where M, CT and KT are the mass, tangent damping and tangent stiffness matrices, respectively, defined by

$$M_{ij}=\left.\frac{\partial R_i}{\partial \ddot{q}_j}\right|_0 \quad CT_{ij}=\left.\frac{\partial R_i}{\partial \dot{q}_j}\right|_0 \quad KT_{ij}=\left.\frac{\partial R_i}{\partial q_j}\right|_0 \tag{3.37}$$

with R_i the ith equation of motion derived from Equation 3.35 and defined in the form of a residual

$$\underline{R}\left(\underline{q},\underline{\dot{q}},\underline{\ddot{q}},t\right)=0 \tag{3.38}$$

3.3.2 Finite Element Analysis

The first application cases in finite element analysis come from structural analysis. The aim of this section is not to describe the finite element method (FEM) extensively, but to give an overview and remind its theoretical basis. Several books abound in the literature and can complete and describe the background of the method.

As the name suggests, the domain studied is divided into (very) small elements. In each of them, some interesting quantities, typically the displacement field $\underline{u}$, are approximated using form functions N and other selected parameters, such as the displacement values $\underline{a}^e$ at the nodes, are defined as

$$\underline{u} = \mathrm{N}\underline{a}^e \tag{3.39}$$

The form functions, often defined as linear, are selected in order to uniquely define the displacement field into each element as a function of the displacement at the selected nodes.

These functions allow calculating the deformation inside the element with respect to the sole nodal displacements. By taking into account the mechanical behavior of the material, these deformations, possibly associated to initial deformations, define the state of strain in any point of the elements, and, hence, on the border of the model.

Subsequently, a set of forces (or pressure) concentrated at the nodes are determined and balance the strain acting on the boundaries and the possible body and surface forces acting on the model. The solving process can therefore carry on a classical way by establishing equilibrium equations at the nodes and then by introducing the boundary conditions before solving the system of differential equations of the following form

$$[\mathrm{M}]\{\ddot{\underline{q}}\} + \{\underline{Q}(\underline{q}, \dot{\underline{q}}, t)\} = \{\underline{f}\} \tag{3.40}$$

where the degrees of freedom of all the nodes are defined in the vector $\underline{q}$ of the configuration parameters. Vector $\underline{f}$ represents all the forces acting on the nodes. When the system is defined as linear, which is an obvious condition in modal analysis, the system to solve becomes

$$[\mathrm{M}]\{\ddot{\underline{q}}\} + [\mathrm{C}]\{\dot{\underline{q}}\} + [\mathrm{K}]\{\underline{q}\} = \{\underline{f}\} \tag{3.41}$$

by introducing a mass matrix M, a stiffness matrix K and, when possible, a damping matrix C.

It is obvious that the aforementioned described process introduces a certain number of approximations. First, it is difficult for the form functions to satisfy the conditions of displacement continuity between elements. These compatibility conditions at the boundary could be eventually violated. On the other hand, by concentrating the external forces at the nodes (and creating equivalent forces), the equilibrium conditions are globally satisfied. Local violations of the equilibrium condition could occur inside the elements and at their boundaries. For each particular case, the selection of the form functions and of the element types is left to the user: the accuracy of this approximation strongly depends on these choices.

One of the difficulties of understanding the finite element method may lie in the mathematical formalism required by its implementation in practice, less intuitive that the aforementioned description. Indeed, taking into account the mathematical complexity of the models, it is necessary to build upon the results of functional analysis, elaborated to formulate this method of approximation. The variational method is a method that approximates the solution of partial differential equations built from a formulation equivalent to the problem to solve. Another general method exists, which is based on an approach introducing two weighted residuals:

- one residue function related to the differential equations representing the mechanical system studied

$$R(\underline{u}) = \mathcal{D}\underline{u} + \underline{f_V} = 0 \tag{3.42}$$

  with $\mathcal{D}(\underline{u})$ the contribution of internal and inertial forces and $\underline{f_V}$ the vector of external body forces,
- the other residue function is associated to the boundary conditions

$$B(\underline{u}) = \mathcal{R}\underline{u} - \underline{f_S} = 0 \tag{3.43}$$

  where $\mathcal{R}(\underline{u})$ represents the contribution of strain at the boundary and $\underline{f_S}$ the vector of contact surface forces.

The method involves both residues in an integral form where they appear as weighted:

$$W(\underline{u}) = \int_V \Psi^T R(\underline{u})\, dV + \int_S \Psi'^T B(\underline{u})\, dS = 0 \tag{3.44}$$

When the weighting functions Ψ^T and Ψ'^T are identical to the form functions, this method is called Galerkin's method, that often induces a symmetrical stiffness matrix K. This property can explain why Galerkin's method is mostly used in the finite element calculation. This method provides powerful numerical solutions to differential equations, as well as modal analysis.

3.4 ANALYSIS AND POST-PROCESSING OF EXPERIMENTAL DATA

The use of experiments presents undeniable advantages:

1. It allows the researcher to develop a modal model by using experimental data only.
2. It gives a rapid means of identifying the resonance frequencies and the associated parameters (damping and mode shapes).
3. It may be utilized to calibrate/update numerical (or theoretical) models.

By fixing sensors (typically accelerometers, see Chapter 1), the response of the mechanical system at these locations can be assessed. It can be pointed out that fixing the position of the sensors is a major concern. For example, it is judicious to place sensors at vibration antinodes (where the vibration amplitude of the studied mode is maximum) and to avoid fixing them at vibration nodes whenever possible, at the risk of missing important information for some modes. An extensive description of equipment selection and mounting of sensors can be found in a number of books and technical documents (e.g. Piersol and Paez [2009]). In addition, in the case of a numerical model calibration, the number of coordinates used in FRF measurement is usually far less than the number of degrees of freedom used in the numerical modal, implying an incomplete spatial description. This does not bring specific issues except for higher frequency modes where some kind of spatial aliasing could occur if the number of coordinates is not sufficient to describe the mode shapes.

The nature of excitation is of great importance and affects the choice of the experimental procedure:

- By exciting the mechanical system with a known force (using a calibrated impact—hammer test—or generating an excitation using a shaker) and measuring the force acting on it, a suitable set of FRFs can be obtained (see Chapter 1). The associated identification procedure is called experimental modal analysis (EMA) which identifies modal parameters from both measurements of the applied force and the vibration response.
- If the control and the measurement of the excitation is complex (or impossible), a complementary testing procedure, called operational modal analysis (OMA), yields estimates of the modal parameters from measurements of the vibration response only. This is based on the natural and freely available excitation due to ambient forces and operational loads.

Starting from these considerations, this section presents the philosophy behind both techniques. Special attention is devoted to the implementation of these techniques which follows similar steps:

- Choose the set of FRFs.
- Identify the possible modes in order to focus on a given frequency range.
- Apply an identification method able to determine natural frequencies, damping ratios and residues for various modes for each FRF.
- Compute mode shape vectors.
- Validate the results obtained.

3.4.1 Preliminary Identification of Possible Modes

From a practical consideration, the number of modes that can be determined and the corresponding frequency band are useful information that could prevent useless analysis and incorrect interpretation of FRFs. To this end, it is well admitted that the calculation of mode indicators, such as the summation function indicators or the mode indicator functions, are of interest. For instance, the SUM function

$$I_{\text{SUM}} = \sum_i \sum_j \left| H_{ij}(\omega) \right| \tag{3.45}$$

takes advantage of the property of FRF magnitude reaching a peak in the region of a mode of the system, in a single curve (instead of considering the visualization of all the FRFs considered). Using more complex interpretation, other characteristics, such as the model order for a set of representative data can be estimated. Techniques based on eigenvalue problems have been developed as guides to users. For example, the complex mode indication function (CMIF) is based on singular value decomposition (SVD) methods applied to multiple reference FRF measurements assembled in a single matrix $H(\omega)$, defined by

$$\left[H(\omega)\right] = \left[U(\omega)\right]\left[S(\omega)\right]\left[V(\omega)\right]^H \tag{3.46}$$

where:

- U and V are unitary singular matrices
- S a diagonal singular value matrix which provides the existence of modes and the corresponding frequencies
- $[\ ]^H$ the Hermitian (conjugate transpose) matrix operator

3.4.2 Experimental Modal Analysis (EMA)

EMA is a fundamental technique which has experienced growing interest and rapid development over the last 30 years. It embraces a wide range of disciplines and has been considered from the start as a useful tool to study and illustrate the dynamic characteristics of a structure, given its substantial scope of application (Ewins, 1986). It is now generally included in the process of structural design and modification, as illustrated in Figure 3.12. The term "experimental modal analysis" is therefore prevalent. It is defined as the study of the dynamic characteristics of a mechanical structure. The theory not only emphasizes modal identification techniques, but also the associated signal processing and the FRF testing (Ewins, 1991).

Many experimental modal analysis software packages have been developed (Maia, 1997). Initially composed by simple SDOF methods (peak picking/mode picking, circle-fit or line-fit), commercial software packages have been updated with new complex methods, allowing rapid and efficient analyses. Indeed, simple methods

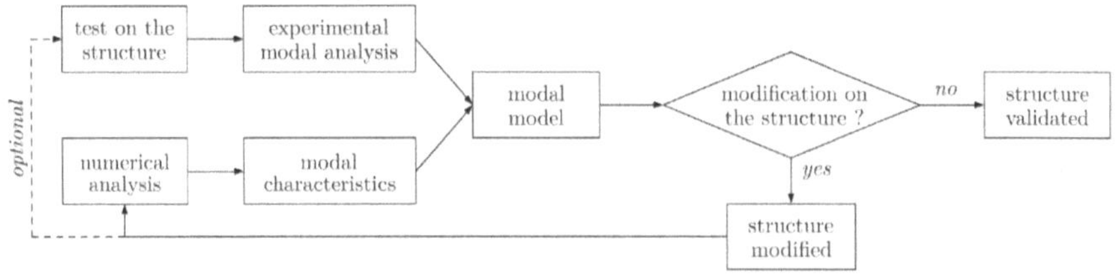

FIGURE 3.12 Scheme of the structural dynamics modification process.

have been discarded in favor of MDOF ones. Commercial tools are developed with the concerns of calculation power, user-friendliness and efficiency but represent a human and financial investment that the industry cannot afford if the application is not inside its core business. Nonetheless, these simple methods are very interesting for the understanding of EMA and allow a progressive approach in the modal analysis fundamentals teaching. Afterwards, the understanding of complex methods, like rational fraction polynomial method, Ibrahim time domain method (ITD), random decrement technique, ARMA time series method, least-square complex exponential (LSCE) method or PolyMax estimator (Guillaume et al., 2003), which are considered as "industry standard" estimation methods, is facilitated, by pointing out their main advantages. To present the EMA techniques, a classification approach is thus preferred in this chapter, from simple to recent algorithms, based on some important main properties of identification methods.

3.4.2.1 Time versus Frequency Domain

Historically, the identification was only performed in the frequency domain: the use of FRFs and their graphical properties motivated the development of methods directly using frequency data. Afterwards, time domain methods were developed presenting some interest for lightly damped structures and systems. Indeed, with a limited frequency resolution (imposed by the experimental test), frequency domain techniques were limited to a few points around the resonance implying a non-negligible variation in estimating the modal parameters. By transforming an FRF into an impulse response function (IRF, see Equation 3.17), a larger set of discrete points could be available for an accurate description of the lightly damped modes. Figure 3.13 shows this property where, although discrete data covers the resonance damping in the frequency domain (the half energy of a resonance magnitude sparse lies within a range of 3 dB!), the time decay—reflecting damping—clearly appears in the time domain. The contrary also holds: highly damped modes are more visible in the frequency domain (with discrete sample of data). The first classification that can be done is therefore based on the use of FRF (frequency domain techniques) or IRF (time domain techniques).

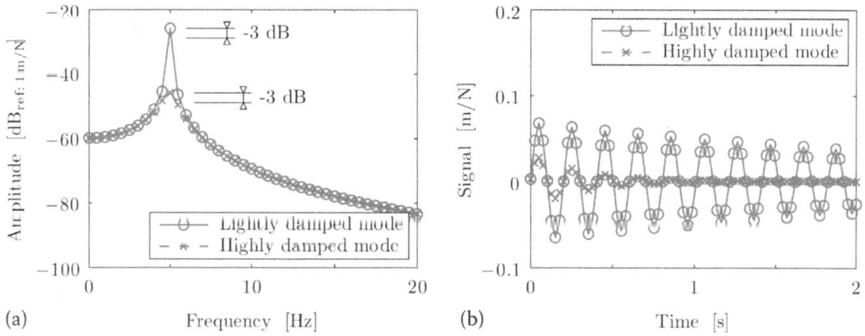

(a) (b)

FIGURE 3.13 Illustration of the interest of time and frequency domain methods for low and high damping: (a) Bode representation of FRFs and (b) time history of IRFs.

3.4.2.2 SDOF versus MDOF Methods

SDOF methods assume that, in a given frequency band, only one mode dominates the vibration response and the parameters of this mode can be determined solely. This assumes that the effect of neighboring modes can be neglected (and degenerated to residuals in the FRF). This assumption forms the basis of simple techniques such as peak picking, mode picking, circle-fit or line-fit methods.

On the contrary, MDOF methods are free from this decoupling approach and can be used in a model that includes several modes in the studied frequency band. They are however more complex than SDOF methods and require important calculation efforts.

3.4.2.3 SISO versus MIMO Methods

In SISO (single input/single output) methods, each data record (FRF or IRF) is analyzed individually, and a potential different estimation of modal parameters (natural frequency and damping ratio) is found each time. While the modal constant is unique for each data record, many estimates of natural frequency and damping ratio of a mode are therefore produced by such techniques and it is left to the user to decide which value is the best or to somehow calculate the best average of all the estimates.

In MIMO (multi input/multi output) methods, all the data records are analyzed simultaneously in order to globally estimate the modal characteristics of the system. Some methods however make the distinction if one reference is used or one response is recorded with multiple excitations: they are called single input/multiple output (SIMO) or multiple input/single output (MISO). The distinction between them is not important due to Betti–Maxwell reciprocity theorem (excitation can be seen as a response and inversely). Technically speaking, the distinction between SIMO and MIMO methods relies on the number of signals used to excite the system under test. LSCE method is probably the most powerful and well-known SIMO method.

An extension of the LSCE method to a MIMO version is the polyreference complex exponential method (PRCE). It includes information not only from several output locations, but also from several input reference points on the mechanical system. Apart from being a more general and automatic way of analyzing a structure dynamically, this overcomes the problem that may occur when using a SIMO method, where one mode of vibration is not excited due to the excitation being located close to a node of the structure. In addition to providing a more accurate modal representation of the test structure, this method can determine multiple roots or closely spaced modes of a structure. The time required for the analysis is reduced and the accuracy in the results increased.

3.4.3 Operational Modal Analysis (OMA)

Most OMA techniques are derived from traditional EMA modal identification procedures but refer to a different mathematical framework. For example, traditional processing of operational data is based on a peak-picking technique applied to the auto- and cross-powers of the operational responses. It is therefore important to consider a proper set of data, while defining the same reference for each measurement location at the risk of missing the phase between each response.

Over recent years, several modal parameter estimation techniques have been proposed and studied for modal parameter extraction from output-only data. Examples include auto-regressive moving averaging models (ARMA), natural excitation technique (NExT) or stochastic subspace methods. For example, the underlying principle of the NExT technique is that correlation functions between the responses can be expressed as a sum of decaying sinusoids. Each decaying sinusoid has a damped natural frequency and damping ratio that is identical to the one of the corresponding structural mode. Consequently, conventional modal parameter techniques such as PRCE can be used for output-only system identification.

Widely used in civil structure due to the uncertainty or ignorance regarding excitation, OMA techniques become interesting in extracting a relevant model from operating data. This is why major references to OMA are available in such discipline (e.g. Rainieri and Fabbrocino [2014]).

3.4.4 On the Use of EasyMod

As an alternative to available commercial software packages, free tools and computer programs are proposed to understand the modal analysis. Although commercial tools provide a support for illustrating these mathematical concepts, they often mask a number of features which are important for a good understanding. It is therefore difficult to estimate the importance of measurement methods, the choice of a given frequency range or the selection of relevant coordinates. For example, the in-house MATLAB® toolbox EasyMod fills this gap and offers a robust platform to establish a complete modal analysis step by step from experimental data (Kouroussis et al., 2012a,b). It consists in several functions developed for various applications in structural dynamics:

- reading and writing files in a universal format readable by many other commercial software solutions (UFF—universal file format),
- generating FRFs from mass M, stiffness K, damping C matrices defined by the user,
- mode indicators (sum of FRFs I_{SUM}, sum of FRFs real part $I_{S\,Re}$ and sum of FRFs imaginary part $I_{S\,Im}$) and their visualization,
- modal identification methods (SDOF and MDOF methods, SISO and SIMO methods, time or frequency domain analysis),
- validation (modal assurance criterion and modal collinearity for a comparison of two sets of analysis).

The functioning and the possibilities of EasyMod are illustrated in Figure 3.14. The toolbox is freely available from the following website: http://mecara.fpms.ac.be/EasyMod.

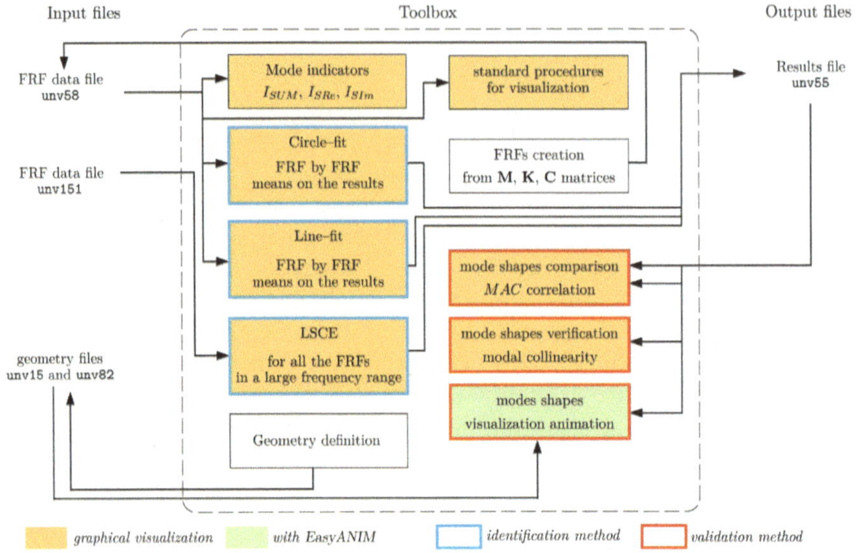

FIGURE 3.14 Schematic operating diagram of toolbox EasyMod (Kouroussis et al., 2012b).

3.4.5 Examples

The human body is most often in interaction with other structures and the knowledge of the dynamic characteristics of such equipment remains important. Two illustrative examples are retained to show the interest and the applicability of modal analysis.

Figure 3.15 displays an example of correlation between a numerical model, here using the finite element approach, with experimental data, based on an alpine ski. Measurement has been made according to ISO6267:1980 standards (International Organization for Standardization, 1980) related to measurement of bending vibrations on this kind of equipment. To perform statistical comparisons between any two mode shapes from numerical and experimental analyses, the correlation based on the natural frequency may not be sufficient and could induce important errors on results interpretation. To prevent it, the correspondence between mode shapes is necessary. The use of the modal assurance criterion (MAC), defined by a matrix of elements given by

$$\mathrm{MAC}_{ij} = \frac{\left(\left\{\underline{\psi}_i^{\text{meas.}}\right\}^T \left\{\underline{\psi}_j^{\text{sim.}}\right\}^*\right)^2}{\left\|\underline{\psi}_i^{\text{meas.}}\right\|^2 \left\|\underline{\psi}_j^{\text{sim.}}\right\|^2} \tag{3.47}$$

allows direct comparison between experimental mode shapes $\underline{\psi}_i^{\text{meas.}}$ and numerical counterparts $\underline{\psi}_j^{\text{sim.}}$. MAC value is comprised between 0 and 1. If the coefficient is

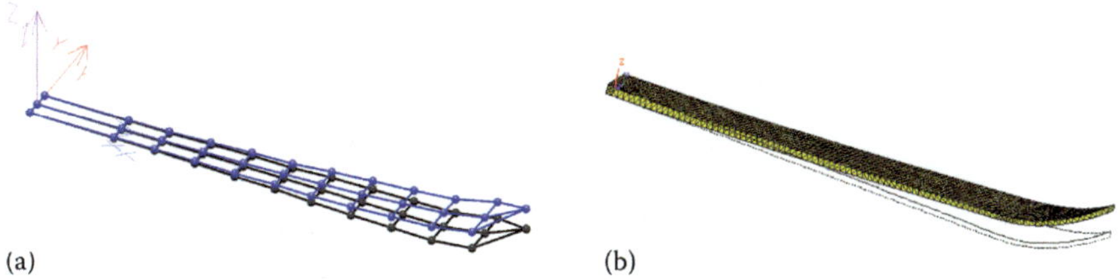

(a) (b)

FIGURE 3.15 First bending mode of an alpine ski at 10.8Hz: (a) Experimental mode shape, (b) Numerical mode shape.

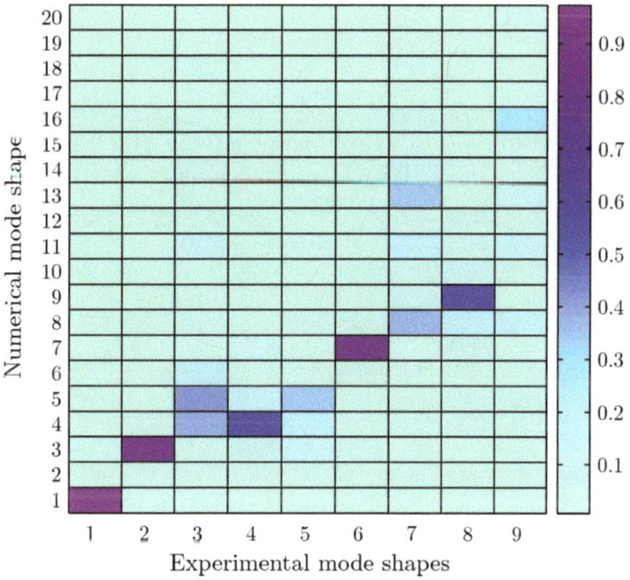

FIGURE 3.16 Modal assurance criterion applied on alpine ski analysis.

equal to unity, then the two shapes are perfectly correlated. In practice, any value between 0.8 and 1 is considered as a good correlation. This good correlation is retained for some mode shapes by visualizing the MAC matrix (e.g. Figure 3.16).

The second example illustrates the interest of OMA by evaluating the rider's comfort during cycling through the vibrational response of the bicycle frame. A number of works demonstrated that classical EMA identification methods fail in the presence of a rider, while OMA techniques are able to take such an external condition into account, such as the activation of some muscles to ensure the grip and the stability of the bike. Figure 3.17 shows a typical test performed on a bike. This offers a way to evaluate the effects of cycling speed and mass of the subject on the operational modal analysis. Detailed findings are available in (Chiementin et al., 2016).

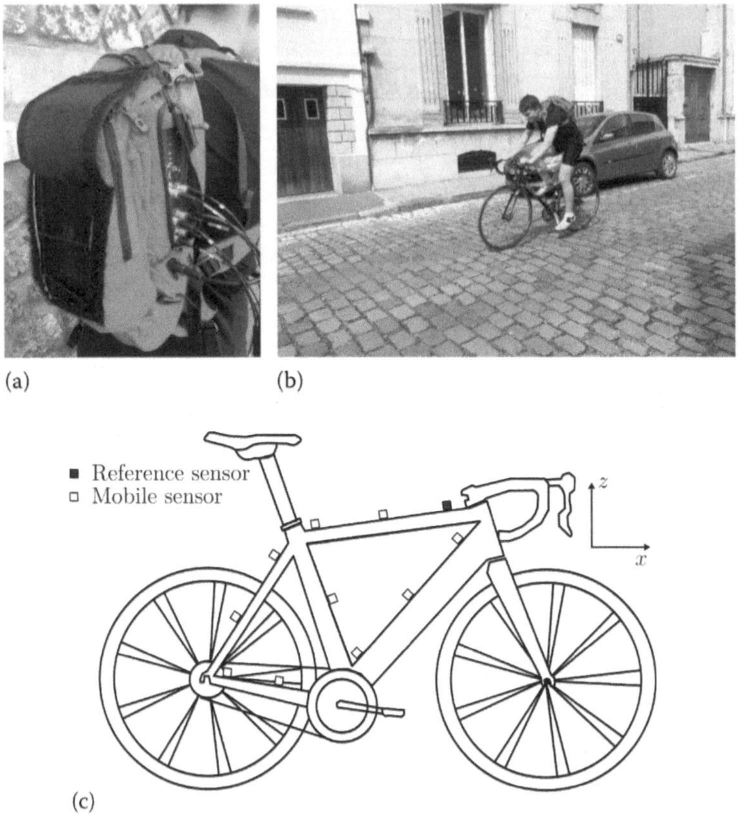

FIGURE 3.17 Operational modal analysis during cycling: (a) backpack with the data collector and tablet computer, (b) test on cobblestone road and (c) accelerometer positions on the bicycle frame (Crequy et al., 2017).

3.5 CONCLUSION AND GENERAL REMARKS

The theory of mechanical vibrations described in this chapter could be applied in the case of the human body, and the tools presented could help the user (industrials and academics) in the analysis of the dynamic response of the human body. Having in mind that the modal basis is a powerful theory to diagnose the mechanical behavior of any kind of physical system, and more particularly the human body, the knowledge of SDOF model theory allows the user to insist on important issues. The multibody system or finite element approaches can be then used to establish a model of the human body. Experimental and operational modal analysis are finally powerful tools to identify the dynamic properties of the human body.

Both experimental and numerical approaches are complementary. The best is to have the possibility to have an available numerical model, calibrated and completed by experimental data, in order to use it as a virtual prototype to assess the mechanical characteristics of the human body model devoted for the studied application. Such analysis offers a way to systematically control the vibration induced in human bodies without making fastidious and complicated measurement. However, in practice, either the lack of measurement or of model data restricts this unified topic and it is left to the user to properly exploit these existing techniques in a responsible manner, knowing *a priori* the limitations of each approach.

REFERENCES

Avitabile, P. (2001). Experimental modal analysis – A simple non-mathematical presentation. *Sound & Vibration*, 1:1–11.

Avitabile, P. (2017). Beware of these top-10 issues in modal testing. *Sound & Vibration*, 1:48–52.

Carfagni, M., Lenzi, E., and Pierini, M. (1998). The loss factor as a measure of mechanical damping. *IMAC XVI, 16th International Modal Analysis Conference*, Santa Barbara, California.

Chiementin, X., Crequy, S., and Kouroussis, G. (2016). Vibration evaluation in cycling using operational modal analysis. *Proceedings of the 23rd International Congress on Sound and Vibration*, Athens.

de Silva, C. (2006). *Vibration: Fundamentals and Practice*. CRC Press, New York. 2nd edition.

Ewins, D. (1986). Modal analysis as a tool for studying structural vibration. In Braun, S., ed., *Mechanical Signature Analysis*. Academic Press, London.

Ewins, D. (1991). *Modal Testing: Theory and Practice*. John Wiley & Sons, Great Yarmouth (UK).

Fisette, P. and Samin, J. C. (2007). Engineering education in multibody dynamics. In Orden, J. C. G., Goicolea, J. M., and Cuadrado, J. eds., *Multibody Dynamics*, 4:159–178. Springer, Netherlands.

Fu, Z. and He, J. (2001). *Modal Analysis*. Butterworth-Heinemann, Oxford.

Guillaume, P., et al. (2003). A polyreference implementation of the least-squares complex frequency-domain estimator. *IMAC 21, The International Modal Analysis Conference*, Kissimmee (USA).

International Standards Organization (1980). Alpine skis — Measurement of bending vibrations, ISO 6267. Geneva.

Kouroussis, G., et al. (2012a). EasyMod: du développement d'un toolbox sous MatLab vers l'enseignement des bases de l'analyse modale expérimentale. *3ième Colloque "Analyse vibratoire Expérimentale*". Blois, France.

Kouroussis, G., et al. (2012b). EasyMod: A MatLab/SciLab toolbox for teaching modal analysis. *Proceedings of the 19th International Congress on Sound and Vibration*. Vilnius, Lithuania.

Lang, G. (2016). How heavy are your mode shapes? *Sound & Vibration*, 11:4–5.

Lipinski, K., et al. (2012). Mechanical engineering education via projects in multibody dynamics. *Computer Applications in Engineering Education*, 20:529–539

Maia, N. (1997). *Theoretical and Experimental Modal Analysis*. John Wiley & Sons, Exeter (UK).

Masarati, P. and Morandini, M. (2008). An ideal homokinetic joint formulation for general-purpose multibody real-time simulation. *Multibody System Dynamics*, 20:251–270.

Piersol, A. and Paez, T. L. (2009). *Harris' Shock and Vibration Handbook*. McGraw-Hill, New York. 6th edition.

Popp, K. and Schiehlen, W. (2010). *Ground Vehicle Dynamics*. Springer, Berlin.
Rainieri, C. and Fabbrocino, G. (2014). *Operational Modal Analysis of Civil Engineering Structures — An Introduction and Guide for Applications*. Springer, New York.
Verlinden, O., Fekih, L., and Kouroussis, G. (2013). Symbolic generation of the kinematics of multibody systems in EasyDyn: From MuPAD to Xcas/Giac. *Theoretical & Applied Mechanics Letters*, 3(1):013012.

4 Effect of Mechanical Vibration on Performance

Pedro J. Marín, Marcela Múnera, Maria Teresa García-Gutiérrez, and Matthew R. Rhea

CONTENTS

4.1 INTRODUCTION

Vibration training has been investigated throughout the last decade as an alternative or complementary mode of training to traditional resistance programs for fitness improvements. Whole body vibration (WBV) as a mode of exercise training has become increasingly more popular in health, physical therapy, rehabilitation, professional sports, and wellness applications because of its effects on the neuromusculoskeletal system (Cochrane, 2011). The main benefits of WBV training are increased maximal power output (Cochrane, 2011; Marín and Rhea, 2010a), strength gains (Marín and Rhea, 2010b; Marín et al., 2012), and increased muscle activity as evaluated with surface electromyography (EMG) (Cardinale and Lim, 2003; Hazell et al., 2010; Pollock et al., 2012).

WBV has been the most often studied method of vibration training. Typically, whole body vibration exercise involves intermittent exposures to a vibration stimulus while performing traditional body weight exercises such as squats or lunges on a ground-based platform. As stated, WBV evokes improvements in muscular strength and power attainable in a short period of time (Adams et al., 2009). Numerous investigators have studied performance effects that vibratory platform exposure produce on the lower body. This exposure has been shown to result in improvements in jump height (Ronnestad, 2009), improved sprint performance (Delecluse et al., 2005), metabolic (Rittweger et al., 2002), and hormonal changes (Bosco et al., 2000), and neuromuscular performance, both acute (Mileva et al., 2006) and chronic (Verschueren et al., 2003).

Exercises performed on ground-based WBV platforms have demonstrated significant increases in lower body muscle activity compared to the same exercise with no WBV (Cardinale and Lim, 2003; Hazell et al., 2010; Abercromby et al., 2007a; Marín et al., 2009). The magnitude of this response depends on the amplitude (size of each deflection, measured in mm) and the frequency (number of deflections per second, measured in Hz) of the stimulus (Ritzmann et al., 2010, 2013). Higher vibration frequencies and amplitudes induce greater muscle activity than lower frequencies and/or amplitudes during a regular isometric squat (Ritzmann et al., 2013; Hazell et al., 2007; Marín et al., 2009). Furthermore, the combination of high frequency and amplitude vibration has been investigated by several studies which demonstrated that higher frequencies of vibration on a platform combined with higher amplitudes result in even greater muscle activity (Hazell et al., 2007, 2010; Ronnestad, 2009). Besides, WBV using a platform as the stimuli source has been used as a laboratory method to study the negative effects of vibration in sports that have an intrinsic presence of vibrations. Endurance sports like cycling show negative performance effects of vibration, produced by the uneven surface on the road (Chiementin et al., 2013; Lépine et al., 2012; Srinivasan and Balasubramanian, 2007), suggesting higher muscle activity than would be needed in the absence of such vibrations.

These facts suggest that there is an increased muscle activity due to vibration stimulus. Because of increasing popularity, much research has been performed to understand both if WBV exercise is beneficial for users and, if so, why WBV achieves these benefits.

The observed strength improvements, in the first few weeks of a training program, have been attributed to neural performance aspects, because changes in the morphology, architecture, and size of muscle tissue occur at a later stage (Goodwill and Kidgell, 2012; Preatoni et al., 2012). EMG activity increases in parallel with the levels of force used for training and reaches maximum with loads near maximal voluntary contraction (Preatoni et al., 2012). Vibration training increases EMG activity on lower body (Abercromby et al., 2007b; Hazell et al., 2007) as well as on upper body muscles (Moran et al., 2007; Bosco et al., 1999a) during exercise.

Several theories have been proposed on how vibration stimuli can have such effects on the neuromuscular system (Cardinale et al., 2003; Ronnestad, 2004), such as a stimulation homonymous a-motor neurons and/or perturbation of the gravitational field during the time course of intervention (Cardinale and Bosco, 2003) (see Figure 4.1).

This early theory suggested WBV exercise improved neuromuscular function (Cochrane, 2011; Cardinale and Wakeling, 2005; Dolny and Reyes, 2008; Issurin, 2005), though whether this includes increases in muscle strength, remains equivocal (Cochrane, 2011; Rittweger, 2010). However, WBV exercise has demonstrated acute increases in strength following a WBV exposure suggesting a degree of neuromuscular potentiation (Armstrong et al., 2008; Bosco et al., 1999b; Cochrane and Stannard, 2005). Performing exercises on a synchronously (uniform vertical vibrations) oscillating platform induces short and rapid changes in muscle fiber length stimulating

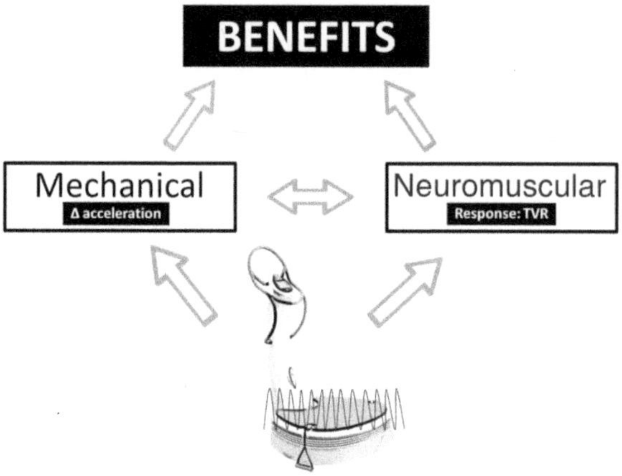

FIGURE 4.1 Schematic diagram illustrating the potential mechanisms of the vibration training.

reflexive muscle contractions in a response akin to monosynaptic reflexes (Pollock et al., 2012; Ritzmann et al., 2010; Hagbarth and Eklund, 1966). This response results in an increase in muscle activity (Hazell et al., 2010; Abercromby et al., 2007a; Marín et al., 2009), enhanced excitability of the cortical motor pathway, as well as modulation of intracortical circuits (Mileva et al., 2009). The mechanical vibration stimulus may also affect skin and joint receptors that provide sensory input to the gamma motor system increasing the sensitivity and responsiveness of the muscle spindle to further mechanical perturbations (Cardinale and Bosco, 2003; Eklund and Hagbarth, 1966).

Enhanced skeletal muscle activity increases the metabolic demand of activity as evidenced by increases in oxygen consumption/energy expenditure due to the WBV exposure (Rittweger et al., 2002; Da Silva et al., 2007; Hazell and Lemon, 2012; Rittweger et al., 2001), though the magnitude of this increased demand is much less than traditional resistance training (Hazell et al., 2010). Nevertheless, the potential for the WBV stimulus to augment different exercises being performed on the WBV platform is intriguing, as many researchers and health professionals have sought methods to further increase the training stimulus using WBV platforms.

In addition, the magnitude of the WBV-induced increase in muscle activity during a dynamic squat was demonstrated to be either the same or increased with the addition of an external load (~30% of body mass) compared to an unloaded condition with WBV (Hazell et al., 2010; Ritzmann et al., 2010). These results demonstrate the associated increase in skeletal muscle activity with exposure to WBV is not dissipated with the addition of an external load suggesting the potential effectiveness of using external loads with WBV exercise. However, the addition of external loads to increase the intensity of effort is not always feasible in terms of interest, accessibility, or practicality (i.e. rehabilitation from injuries, cardiovascular diseases, older adults). One possible solution could be the use of unstable surfaces. That form of instability training has become popular (Saeterbakken and Fimland,

2013), and the benefits are that unstable surfaces can induce similar levels of muscle activation while using less external load (Behm et al., 2002; Anderson and Behm, 2005; McBride et al., 2006).

4.2 VIBRATION DEVICES

The vibration stimulus can be applied locally by two main ways: (i) directly to the muscle belly (Jackson and Turner, 2003) or the tendon (Moran et al., 2007; Luo et al., 2005), or (ii) indirectly applied by gripping a vibration system (Humphries et al., 2004), dumbbell (Cochrane and Hawke, 2007), bar (Poston et al., 2007), or pulley system (Kaeding et al., 2017). Vibration stimulus that is applied via the feet while standing on a vibration platform (Kaeding et al., 2017) is known as whole body vibration. Recently, research has shown that a specific wearable vibration device increased elbow flexor peak power in master athletes (Cochrane, 2016). Such a device that is battery powered, portable, and can be self-applied provides an advantage to current vibration training equipment. In addition, other research has shown that, in the short-term, a specific wearable vibration device was able to significantly reduce the level of biceps brachii pain at 24 hrs and 72 hrs, enhance pain threshold at 48 hrs and 72 hrs, improve range of motion at 24 hrs, 48 hrs, and 72 hrs, and significantly reduced creatine kinase at 72 hrs compared to control (Cochrane, 2017) (Figure 4.2).

4.3 VIBRATION AND SPORTS

Negative effects of vibration can appear under uncontrolled conditions in work and at sports. Those negative effects could be physiological, pathological, psychological, and biodynamic, and can be classified as acute and/or chronic (Griffin, 1990). Among those negative effects include musculoskeletal disorders, such as tendinitis (Chetter et al., 1998), muscular strength decrease (Gurram, 1993), joint wear (Boshuizen et al., 1990; Carlsöö, 1982), fatigue increase, and pain (Hill et al., 2009). These disorders have a direct relation to health, comfort, and performance in sports; and depend on characteristics of the vibration source, but also in its transmission through mechanical structures or the human body.

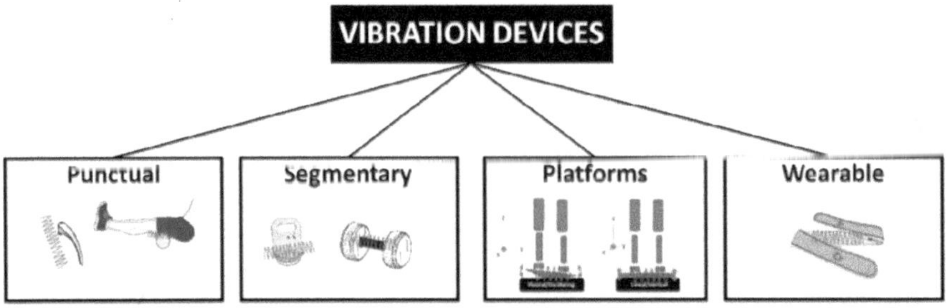

FIGURE 4.2 The vibration devices marketed by several producers and often used in clinical studies.

In sports, vibration presence is less studied than in the industrial field, even if this presence is not negligible (Munera et al., 2015a). Vibration could be caused as a reaction to shock in sports such as golf (Roberts et al., 2005; Osis and Stefanyshyn, 2010), tennis (Stroede et al., 1999), and table tennis (Manin et al., 2012); or as an external excitation in sports like cycling (Chiementin et al., 2013). There are also vibrations in sports like sailing, surfing, downhill skiing, and horse riding (Mester et al., 1999).

When the vibration is caused by a shock, studies focus primarily on the measurement of this vibration at the entrance of the hand–arm system and on the development of systems to decrease this transmission or modify its frequency.

In cycling, research has been carried out on a "transmissibility phenomenon" (Chiementin et al., 2013; Munera, 2014), the physiological response to vibrations (Munera et al., 2015b; Sperlich and Kleinoeder, 2009) and the muscular activation response measured by EMG (Srinivasan and Balasubramanian, 2007; Munera et al., 2016). Aiming to decrease the vibration that enters the human body in cycling practice, some studies search to measure the vibration taking into account the type of bicycle (Parkin and Sainte Cluque, 2014; Peretti et al., 2009), its structural characteristics (Giubilato and Petrone, 2012; Olieman et al., 2012), the surface rough, the speed (Giubilato and Petrone, 2012), the frequencies of the vibration exposure (Chiementin et al., 2013; Munera et al., 2015a), and its amplitude (Munera et al., 2015b).

Nonetheless, the human–machine interaction behavior in those sports under the presence of vibration depends on a great number of factors in the equipment and the human body, and in most cases has a non-linear response. For that reason, some of the studies found contradictory results and there is not a general group of parameters to decrease the vibration exposure and its effects.

4.4 VIBRATION AND MOTION

Existing studies about the assessment of vibrations in sports show different kinds of measures and interpretations of its effects based on dynamic and physiological response of the human body to vibrations. However, most of those studies are conducted in static conditions.

Motion is a fundamental parameter in sports practice and it can affect the vibration transmission through the human body, as well as its eventual effects. If vibration training is performed during a moving exercise, it can be assumed that there is an influence of the motion on the vibratory response. This influence can have an impact on the value of the transmissibility of the vibrations through the joints, as well as on the muscular activation of the lower limb. This hypothesis is established through two observations. The first is based on the dependence between the position on the platform and the transmissibility (Kiiski et al., 2008; Harazin and Grzesik, 1998; Crewther et al., 2004; Avelar et al., 2013). The second is that motion is a continuous change of position causing variations in the rigidity and absorption of energy in the segments involved.

Cycling is an example of this coupling between the motion of the rider and the vibration produced by the road. Although pedaling is limited by the circular trajectory of the pedal, it is not a simple motion. In cycling, the human body works in association with the bicycle to produce a displacement from a rotational motion of the lower limbs of the cyclist (Grappe, 2009).

To study the influence of motion on the vibration transmission, two studies have been developed on the lower limb. In the first study, the evaluation of the influence of motion on the response to vibrations is made for a simple motion and without the interaction of a structure (Munera et al., 2016). In the second study, the vibration and motion coupling is studied during a cycling exercise, where the interaction with the bike will be an important parameter (Munera, 2014).

For the first study, the chosen motion was a half-squat movement. The purpose was to assess the transmissibility of vibration and muscle activity in the lower limb. For this, 15 healthy males were exposed to vertical sinusoidal vibration at different frequencies (20–60 Hz) over a vibrating platform, while performing a dynamic half-squat exercise (Figure 4.3).

The measurements in the dynamic conditions allow for the computing of mean transmissibility in several parts of the movement at the ankle (Figure 4.4a), the knee (Figure 4.4b), and the hip (Figure 4.4c).

The results showed statistically significant differences ($P<0.05$) in the transmissibility values caused by the frequency, the position, and to the presence of the movement and its direction at the different conditions. Furthermore, higher exposure for each joint was not always at the same angle. This fact suggests that one position

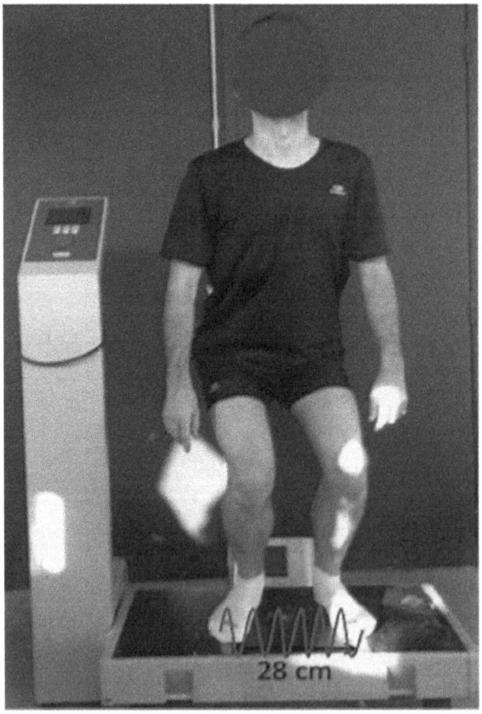

FIGURE 4.3 Experimental set-up for a half-squat movement.

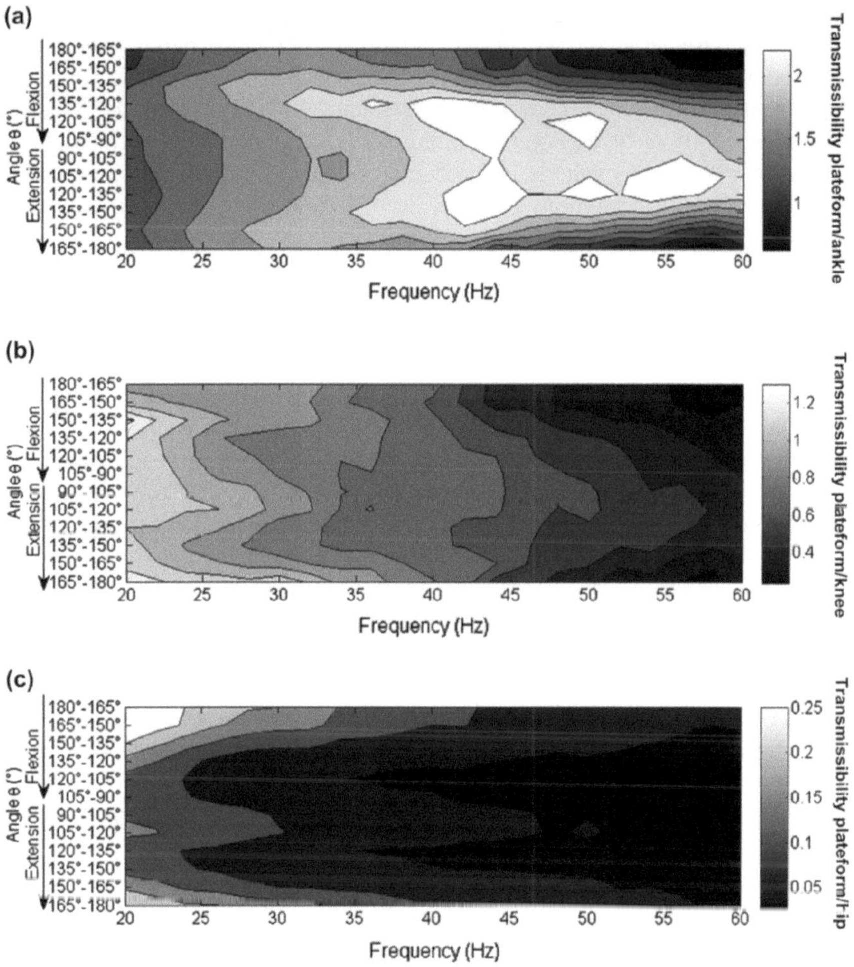

FIGURE 4.1 Transmissibility curve of vibration in dynamic conditions for the ankle (a), knee (b), and hip (c).

could have lower exposure for a specific joint, but a higher exposure for a different joint. Besides, the frequency effects on the transmissibility behavior were not the same for the three joints. The most important original aspect of this work is the consideration of the movement for the transmissibility assessments. The results showed that motion and its direction have an influence on the vibration dose experienced by each joint; this could be due to the movements of flexion and extension that represent an active lengthening and shortening of the different muscles of the lower limb. These movements could change the dynamic properties of the musculature, affecting the vibration transmission. The method proposed could be integrated for comfort and safety estimations in different sports where motion and vibration have important presence as cycling, rolling, or skating.

For the second study, 15 participants performed eight, six-minute, sub-maximal pedaling exercises at a constant power output (150 W) and pedaling cadence (80 RPM) being exposed to vibration at different frequencies (20, 30, 40, 50, 60, 70 Hz) or without vibration (Figure 4.5). Oxygen uptake (VO2), heart rate (HR), surface EMG activity of seven lower limb muscles (gluteus maximus, rectus femoris, biceps femoris, vastus medialis, gastrocnemius, soleus, and tibialis anterior) and three-dimensional accelerations at ankle, knee, and hip were measured during the exercises.

To analyse the dynamic response, the influence of the pedaling motion was taken into account. The results show that there was not a significant influence of vibrations on HR and VO2 during this pedaling exercise. However, muscular activity significantly increased with the presence of vibration with the effect being influenced by frequency. In addition, the dynamic response showed an influence of the frequency as well as an influence of the different parts of the pedaling cycle (Figure 4.6). Those results could explain the effects of vibration on the human body and the influence of the rider/bike interaction in those effects.

An influence of the different stages of the pedaling cycle on the measure of acceleration at the pedal and joints was evidenced for all frequencies, confirming the hypothesis that movement has an influence on the vibration transmission. Acceleration measures at the pedal represent the dynamic response of the bike. This response confirms the hypothesis of the influence of two parameters: the frequency and the pedaling movement. The movement not only influences the biological structures such as the lower limb, but also the mechanical structure, the bike. This confirms that movement is an important parameter to take into account in the dynamic response of the bike, especially for the evaluation of the dose of vibration in the pedals.

FIGURE 4.5 Pedaling exercises under vibration exposure.

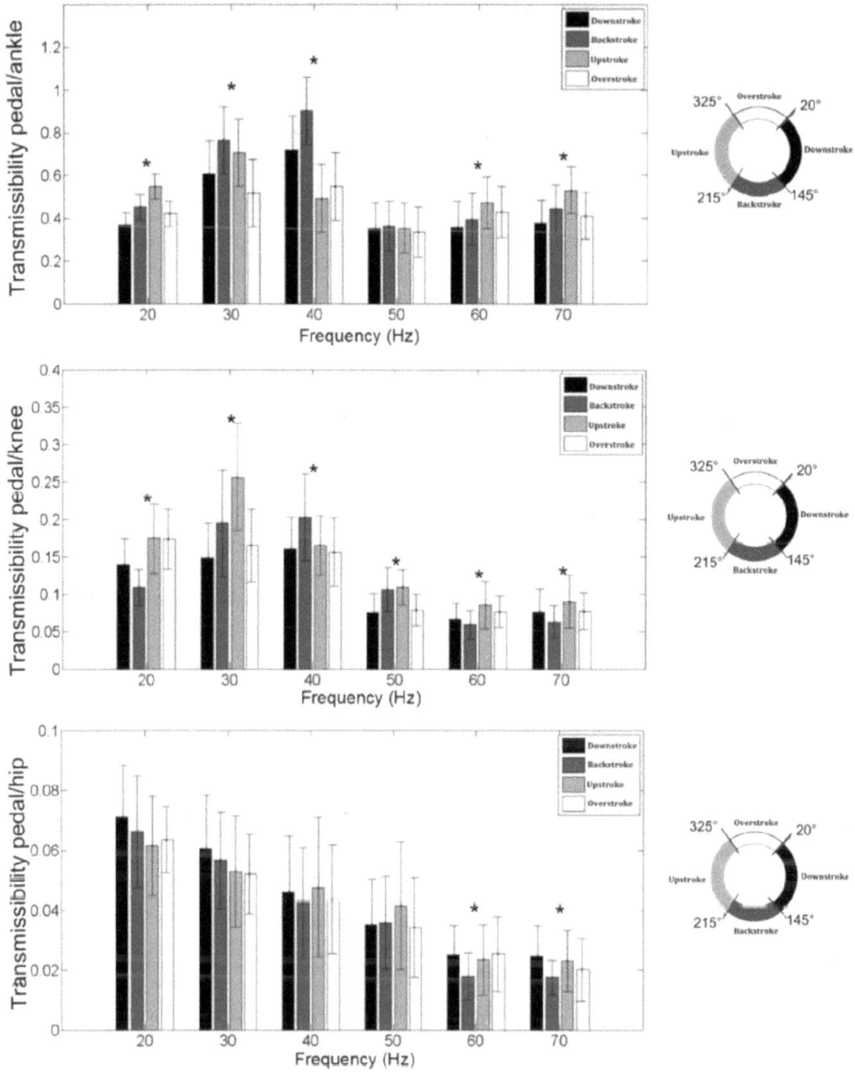

FIGURE 4.6 Transmissibility pedal/joint for the ankle, knee and hip. (*) Significant difference between the parts of the pedaling cycle ($P<0.05$).

The transmissibility values computed for the ankle, knee, and hip joints have decreased when measured at a more proximal position in the lower limb. These results agree with several studies on whole body vibration that show that the transmissibility is lower at a more proximal localization in the lower limb (Kiiski et al., 2008; Harazin and Grzesik, 1998; Bressel et al., 2010). This indicates that the damping properties of tissues are also preserved when pedaling at a rider/bike interaction. However, transmissibility values are lower than those measured in the half-squat study. This decrease is related to the fact that the member is in contact with the structure and this can absorb some of the vibrations during the exercise. The vibration is dampened and modified by means of the pedal–foot interface. This proves that the structure (bicycle) has a major influence on the vibration propagation. Thus, the structure reduces the probability of risk.

Some studies showed that higher acceleration values could produce higher risks of joints injury (Carlsöö, 1982; Griffin, 1990). In contrast, there are studies that reported potential benefits of the vibration training for athletes (Delecluse et al., 2005; Marín et al., 2011). Nevertheless, the connection between potential benefits and possible harmful consequences is still unknown. This lack of knowledge can justify the importance to determine the vibration dose in the joints when there is vibration exposure. Subsequently, transmissibility could help to predict the vibration dose using the input acceleration value. The vibration inherent to some sport practices could have one of these effects, and it could be controlled by using the experimental measurements and taking into account the motion effect on the transmission.

4.5 STRENGTH

The use of platforms represents the most common form of vibration exercise when studied for strength adaptations. As we wrote earlier, there are basically two types of vibration platforms: platforms that vibrate in a predominantly vertical direction (vertical platform) and the platform that vibrates through rotation about a horizontal axis such that distances farther from the axis of rotation result in larger amplitude vibrations (oscillating platform).

(Lamont et al., 2008) found a significant augment to strength in an isometric squat test after resistance training was supplemented with WBV.

(Marín and Rhea, 2010a) published a meta-analysis in which they analyzed the magnitude of the strength gains elicited by acute and chronic vibration training and examined specific factors that influence the treatment effects. The findings of the analysis identified differences in both acute and chronic changes in muscle strength when vertical vibration platforms were compared with oscillating platforms. Vertical platforms elicit a significantly larger treatment effect for chronic adaptations (ES = 1.24) compared with oscillating platforms (ES = −0.13). However, oscillating platforms elicit a greater treatment effect for acute effects (ES = 0.24) compared with vertical platforms (ES = −0.07).

The data also showed that age (younger people ES = 1.18 and older people ES = 1.83), gender (women experienced a much greater increase in strength with vibration training compared with men), training status, and exercise protocol (the data demonstrate that a combination of such contractions results in nearly twice the strength adaptation compared with solely isometric contractions) are moderators of the response to vibration exercise for strength development (vertical platforms). Moreover, apparently vibration exercise can be effective at eliciting chronic muscle strength adaptations (Marín and Rhea, 2010b).

Thus vibration exercise can be used by exercise professionals to enhance muscular strength.

4.6 POWER

In relation to power, it has been reported that following acute WBV power and strength performance is enhanced in bilateral countermovement vertical jump (Cochrane and Stannard, 2005; Cormie et al., 2006; Torvinen et al., 2002), bilateral squat power performance (Rhea and Kenn, 2009), unilateral knee isokinetic torque (Jacobs and Burns, 2009; Stewart et al., 2009), and unilateral isometric knee extension force (Torvinen et al., 2002). Significant increases (5%–7%) in velocity have also been observed during single-leg press loads (70–130 kg) after unilateral WBV exposure (10×60 seconds) (Bosco et al., 1999a); which is consistent with our muscle velocity enhancement (3%–5%) of the stimulated and non-stimulated leg following 50 Hz-High.

García-Gutiérrez et al. (2016) found that performing dynamic decline bench press exercise with WBV (applied through a hamstring bridge exercise) and a high load (70% of one-repetition maximum) produced higher peak power values compared to the same exercise condition without WBV (Figure 4.7). In addition, while peak acceleration was higher at lower loads as expected, it was increased with WBV exposure. These effects of WBV on peak power and peak acceleration are partly explained by increases in triceps brachii muscle activity.

4.7 CROSS EDUCATION

Cross education represents the changes evoked in one limb that is not being stimulated, while the opposite or contralateral limb is stimulating by a stimulus such as a conventional training or vibration stimulus.

It has been well documented that "cross education", "cross training effect", or "cross-transfer" enhances strength in the untrained contralateral limb from unilateral resistance training that relies on training specificity of the homologous muscles (Carroll et al., 2006; Lee and Carroll, 2007). Many cross-education studies have

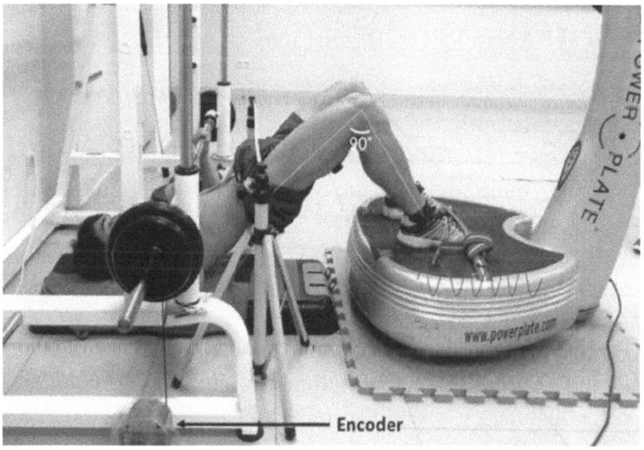

FIGURE 4.7 Dynamic decline bench press exercise with WBV applied through a hamstring bridge exercise.

used various unilateral resistance training methods and loading parameters to elicit changes in contralateral strength (Munn et al., 2004), which have been attributed to neural adaptations (Lee and Carroll, 2007). It has been suggested that WBV elicits a neural potentiation effect similar to that of resistance training (Delecluse et al., 2003) due to its ability to increase acute strength parameters (Armstrong et al., 2008; Bosco et al., 1999a,b; Cochrane and Stannard, 2005). During WBV, the loading parameter is altered by adjusting the acceleration (vibration frequency and amplitude), which differs from conventional strength training, the weight lifted is altered to modify the neuromuscular effects.

It has been demonstrated that three weeks of unilateral training (three sessions per week) with the addition of WBV (35 Hz, 2.5 mm) did not further augment the cross-transfer of strength compared to the same training without WBV (Goodwill and Kidgell, 2012). While this suggests that nine sessions over three weeks was not sufficient to induce an adaptation, the researchers only employed one vibration frequency over the three weeks and utilized large external loads that may have diminished the stimulus generated by the WBV platform.

Marín et al. (2014) found an acute WBV (50 Hz) bout of 30 seconds augments cross-transfer in neuromuscular performance of explosive power parameters (Figure 4.8). To optimize muscle power from synchronous vibration platforms, such as the one used in this study, it has been advocated that higher WBV frequencies of 45–50 Hz at a fixed amplitude can significantly increase EMG during exposure and acute power characteristics post-exposure compared to lower WBV frequencies of 20–35 Hz (Hazell et al., 2007; Ronnestad, 2009; Hazell et al., 2010).

The practical importance of the effects of WBV on the cross-transfer of strength are also clinically important in rehabilitation settings for individuals with compromised capacity to use/train one limb due to injury or limb immobilization following surgery (Goodwill and Kidgell, 2012; Hendy et al., 2012; Pearce et al., 2013).

4.8 PRACTICAL APPLICATIONS: PROGRAM VARIATION AND PROGRESSIONS

Vibration exercise can be performed alone (i.e. full workout with vibration) or combined vibration plus other training methods (i.e. before conventional workout similar to a warm-up) (Bunker et al., 2011), between sets like complex training (Marín et al., 2011) (see Figure 4.9), and/or another great option is after workouts like a cool down up to improve recovery (Rhea et al., 2009).

In general, the first way in which to challenge a client with vibration platform is with movement progressions. The following progression is recommended:

1. Static exercise (i.e. static squat)
2. Dynamic movement (i.e. dynamic squat)
3. Integrated movement (i.e. one leg squat with lunge back)

When the client is ready for progression to other modalities, a trainer must consider that gravitational forces increase quickly as frequency and amplitude are increased.

FIGURE 4.8 Benefits of cross-transfer.

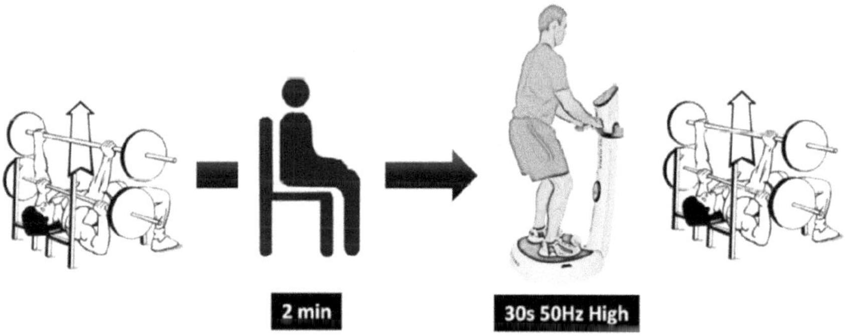

FIGURE 4.9 Example of combined vibration plus other training methods between sets like complex training.

Therefore, it is prudent to change the above movement variables first, before making changes to other modality variables. When the client is ready to progress to greater gravitational forces, it is then recommended to progress the following variables in this order: (i) movement variables; (ii) duration; (iii) number of sets; (iv) work rest ratio; (v) frequency; (vi) amplitude; and (vii) extra-load, for example, vest weight (Figures 4.10 and 4.11).

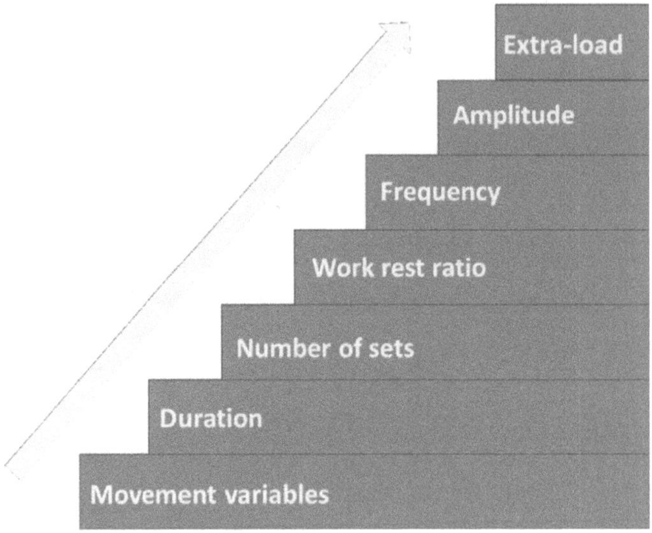

FIGURE 4.10 Vibration training progression.

	Step	Frequency	Amplitude	Mode
Vibration Training Progression	12	45-50 Hz	High	without soft mat
	11	35-40 Hz	High	without soft mat
	10	25-30 Hz	High	without soft mat
	9	45-50 Hz	High	with soft mat
	8	35-40 Hz	High	with soft mat
	7	25-30 Hz	High	with soft mat
	6	45-50 Hz	Low	without soft mat
	5	35-40 Hz	Low	without soft mat
	4	25-30 Hz	Low	without soft mat
	3	45-50 Hz	Low	with soft mat
	2	35-40 Hz	Low	with soft mat
	1	25-30 Hz	Low	with soft mat
	0	Bodyweight (without vibration)		

FIGURE 4.11 Vibration stimulus micro-progression for muscle strength on vertical vibration platform.

In summary, within the training session, the following variables can be manipulated to find the optimal training formula for the client:

- Duration: 30, 45, 60, 90 seconds.
- Number of sets: 1, 2, 3, etc.
- Work to rest ratio: 1:3, 1:2, 1:1, etc.
- Type of movement: Static, dynamic.
- External load: Load, no load.
- Frequency: 25, 30, 35, 40 Hz, etc.
- Amplitude: Low vs high.
- Soft mat: With vs without.
- Integrated: Cable, floor, weight, etc.

REFERENCES

Abercromby, A. F. J., et al. (2007a). Variation in neuro-muscular responses during acute whole-body vibration exercise. *Medicine & Science in Sports & Exercise*, 39(9):1642–1650.

Abercromby, A. F. J., et al. (2007b). Vibration exposure and biodynamic responses during whole-body vibration training. *Medicine & Science in Sports & Exercise*, 39(10):1794–1800.

Adams, J. B., et al. (2009). Optimal frequency, displacement, duration, and recovery patterns to maximize power output following acute whole-body vibration. *Journal of Strength and Conditioning Research*, 23(1):237–245.

Anderson, K. and Behm, D. G. (2005). Trunk muscle activity increases with unstable squat movements. *Canadian Journal of Applied Physiology = Revue canadienne de physiologie appliquee*, 30(1):33–45.

Armstrong, W. J., et al. (2008). The acute effect of whole-body vibration on the Hoffmann reflex. *Journal of Strength and Conditioning Research*, 22(2):471–476.

Avelar, N. C., et al. (2013). Influence of the knee flexion on muscle activation and transmissibility during whole body vibration. *Journal of Electromyography and Kinesiology*, 23(4):844–850.

Behm, D. G., Anderson, K., and Curnew, R. S. (2002). Muscle force and activation under stable and unstable conditions. *Journal of Strength and Conditioning Research*, 16(3):416–422.

Bosco, C., Cardinale, M., and Tsarpela, O. (1999a). Influence of vibration on mechanical power and electromyogram activity in human arm flexor muscles. *European Journal of Applied Physiology*, 79(4):306–311.

Bosco, C., et al. (1999b). Adaptive responses of human skeletal muscle to vibration exposure. *Clinical Physiology*, 19(2):183–187.

Bosco, C., et al. (2000). Monitoring strength training: Neuromuscular and hormonal profile. *Medicine & Science in Sports & Exercise*, 32(1):202–208.

Boshuizen, H. C., Bongers, P. M., and Hulshof, C. T. (1990). Back disorders and occupational exposure to whole-body vibration. *International Journal of Industrial Ergonomics*, 6(1):55–59.

Bressel, E., Smith, G., and Branscomb, J. (2010). Transmission of whole body vibration in children while standing. *Clinical Biomechanics*, 25(2):181–186.

Bunker, D. J., et al. (2011). The use of whole-body vibration as a golf warm-up. *Journal of Strength and Conditioning Research*, 25(2):293–297.

Cardinale, M. and Bosco, C. (2003). The use of vibration as an exercise intervention. *Exercise and Sport Sciences Reviews*, 31(1):3–7.

Cardinale, M. et al. (2003). The effects of whole body vibration on humans: dangerous or advantageous? *Acta Physiologica Hungarica*, 90(3): 195-206.
Cardinale, M. and Lim, J. (2003). Electromyography activity of vastus lateralis muscle during whole-body vibrations of different frequencies. *Journal of Strength and Conditioning Research*, 17(3):621–624.
Cardinale, M. and Wakeling, J. (2005). Whole body vibration exercise: Are vibrations good for you? * Commentary. *British Journal of Sports Medicine*, 39(9):585–589.
Carlsöö, S. (1982). The effect of vibration on the skeleton, joints and muscles: A review of the literature. *Applied Ergonomics*, 13(4):251–258.
Carroll, T. J., et al. (2006). Contralateral effects of unilateral strength training: Evidence and possible mechanisms. *Journal of Applied Physiology*, 101(5):1514–1522.
Chetter, I. C., Kent, P. J., and Kester, R. C. (1998). The hand arm vibration syndrome: A review. *Cardiovascular Surgery*, 6(1):1–9.
Chiementin, X., et al. (2013). Hand–arm vibration in cycling. *Journal of Vibration and Control*, 19(16):2551–2560.
Cochrane, D. J. (2011). Vibration exercise: The potential benefits. *International Journal of Sports Medicine*, 32(02):75–99.
Cochrane, D. (2016). Does muscular force of the upper body increase following acute, direct vibration? *International Journal of Sports Medicine*, 37(07):547–551.
Cochrane, D. J. (2017). Effectiveness of using wearable vibration therapy to alleviate muscle soreness. *European Journal of Applied Physiology*, 117(3):501–509.
Cochrane, D. J. and Hawke, E. J. (2007). Effects of acute upper-body vibration on strength and power variables in climbers. *The Journal of Strength and Conditioning Research*, 21(2):527.
Cochrane, D. J. and Stannard, S. R. (2005). Acute whole body vibration training increases vertical jump and flexibility performance in elite female field hockey players. *British Journal of Sports Medicine*, 39(11):860–865.
Cormie, P., et al. (2006). Acute effects of whole-body vibration on muscle activity, strength, and power. *The Journal of Strength and Conditioning Research*, 20(2):257.
Crewther, B., Cronin, J., and Keogh, J. (2004). Gravitational forces and whole body vibration: Implications for prescription of vibratory stimulation. *Physical Therapy in Sport*, 5(1):37–43.
Da Silva, M. E., et al. (2007). Influence of vibration training on energy expenditure in active men. *The Journal of Strength and Conditioning Research*, 21(2):470.
Delecluse, C. et al. (2003). Strength increase after whole-body vibration compared with resistance training. *Medicine & Science in Sports & Exercise*, 35(6): 1033–1041.
Delecluse, C., et al. (2005). Effects of whole body vibration training on muscle strength and sprint performance in sprint-trained athletes. *International Journal of Sports Medicine*, 26(8):662–668.
Dolny, D. G. and Reyes, G. F. C. (2008). Whole body vibration exercise. *Current Sports Medicine Reports*, 7(3):152–157.
Eklund, G. and Hagbarth, K. E. (1966). Normal variability of tonic vibration reflexes in man. *Experimental Neurology*, 16(1):80–92.
García-Gutiérrez, M. T., Hazell, T. J., and Marín, P. J. (2016). Effects of whole-body vibration applied to lower extremity muscles during decline bench press exercise. *Journal of Musculoskeletal & Neuronal Interactions*, 16(3):204–210.
Giubilato, F. and Petrone, N. (2012). A method for evaluating the vibrational response of racing bicycles wheels under road roughness excitation. *Procedia Engineering*, 34:409–414.
Goodwill, A. M. and Kidgell, D. J. (2012). The effects of whole-body vibration on the cross-transfer of strength. *The Scientific World Journal*, 2012:1–11.
Grappe, F. (2009). *Cyclisme Optimisation de la performance, Frédéric Grappe.*
Griffin, M. J. (1990). *Handbook of Human Vibration*. Academic Press.

Gurram, R. (1993). A study of vibration response characteristics of the human hand-arm system. *Technical Report*, Concordia University.

Hagbarth, K. E. and Eklund, G. (1966). Tonic vibration reflexes (TVR) in spasticity. *Brain Research*, 2(2):201–203.

Harazin, B. and Grzesik, J. (1998). The transmission of vertical whole-body vibration to the body segments of standing subjects. *Journal of Sound and Vibration*, 215(4):775–787.

Hazell, T. J. and Lemon, P. W. R. (2012). Synchronous whole-body vibration increases VO2 during and following acute exercise. *European Journal of Applied Physiology*, 112(2):413–420.

Hazell, T. J., Jakobi, J. M., and Kenno, K. A. (2007). The effects of whole-body vibration on upper- and lower-body EMG during static and dynamic contractions. *Applied Physiology, Nutrition, and Metabolism*, 32(6):1156–1163.

Hazell, T. J., Kenno, K. A., and Jakobi, J. M. (2010). Evaluation of muscle activity for loaded and unloaded dynamic squats during vertical whole-body vibration. *Journal of Strength and Conditioning Research*, 24(7):1860–1865.

Hendy, A. M., Spittle, M., and Kidgell, D. J. (2012). Cross education and immobilisation: Mechanisms and implications for injury rehabilitation. *Journal of Science and Medicine in Sport*, 15(2):94–101.

Hill, T. E., Desmoulin, G. T., and Hunter, C. J. (2009). Is vibration truly an injurious stimulus in the human spine? *Journal of Biomechanics*, 42(16):2631–2635.

Humphries, B., et al. (2004). The influence of vibration on muscle activation and rate of force development during maximal isometric contractions. *Journal of Sports Science & Medicine*, 3(1):16–22.

Issurin, V. B. (2005). Vibrations and their applications in sport. A review. *The Journal of Sports Medicine and Physical Fitness*, 45(3):324–36.

Jackson, S. W. and Turner, D. L. (2003). Prolonged muscle vibration reduces maximal voluntary knee extension performance in both the ipsilateral and the contralateral limb in man. *European Journal of Applied Physiology*, 88(4):380–386.

Jacobs, P. L. and Burns, P. (2009). Acute enhancement of lower-extremity dynamic strength and flexibility with whole-body vibration. *Journal of Strength and Conditioning Research*, 23(1):51–57.

Kaeding, T. S., et al. (2017). Whole-body vibration training as a workplace-based sports activity for employees with chronic low-back pain. *Scandinavian Journal of Medicine & Science in Sports*, 27(12):2027–2039.

Kiiski, J., et al. (2008). Transmission of vertical whole body vibration to the human body. *Journal of Bone and Mineral Research*, 23(8):1318–1325.

Lamont, H. S., et al. (2008). Effects of 6 weeks of periodized squat training with or without whole-body vibration on short-term adaptations in jump performance within recreationally resistance trained men. *Journal of Strength and Conditioning Research*, 22(6):1882–1893.

Lee, M. and Carroll, T. J. (2007). Cross education: Possible mechanisms for the contralateral effects of unilateral resistance training. *Sports Medicine*, 37(1):1–14.

Lépine, J., Champoux, Y., and Drouet, J.-M. (2012). Technique to measure the dynamic behavior of road bike wheels. In *Topics in Modal Analysis II*, Chapter 47, pp. 465–470. Springer, New York.

Luo, J., McNamara, B. P., and Moran, K. (2005). A portable vibrator for muscle performance enhancement by means of direct muscle tendon stimulation. *Medical Engineering & Physics*, 27(6):513–522.

Manin, L., Poggi, M., and Havard, N. (2012). Vibrations of table tennis racket composite wood blades: Modeling and experiments. *Procedia Engineering*, 34:694–699.

Marín, P. J. and Rhea, M. R. (2010a). Effects of vibration training on muscle power: A meta-analysis. *Journal of Strength and Conditioning Research*, 24(3):871–878.

Marín, P. J. and Rhea, M. R. (2010b). Effects of vibration training on muscle strength: A meta-analysis. *Journal of Strength and Conditioning Research*, 24(2):548–556.

Marín, P. J., et al. (2009). Neuromuscular activity during whole-body vibration of different amplitudes and footwear conditions: Implications for prescription of vibratory stimulation. *Journal of Strength and Conditioning Research*, 23(8):2311–2316.

Marín, P. J., et al. (2011). Effects of different vibration exercises on bench press. *International Journal of Sports Medicine*, 32(10):743–748.

Marín, P. J., et al. (2012). Whole-body vibration increases upper and lower body muscle activity in older adults: Potential use of vibration accessories. *Journal of Electromyography and Kinesiology*, 22(3):456–462.

Marín, P. J., et al. (2014). Acute unilateral leg vibration exercise improves contralateral neuromuscular performance. *Journal of Musculoskeletal & Neuronal Interactions*, 14(1):58–67.

McBride, J. M., Cormie, P., and Deane, R. (2006). Isometric squat force output and muscle activity in stable and unstable conditions. *The Journal of Strength and Conditioning Research*, 20(4):915.

Mester, J., et al. (1999). Biological reaction to vibration – Implications for sport. *Journal of Science and Medicine in Sport*, 2(3):211–226.

Mileva, K. N., et al. (2006). Acute effects of a vibration-like stimulus during knee extension exercise. *Medicine & Science in Sports & Exercise*, 38(7):1317–1328.

Mileva, K. N., Bowtell, J. L., and Kossev, A. R. (2009). Effects of low-frequency whole-body vibration on motor-evoked potentials in healthy men. *Experimental Physiology*, 94(1):103–116.

Moran, K., McNamara, B., and Luo, J. (2007). Effect of vibration training in maximal effort (70% 1RM) dynamic bicep curls. *Medicine & Science in Sports & Exercise*, 39(3): 526–533.

Munera, M. (2014). *Analyse vibro-biomécanique et dynamique en sport/santé*. PhD thesis, Université de Reims Champagne-Ardenne.

Munera, M., et al. (2015a). Model of the risk assessment of hand-arm system vibrations in cycling: Case of cobblestone road. *Proceedings of the Institution of Mechanical Engineers, Part P: Journal of Sports Engineering and Technology*, 229(4):231–238.

Munera, M., et al. (2015b). Physiological and dynamic response to vibration in cycling: A feasibility study. *Mechanics & Industry*, 16(5):503.

Munera, M., et al. (2016). Transmission of whole body vibration to the lower body in static and dynamic half-squat exercises. *Sports Biomechanics*, 15(4):409–428.

Munn, J., Herbert, R. D., and Gandevia, S. C. (2004). Contralateral effects of unilateral resistance training: A meta-analysis. *Journal of Applied Physiology*, 96(5):1861–1866.

Olieman, M., Marin-Perianu, R., and Marin-Perianu, M. (2012). Measurement of dynamic comfort in cycling using wireless acceleration sensors. *Procedia Engineering*, 34:568–573.

Osis, S. T. and Stefanyshyn, D. J. (2010). Vibration at the wrist and elbow joints during the golf swing reveals shaft-specific swing kinematics. *Procedia Engineering*, 2(2):2637–2642.

Parkin, J. and Sainte Cluque, E. (2014). The impact of vibration on comfort and bodily stress while cycling. In *UTSG 46th Annual Conference*. Newcastle University.

Pearce, A. J., et al. (2013). Corticospinal adaptations and strength maintenance in the immobilized arm following 3 weeks unilateral strength training. *Scandinavian Journal of Medicine & Science in Sports*, 23(6):740–748.

Peretti, A., et al. (2009). Vibrazioni su biciclette da corsa e da citta. *Giornale degli Ignienisti Industriali*, 34(3):283–293.

Pollock, R. D., et al. (2012). Effects of whole body vibration on motor unit recruitment and threshold. *Journal of Applied Physiology*, 112(3):388–395.

Poston, B., et al. (2007). The acute effects of mechanical vibration on power output in the bench press. *Journal of Strength and Conditioning Research*, 21(1):199–203.

Preatoni, E., et al. (2012). The effects of whole-body vibration in isolation or combined with strength training in female athletes. *Journal of Strength and Conditioning Research*, 26(9):2495–2506.

Rhea, M. R. and Kenn, J. G. (2009). The effect of acute applications of whole-body vibration on the iTonic platform on subsequent lower-body power output during the back squat. *Journal of Strength and Conditioning Research*, 23(1):58–61.

Rhea, M. R., et al. (2009). Effect of iTonic whole-body vibration on delayed-onset muscle soreness among untrained individuals. *Journal of Strength and Conditioning Research*, 23(6):1677–1682.

Rittweger, J. (2010). Vibration as an exercise modality: How it may work, and what its potential might be. *European Journal of Applied Physiology*, 108(5):877–904.

Rittweger, J., Schiessl, H., and Felsenberg, D. (2001). Oxygen uptake during whole-body vibration exercise: Comparison with squatting as a slow voluntary movement. *European Journal of Applied Physiology*, 86(2):169–173.

Rittweger, J., et al. (2002). Oxygen uptake in whole-body vibration exercise: Influence of vibration frequency, amplitude, and external load. *International Journal of Sports Medicine*, 23(6):428–432.

Ritzmann, R., et al. (2010). EMG activity during whole body vibration: Motion artifacts or stretch reflexes? *European Journal of Applied Physiology*, 110(1):143–151.

Ritzmann, R., Gollhofer, A., and Kramer, A. (2013). The influence of vibration type, frequency, body position and additional load on the neuromuscular activity during whole body vibration. *European Journal of Applied Physiology*, 113(1):1–11.

Roberts, J., et al. (2005). Evaluation of vibrotactile sensations in the 'feel' of a golf shot. *Journal of Sound and Vibration*, 285(1–2):303–319.

Ronnestad, B. R. (2004). Comparing the performance-enhancing effects of squats on a vibration platform with conventional squats in recreationally resistance-trained men. *The Journal of Strength and Conditioning Research*, 18(4):839.

Ronnestad, B. R. (2009). Acute effects of various whole-body vibration frequencies on lower-body power in trained and untrained subjects. *Journal of Strength and Conditioning Research*, 23(4):1309–1315.

Saeterbakken, A. H. and Fimland, M. S. (2013). Muscle force output and electromyographic activity in squats with various unstable surfaces. *Journal of Strength and Conditioning Research*, 27(1):130–136.

Sperlich, B. and Kleinoeder, H. (2009). Physiological and perceptual responses of adding vibration to cycling. *Journal of Exercise Physiology*, 12(2):40–46.

Srinivasan, J. and Balasubramanian, V. (2007). Low back pain and muscle fatigue due to road cycling—An sEMG study. *Journal of Bodywork and Movement Therapies*, 11(3):260–266.

Stewart, J. A., Cochrane, D. J., and Morton, R. H. (2009). Differential effects of whole body vibration durations on knee extensor strength. *Journal of Science and Medicine in Sport*, 12(1):50–53.

Stroede, C. L., Noble, L., and Walker, H. S. (1999). The effect of tennis racket string vibration dampers on racket handle vibrations and discomfort following impacts. *Journal of Sports Sciences*, 17(5):379–385.

Torvinen, S., et al. (2002). Effect of a vibration exposure on muscular performance and body balance. Randomized cross-over study. *Clinical Physiology and Functional Imaging*, 22(2):145–52.

Verschueren, S. M., et al. (2003). Effect of 6-month whole body vibration training on hip density, muscle strength, and postural control in postmenopausal women: A randomized controlled pilot study. *Journal of Bone and Mineral Research*, 19(3):352–359.

5 Effects of Whole Body Vibration on the Elderly

Maíra Florentino Pessoa, Helga C. Muniz de Souza, Helen K. Bastos Fuzari, Patrícia E. M. Marinho, and Armèle Dornelas de Andrade

CONTENTS

5.1 INTRODUCTION

The growth of the elderly population due to technological advances and improvements in health care is a present reality in the world demographic structure, with reductions in mortality and birth rates, increasing the number of elderly people. The World Health Organization (WHO) states that for developing countries, the elderly are any individuals over the age of 60 (WHO, 2015).

Aging is a progressive and constant process that affects all organs and tissues due to various biological changes related to advancing age. Although all organic systems will be modified by aging, in this chapter we will only focus on the systems and functions that are known to have direct effects from the application of whole body vibration (WBV) in the elderly.

WBV has been shown to be a well-recognized exercise by seniors for gaining muscle strength, especially since it is easy to use, well-tolerated, with fast, effective, safe results and as another tool available to this population (Jordan et al., 2005).

WBV exercise on the vibration platform for the elderly requires some adjustments, such as a comfortable position, but within the acceptable standards to receive the vibration. Based on the chosen position, one of the protocols available in the literature is implemented and adapted to the reality of this population.

Adverse effects have not been reported in the literature to date, provided that the training is properly developed, under the supervision of a qualified professional and considering some contraindications such as: use of pacemaker, bolts or pins in the body, presence of acute migraine, labyrinthitis, cognitive, visual and/or auditory deficits, cardiac surgeries, pregnancy or having body weight higher than what is allowed for platform specifications.

5.2 WBV IN ELDERLY MUSCLE SYSTEM

A loss of muscle mass associated with reduced maximal strength is a common finding throughout the aging process; in most cases it is also associated with physical and functional decline. Aging leads to a decreased number of muscle fibers and in the cross-sectional area size—sarcopenia—being responsible for a loss of 30% of muscle mass and motor units between the age of 20 and 80 years (Fielding et al., 2011).

Changes in the neuromuscular system affect the senile motor unit, structurally manifested as a reduction in the number of secondary motor units due to the loss of motor neurons. This finding leads to a grouping of several types of muscle fibers, due to repeated cycles of denervation and reinnervation, thereby causing instability of the neuromuscular junction, which culminates in the death of motor neurons or traumatic injury to the muscle fiber. These changes are functionally manifested in reduced strength and coordination, which precede loss of muscle mass and contribute to the development of muscular fatigue (Hepple and Rice, 2016).

Whole body vibration training was introduced to create alternative strategies to reduce the action of sarcopenia and physical deconditioning, to improve the performance of the muscular system in the elderly, and thus reflect the type of prescribed exercise in the care due to the installed weakness and/or reduction of the corporal composition (Srinivas-Shankar et al., 2010; Cardinale and Pope, 2003).

The physiological principle of increasing strength through WBV may be mainly related to the neurophysiological mechanisms generated from vibrational stimuli that drive neuromuscular spindles, especially the alpha, leading to muscle contraction and consequently to increased muscle strength; similar to that occurring with tonic vibration reflex.

Studies show that WBV improves bone morphology, promoting muscle toning without overloading the joints, resembling combined aerobic workouts with endurance and strength training (Von Stengel et al., 2012).

According to the American College of Sports Medicine and the American Heart Association (ACSM/AHA), the maintenance of health status should be accomplished by practicing moderate physical activity for 30 minutes, five days per week, or 20 minutes of vigorous activity for three days a week (Nelson et al., 2007). For the elderly, such an exercise program may not be properly tolerated, due to their reduced physical capacity and a sedentary life history adopted by many of them, thus caution is required in prescribing any type of training.

Studies have demonstrated strength gain in the elderly through WBV, suggesting this form of training is an alternative to resistance training, since there is no active contraction and it results in improved strength (Roelants et al., 2004; Verschueren et al., 2004). Resistance training is where all motor units are recruited from the smallest motor units to those with the highest thresholds (type II), and is used for muscular strengthening, muscular hypertrophy and improving the nervous command, with progressive load increase.

WBV has been widely used in recent years and can be considered a form of effective treatment. A new study using WBV in the elderly showed minimal additional effort and stress on the skeletal muscles and cardiovascular system for example, it minimized joint wear compared to conventional exercise programs (Bogaerts et al., 2009). An assessment of muscle strength gain after WBV training is usually performed through dynamometry, which is a functional and low cost method, where the individual performs a maximum or submaximal contraction at the used interface (Kennis et al., 2013).

It should be emphasized that there is no single protocol for training with the elderly, as they may vary: in frequency (according to each equipment), in vibration dosage and amplitude (2 to 10 mm). Another important point regarding the rehabilitation protocol in the elderly using WBV is determining the intervention periods which can be performed two to three times a week, and not exceeding a maximum of 30 minutes of vibration per day for each individual, regardless of age (ISO 2631-1) (ISO 2631-1, 1997). However, a minimum of three months of training is required for strength gain. There are protocols in the literature ranging from six to 24 weeks (Roelants et al., 2004; Tankisheva et al., 2014).

5.3 WBV IN ELDERLY OSTEOARTICULAR SYSTEM

In skeletal aging, bone volume and bone mass decrease in both genders, usually manifesting as osteoporosis and increased risk of fractures. Loss of cartilage thickness in synovial joints with age progression contributes to osteoarthritis, while it decreases the structural integrity of the intervertebral disc (DIV), leading to a decrease in disc height, disc collapse and spinal cord compression (Roberts et al., 2016).

The osseous system consists of osteoclastic cells (which degrade bone tissue), osteoblastic cells (which produce and increase bone mineral tissue) and osteocytes (which maintain bone tissue) that act in constant equilibrium for remodelling and preserving bone strength and volume throughout life. Aging leads to an imbalance between osteoclastic and osteoblastic activity, causing greater bone degradation and a reduction in the rate of osteogenesis, leading to age-related bone reduction.

In addition to changes in the regulation mechanisms of osteogenesis, changes in age-related sexual hormones also affect normal skeletal biology. In women, the decline in estrogen levels after menopause lead to cortical and trabecular bone loss. These results in high rates of bone resorption, which was previously inhibited by estrogen, triggering insufficient bone formation. In men, low levels of androgens also cause remodelling and loss of bone tissue due to reduced levels of estrogen, derived from aromatization of the testosterone hormone (Roberts et al., 2016). In addition to these, cellular responses to growth hormones are also altered with aging, leading to an osteoblastic and osteogenic progenitor cell reduction due to reduced cellular sensitivity to the signalling of this hormone, contributing to bone loss.

WBV training has been a very important and beneficial intervention tool for the elderly population in the prevention of falls, since WBV has been used in treating and alleviating some diseases and comorbidities, as well as in their prevention. Falls occur in this population due to the decrease in balance and muscular strength (among other reasons), which are conditions conducive to intervention with WBV.

Another aspect to be considered is the presence of osteoporosis or even osteopenia, associated with the risk of fractures. Osteoporosis is a metabolic condition identified by progressive decrease in bone density. Therefore, bones become more porous and lose resistance over time; in this context, the use of WBV has also been valuable, acting directly on improving bone density (Slatkovska et al., 2010; Segal et al., 2012).

5.4 WBV IN ELDERLY IMMUNE SYSTEM

The immune system of the elderly undergoes transformations and their immune response patterns are age dependent. Changes that occur from the age of 50 have received special attention due to their clinical impact, and such changes have been globally known as immune senescence or immunosenescence.

Restructuring of the immune system occurs during aging, where all immune cells are generally affected, thus contributing to high vulnerability to infections and increased mortality. This mortality is associated with a growing increase in the basal inflammatory state of the organism, represented by increased serum levels of C-reactive protein (CRP), interleukin-6 (IL-6) and tumor necrosis factor alpha (TNF-α), implicating in numerous inflammatory diseases. Moreover, innate immune system cells (neutrophils, monocytes, macrophages, killer cells, dendritic cells) worsen in their function, deteriorating the body's defense process. Significant decline in adaptive immune response during aging is also observed due to decreased lymphocyte responsiveness.

Progressive deficits of functional T and B lymphocytes have been suggested as the major factors responsible for age-related disorders. Lymphocytes are largely affected during immunosenescence, and the age-related continuous antigenic stress causes a variety of changes in the immune system, exhausting and reducing T lymphocytes, as well as its responsiveness.

The use of WBV may be a viable alternative to attenuate these changes and represents a therapeutic instrument that minimizes the basal inflammatory process present in this population. Some studies performed with the elderly found a reduction in plasma levels of TNF-α and CRP (Rodriguez-Miguelez et al., 2015) after a WBV program, as well as a reduction in plasma concentrations of inflammatory markers TNF-1 and TNF-2 (Simão et al., 2012), suggesting that WBV can improve the basal anti-inflammatory status in these individuals.

By evaluating elderly mice, Lin et al. in 2015 observed that serum levels of lactate, ammonia, alkaline phosphatase and creatinine decreased, and serum levels of albumin and total fasting protein increased according to the vibration dose and frequency applied. These findings suggest that WBV can improve performance during exercise, reduce fatigue and prevent the biochemical changes associated with senescence in elderly mice; however, the mechanism of how WBV regulates these changes needs more clarification.

It appears that the use of WBV to prevent loss of muscle strength associated with sarcopenia in the elderly may be a safe training method that does not induce training-related inflammatory effects. In 2014, Cristi et al. investigated the effects of WBV and observed significant increases in physical fitness measures without changes in inflammation markers in the elderly. The levels of CRP, IL-6, IL-1, TNF-α and IL-10 were analysed, and did not differ from baseline.

The acute effects of WBV on the elderly when compared to exercise without vibration revealed an increase in the circulating levels of insulin-1 growth factor and higher cortisol than that observed in the same exercise protocol conducted without vibration (Cardinale et al., 2010), and no differences were found between testosterone and growth hormone levels in the groups. These findings suggest that perhaps the acute changes resulting from a single WBV session differ from those found in traditional exercises, although further studies are needed regarding these acute effects.

In addition to its application to healthy elderly people, WBV can have beneficial effects when used in elderly individuals with diseases that have an increased inflammatory status, as shown in a study (Simão et al., 2012) that investigated the effects of squatting exercises combined with WBV on plasma inflammatory markers in elderly with knee osteoarthritis. The WBV group showed a reduction in TNF-1 and TNF-2 levels with improved gait quality and balance.

The effects of WBV found suggest that vibration training in the elderly population seems to attenuate the basal inflammatory state, more studies are needed to verify the association between physical, functional and therapeutic improvements, notably on the elderly immune system.

5.5 WBV IN ELDERLY FUNCTIONALITY

Although aging is a physiological and progressive process, the alterations found in the osteomioarticular, cardiopulmonary and immune systems generate impacts on the functionality of this population. A decrease in independence and loss of functional capacity are linked to old age, with aging also being related to declines in mobility, balance and flexibility, which results in the elderly population presenting a decrease in physical activity levels and sometimes gait limitations.

5.5.1 WBV and Functional Capacity

In aging, muscular weakness caused by sarcopenia is predominant in the lower limbs, especially in the quadriceps. A reduction in quadricipital volume generates early muscle fatigue, progressing to reduced levels of physical activity, culminating in a sedentary lifestyle and replacement of muscle fibers by adipose tissue. Whole-body vibration presents similar adaptive responses in skeletal muscles, however in an overall manner involving morphological and neural factors, with minimal modifications to the muscular cross-sectional area and a great increase in neuromodulated response speed (Cardinale and Pope, 2003).

During vibration, muscle spindles are the first sensory structure activated, followed by Golgi tendon organs and type I and type II joint mechanoreceptors. Vibratory stimuli initially act on these three sensory structures, hypersensitizing the gamma (γ) system and leading to a contraction reflex called tonic vibration reflex (TVR), which recruits the motor units through the polysynaptic pathway. Parallel to this recruitment, maintenance of the vibratory stimulus leads motor units that are already in operation to derecruitment through a presynaptic inhibition mechanism with consequent reduction of the monosynaptic reflexes, so that the muscle tends to reduce its reflex stretch, improving contraction by synchronizing the motor units (Bogaerts et al., 2011).

It is the vibration reflex contraction that increases peripheral muscle strength and improves muscle contractile function, since it improves the oxidative capacity at the same time that it reflexively vasodilates muscular blood vessels, allowing rapid inflow of metabolites and catabolite drainage, improving muscle function and functional capacity of trained elderly. Some studies report that improvement has been measured through Katz index scores, and functional tests such as the "10 Minute Walking", the "Endurance Shuttle Walk", the "Timed Up and Go", the "Chair Rising", the "Short Physical Performance Battery" and the "Senior Fitness Tests" (Kessler et al., 2014; Santin-Medeiros et al., 2015).

5.5.2 WBV and Flexibility

According to the National Strength and Conditioning Association (NSCA), the practice of any physical activity or exercise section should be preceded by five to ten minutes of stretching activities, followed by physical activity itself carried out slowly in order to assist in the transition from rest to exercise, increasing blood flow and flexibility in order to maximize performance.

Using vibration as a form of training has shown to be effective in improving flexibility in the elderly. The increase in the range of motion in the joints probably results from a reduction in the rigidity of the musculo-tendinous unit, provoked by the tonic vibration reflex. Bosco et al. in 1999, suggest that the neurophysiological alterations of WBV indicate a reduction in the efficacy between 1a sensory fibers and alpha motoneuron transmission. Another proposition is that the vibration is related to an increase in the sensitivity of the stretch reflex, caused by a blockage of the monosynaptic pathways, and that it inhibits the activation of antagonist muscles through inhibitory Ia neurons, thereby altering muscle coordination patterns and decreasing the locking force around the joints (Gómez-Cabello et al., 2013).

Factors such as gender, age, joint structure and ligament and tendon composition may influence the primary range of motion, as well as the gains after vibration training. Women tend to naturally have greater flexibility generated by hormonal factors, while aging makes the joints more rigid due to elastic ligament tissue loss and the consequent replacement by collagen, in addition to lower water content.

Gains in articular flexibility occur during an acute approach and they do not depend on the continuity of the training, so that chronic training does not produce better results. Di Giminiani et al. demonstrated that 20 ten-second sets of low frequency vibration in addition to stretching provided 16% higher gains in flexibility when WBV was used, which can result in up to seven degrees of flexibility measured by the goniometer when compared to static stretching in a similar position (Di Giminiani et al., 2014).

Still regarding flexibility, after observing the results found and the consistency of the studies, the Center of Evidence Based Medicine found moderate evidence to support the use of WBV for increasing flexibility, recommending its use with an evidence grade of B (Houston et al., 2015).

5.5.3 WBV, Balance and Falls

Balance is achieved when the forces acting on the body allow it to maintain the desired posture, or to move in a controlled manner. However, biomechanical and postural changes may contribute to reduced stability. These structural changes coincide with shortening the anterior muscle groups of the thoracic region and elongation of the posterior chain. This shortening or excessive stretching affects the endurance of these muscle groups, leading to a greater chance of fatigue, and making them less efficient at maintaining upright postures, in addition to changing the body's centre of gravity.

More recent studies also point out that in addition to fiber loss, monosynaptic reflex speed is slowed down and proprioception is impaired, so that neurophysiological changes add up to functional alterations, predisposing the elderly to falls.

Although fall events are associated with the sum of intrinsic (neurophysiological) and extrinsic factors (environmental and situational hazards related to the task that is being performed), intrinsic factors can be greatly minimized when the elderly use vibration platforms.

Vibratory stimuli favour gait coordination by improving balance and proprioception, with a positive effect on the Tinetti Test (Bruyere et al., 2005) and significant improvement in postural control evaluated by the Sensory Organization Test, which measures the ability to respond to visual, vestibular and proprioceptive information to maintain balance (ISO 2631-1, 1997). In 2011, Bogaerts et al. noticed that WBV training concomitant with calcium and vitamin D supplementation was more effective than supplementation alone, increasing walking speed by 10% and improving oscillation speed. Although these increases only occurred in relation to the baseline values, they have clinical relevance since they are important protective factors for falls. When compared to the group that only received supplementation, vibration generated an 18% reduction in the risk of falls in these elderly individuals.

Studies have shown results that indicate improvements in relation to balance and reduced risk of falls with the most varied frequencies offered in vibrating platforms from 10 Hz to 50 Hz, as well as with various training periods with durations ranging from four to 48 weeks. The proposal is that the WBV would act on the root causes of the reduced balance by performing rapid variations in position on a fixed support base which would promote an increase in body balance, activating tactile and articular sensitivity, previously impaired. Moreover, the mechanical vibrations activate the neuromuscular control system and improve the efficiency of the spinal reflex, coordinating the action between synergistic and antagonistic muscles during fast muscle contractions, increasing the speed of muscular response (Bogaerts et al., 2011).

One last theory uses another neural mechanism which may explain how WBV affects balance. The proposal is that vibration would stimulate bone myoregulation reflex (BMR) by acting on the area of slightly bent knees. Thus, by generating a supragravitational stimulus, WBV would increase the extension moment of the knee joint, increasing the mechanical load on the femur, and thus the strength of the quadriceps, which would prevent knee flexion at the falling moment (Bruyere et al., 2005). BMR is the reflex mechanism of muscle activity that is triggered when osteocytes are exposed to cyclical mechanical load, as is the case of vibration stimulus (Bruyere et al., 2005).

5.5.4 WBV and Quality of Life

Quality of life assessment is used as an indicator of the impact of certain diseases or treatments. Although quality of life is accepted as a good predictor of survival, there is no clear definition of its concept, although there is consensus on two aspects: it is a subjective and multidimensional measurement that aggregates several domains of analysis and it has a strong correlation with maintaining functionality and independence in the elderly, which in turn is related to psychological, social and mainly physical aspects such as the ability to move, reducing fatigue or pain, and more subjectively to the sensation of fulfilling activities.

In 2005, Bruyere et al. used WBV for six weeks in institutionalized elderly, presenting gains in the functional capacity domain evaluated through the SF-36 questionnaire. Although the positive results of this study on quality of life are limited to one domain, it is worth emphasizing that this is the domain that best represents gains in relation to measuring the preservation of the capacity to independently carry out activities, allowing for an active lifestyle.

In 2017, Pessoa et al. found benefits in all physical domains (physical aspects, pain, general health and energy) using the SF-36 questionnaire, with the most significant improvement occurring in relation to the functional capacity domain, but without improvements in the mental or social domains. The functional capacity domain is clinically related to a better condition of autonomy and independence, and its maintenance is associated with reduced risk of institutionalization and mortality.

Although WBV is a physical stimulus, it can be assumed that better mobility allows greater independence, consequently positively interfering in psychological aspects such as anxiety and depression.

5.6 WBV IN THE ELDERLY CARDIOPULMONARY SYSTEM

5.6.1 WBV and Cardiac Effectiveness

Cardiopulmonary deconditioning in the elderly occurs due to reduced effectiveness of the cardiac pump and by alterations in the pulmonary volumes, resulting in a decrease in exercise tolerance, with an annual decrease of 1% in maximal O_2

consumption (VO_{2max}), which limits exercise capacity, thereby becoming a progressive process and reducing the quality of life, with inactivity being the main cause of physical decline. Training programs are instituted in order to reverse this process, seeking to increase the effort load so that the VO_2 can adapt.

One of the first studies to verify the use of the platform in order to bring benefits to cardiac function was the study by Rittweger et al. in 2001, who suggested that the platform could be considered aerobic training, since the improvement promotes increased heart rate and oxygen consumption.

The mechanisms which influence the improvement of cardiopulmonary fitness after WBV are not exactly known. We know that this exercise modality increases blood flow to the muscles that are subjected to vibration, facilitating capillary collateral circulation and optimizing vascular function. This leads to a reduction of peripheral vascular resistance and increased blood inflow to the heart. Over time, this would allow eccentric hypertrophy with increased cardiac chamber and increased systolic ejection volume and cardiac output, increasing the VO_2 (Kerschan-Schindl et al., 2001).

Intense aerobic exercises subject the muscles to lipid oxidation as early as five minutes after the beginning of the activity, while the tonic-vibratory reflex produces continuous contraction of large groups, gradually increasing until reaching a plateau at approximately 30 seconds (Oyarzún, 2009). Thus, there is a requirement for continuous oxidative muscle fiber work, increasing the size and number of mitochondria and amount of myoglobin, assisting the enzyme system yield and its ability to generate ATP through lipid oxidative phosphorylation. In the long-term, oxidative changes increase O_2 extracted from the blood, improving the capacity of the trained muscle to use it.

5.6.2 WBV and Blood Flow

Aging leads the vascular system to increase its calcium deposition, which results in a decrease in arterial compliance, making the artery walls more rigid and leading to less peripheral vascularization. This arterial stiffness makes the vascular system less able to contain pressure variations.

Arterial stiffness can be measured in several non-invasive ways, of which pulse wave velocity (PWV) is considered the standard measure. The most common ways to measure PWV are brachial-ankle pulse wave velocity (baPWV), leg pulse wave velocity (lPWV) and carotid-femoral pulse wave velocity (cfPWV). Stiffness is considered a predictor for cardiovascular mortality, since the greater the risk, the worse the cardiac overload will be in order to cause blood to flow through the extremities. Thus, it can contribute to an increase of afterload and left heart failure.

WBV as aerobic training with minimal risk of muscle injury or joint overload is the ideal exercise for seniors wishing to improve their perfusion. According to Herrero et al. in 2011, the mechanism underlying the vasodilatory effect of WBV may be related to an acute increase in the production of substances such as vasodilator metabolites. Another explanation is that the frictional forces applied by the vibrating platform on the skin would cause endothelial cells to vibrate at a cellular level, releasing substances (Lohman et al., 2007). Data from the literature suggest that the effect of vasodilation continues for some time after WBV, and that it can persist throughout the following day, as vibration changes endothelial factors rather than contraction-related vasodilatory metabolites that would be rapidly drained (Herrero et al. in 2011).

5.6.3 WBV and Respiratory Strength

As WBV has been applied to sarcopenia and increases in peripheral force, and as it is a global intervention, only one study has verified the vibration action on the respiratory musculature of the elderly. In 2017, Pessoa et al. verified that vibratory stimulus generated a significant increase in the respiratory force measurements measured by the maximal inspiratory pressure (MIP) and the maximal expiratory pressure (MEP) when compared to the control.

The hypothesis for the MIP increase is that the vibration promoted non-specific training in the accessory musculature and in the diaphragm through overall vibration, in addition to a more intense action exerted in the breathing accessories through the transmission of impulses by the arm. Because the position assumed during WBV requires semi-flexion of the knees, there is a tendency on the part of the elderly to keep their arms outstretched for better support, which would facilitate transmission of impulses through the shoulder girdle axially through the handle of the vibrating platform, intensifying stimulation on this accessory musculature. Concerning MEP, it is already known that vibration is an overall training that increases strength and tone, and that along with the isometric position assumed during the intervention makes the abdominal muscle group a more efficient pressure generator (Pessoa et al., 2017).

The increase of respiratory strength in this population clinically favors the maintenance of their health status and autonomy, considering that higher MIP allows elderly to perform their daily tasks (even those that require a certain overload) without the sensation of dyspnea occurring, as well as increased MEP which will likely help minimize the risk of respiratory infections by improving cough efficacy.

5.6.4 WBV and Lung Capacities

Installation of a hyperinflation pattern will lead to modifications in pulmonary volumes. Residual volume (RV) increases as a function of trapped air, altering functional residual capacity (FRC), reducing volume tidal (VT), vital capacity (VC) and inspiratory capacity (IC) (Oyarzún, 2009).

Only one study (Pessoa et al., 2017) evaluated the action of vibration platforms on the regional distribution of the thoracic cavity. No changes were observed in analyzing the three chest wall compartments (pulmonary rib cage, abdominal rib cage and abdomen) during quiet breathing. However, when the respiratory effort maneuver is required, the elderly who used vibration training reached significantly higher inspiratory capacity values than the other groups. An analysis of thoracic-abdominal kinematics also shows that the greater volume displaced during the inspiratory capacity maneuver can be found in the pulmonary rib cage and abdominal rib cage; regions which are generally related to the accessory musculature, indicating that there was improvement in the function of these muscles as inspiratory pressure generators, a result corroborated by the positive correlation of r = 0.979 between pulmonary rib cage and MIP.

Since IC volume optimization is due to an increase in MIP, which in turn was provoked by the accessory musculature training due to the assumed position, this increase becomes clinically significant, since it allows greater ventilation competence during the effort, which reduces respiratory fatigue; this is the main limiting factor for the elderly to exercise, thereby making them more capable of exercising (Oyarzún, 2009).

5.7 CONCLUSION

Although other types of training programs such as aerobic or resistance exercises are also capable of promoting improvements in the physical fitness of this population segment, WBV can be a very attractive training option for potentially sedentary subjects with reduced muscular strength that need to incorporate an exercise program that does not involve additional physical exertion or joint injury.

REFERENCES

Bogaerts, A. C., et al. (2009). Effects of whole body vibration training on cardiorespiratory fitness and muscle strength in older individuals (a 1-year randomized controlled trial). *Age and Ageing*, 38(4):448–454.

Bogaerts A. C., et al. (2011). Changes in balance, functional performance and fall risk following whole body vibration training and vitamin D supplementation in institutionalized elderly women. A 6 month randomized controlled trial. *Gait Posture*, 33(3):466–472.

Bosco, C., et al. (1999). Adaptive responses of human skeletal muscle to vibration exposure. *Clinical Physiology*, 19: 183–187.

Bruyere, O., et al. (2005). Controlled whole body vibration to decrease fall risk and improve health-related quality of life of nursing home residents. *Archives of Physical Medicine and Rehabilitation*, 86(2):303–307.

Cardinale, M. and Pope, M. H. (2003). The effects of whole body vibration on humans: Dangerous or advantageous? *Acta Physiologica Hungarica*, 90:195–206.

Cardinale, M., et al. (2010). Hormonal responses to a single session of whole body vibration exercise in older individuals. *British Journal of Sports Medicine*, 44(4):284–288.

Cristi, C., et al. (2014). Whole-body vibration training increases physical fitness measures without alteration of inflammatory markers in older adults. *European Journal of Sport Science*, 14(6):611–619.

Di Giminiani, R., et al. (2014). Hormonal and neuromuscular responses to mechanical vibration applied to upper extremity muscles. *PLoS ONE*, 9(11).

Fielding, R. A., et al. (2011). Sarcopenia: an undiagnosed condition in older adults. Current consensus definition: Prevalence, etiology, and consequences. International working group on sarcopenia. *Journal of the American Medical Directors Association*, 12(4):249–256.

Gómez-Cabello, A., et al. (2013). Effects of a short-term whole body vibration intervention on physical fitness in elderly people. *Maturitas*, 74(3):276–278.

Herrero, A. J., et al. (2011). Whole-body vibration alters blood flow velocity and neuromuscular activity in Friedreich's ataxia. *Clinical Physiology and Functional Imaging*, 31:139–144.

Hepple, R. T. and Rice, C. L. (2016). Innervation and neuromuscular control in ageing skeletal muscle. *The Journal of Physiology*, 594(8):1965–1978.

Houston, M. N., et al. (2015). The effectiveness of whole-body-vibration training in improving hamstring flexibility in physically active adults. *Journal of Sport Rehabilitation*, 24(1):77–82.

International Standards Organization (1997). Mechanical Vibration and Shock – Evaluation of Human Exposure to Whole Body Vibration. Part 1: General Requirements, ISO 2631-1. Geneva.

Jordan, M. J., et al. (2005). Vibration training: an overview of the area, training consequences, and future considerations. *Journal of Strength and Conditioning Research*, 19(2):459–466.

Kenni, S. E., et al. (2013). Effects of fitness and vibration training on muscle quality: A 1-year postintervention follow-up in older men. *Archives of Physical Medicine and Rehabilitation*, 94(5):910–918.

Kerschan-Schindl, K., et al. (2001). Whole-body vibration exercise leads to alterations in muscle blood volume. *Clinical Physiology*, 21:377–382.

Kessler, J., et al. (2014). Effect of stochastic resonance whole body vibration on functional performance in the frail elderly: A pilot study. *Archives of Gerontology and Geriatrics*, 59(2):305–311.

Lin, C. I., et al. (2015). Effect of whole-body vibration training on body composition, exercise performance and biochemical responses in middle-aged mice. *Metabolism*, 64(9):1146–1156.

Lohman E. B., et al. (2007). The effect of whole body vibration on lower extremity skin blood flow in normal subjects. *Medical Science Monitor*, 13(2):71–76.

Nelson, M. E., et al. (2007). Physical activity and public health in older adults: Recommendation from the American College of Sports Medicine and the American Heart Association. *Circulation*, 166(9):1094–1105.

Oyarzún, M. (2009). Pulmonary function in aging. *Revista Médica de Chile*, 137(3):411–418.

Pessoa, M. F., et al. (2017). Vibrating platform training improves respiratory muscle strength, quality of life, and inspiratory capacity in the elderly adults: A randomized controlled trial. *The Journals of Gerontology. Series A, Biological Sciences and Medical Sciences*, 72(5): 683–688.

Rittweger J., Beller, G., and Felsenberg, D. (2001). Oxygen uptake during whole-body vibration exercise: Comparison with squatting as a slow voluntary movement. *European Journal of Applied Physiology*, 86:169–173.

Roberts, S., et al. (2016). Ageing in the musculoskeletal system: Cellular function and dysfunction throughout life. *Acta Orthopaedica*, 87(363):15–25.

Rodriguez-Miguelez, P., et al. (2015). Whole-body vibration improves the anti-inflammatory status in elderly subjects through toll-like receptor 2 and 4 signaling pathways. *Mechanisms of Ageing and Development*, 150:12–19.

Roelants, M., Delecluse, C., and Verschueren, S. M. (2004). Whole-body-vibration training increases knee-extension strength and speed of movement in older women. *Journal of the American Geriatrics Society*, 52(6):901–908.

Santin-Medeiros, F., et al. (2015). Effects of eight months of whole body vibration training on hip bone mass in older women. *Nutrición Hospitalaria*, 31(4):1654–1659.

Segal, N. A., et al. (2012). Vibration platform training in women at risk for symptomatic knee osteoarthritis. *PM and R*, 53.

Simão, A. P., et al. (2012). Functional performance and inflammatory cytokines after squat exercises and whole-body vibration in elderly individuals with knee osteoarthritis. *Archives of Physical Medicine and Rehabilitation*, 93(10):1692–1700.

Slatkovska, L., et al. (2010). Effect of whole-body vibration on BMD: A systematic review and meta-analysis. *Osteoporosis International*, 21:1969–1980.

Srinivas-Shankar, U., et al. (2010). Effects of testosterone on muscle strength, physical function, body composition, and quality of life in intermediate-frail and frail elderly men: A randomized, double-blind, placebo-controlled study. *Journal of Clinical Endocrinology and Metabolism*, 95(2):639–650.

Tankisheva, E., et al. (2014). Effects of intensive whole-body vibration training on muscle strength and balance in adults with chronic stroke: A randomized controlled pilot study. *Archives of Physical Medicine and Rehabilitation*, 95(3):439–446.

Verschueren, S. M., et al. (2004). Effect of 6-month whole body vibration training on hip density, muscle strength, and postural control in postmenopausal women: A randomized controlled pilot study. *Journal of Bone and Mineral Research*, 19(3):252–359.

Von Stengel, S., et al. (2012). Effect of whole-body vibration on neuromuscular performance and body composition for females 65 years and older: A randomized-controlled trial. *Scandinavian Journal of Medicine and Science in Sports*, 22(1):119–127.

WHO (2015). World aging report. Retrieved from http://www.who.int/ageing/events/world-report-2015-launch/en/ [accessed 2 October 2018].

6 Effects of WBV in Individuals with Diabetes and Diabetic Neuropathy

Caroline Cabral Robinson

CONTENTS

6.1 DIABETES

Diabetes mellitus (DM) is a syndrome of multiple etiology, due to lack of insulin or the inability of insulin to adequately perform its metabolic effects (ADA, 2016). It is characterized by chronic hyperglycemia, frequently associated with dyslipidemia, arterial hypertension and endothelial dysfunction (American Diabetes Association [ADA], 2016).

According to the ADA, an international reference for the diagnosis and treatment of this condition, the classification of diabetes includes four clinical classes, three of which are the main ones:

Type 1 diabetes (T1DM): Occurs by destruction of β cells of the pancreas, generally leading to absolute insulin deficiency. It is autoimmune or idiopathic and is usually diagnosed in childhood;

Type 2 diabetes (T2DM): Occurs due to insulin resistance and/or insulin deficiency, caused by the interaction of genetic and environmental factors. Among the associated environmental factors are sedentary lifestyle, high fat diets and high consumption of tobacco and alcohol. It can occur at any age, but is usually diagnosed in the fourth decade of life;

Gestational diabetes (GDM): Occurs from any glucose intolerance of variable magnitude, with onset or diagnosis during gestation. In most cases, there is reversal to normal tolerance after pregnancy, but there is a 10%–63% risk of developing T2DM within 5–16 years of pregnancy.

There are other specific types of diabetes from other causes such as genetic defects in the functioning of pancreatic β cells or insulin action, exogenous or endogenous diseases damaging the pancreas, induction of diabetes by traumatic stress, drugs, chemicals or infections, but its incidence represents less than 1% of the overall incidence (ADA, 2016).

The frequency of DM is assuming epidemic proportions. It was estimated that by the end of 2015, about 415 million individuals between 29 and 79 years of age lived with DM in the world, 90% of whom were T2DM (International Diabetes Federation [IDF], 2015). Developing countries contribute about 80% of these estimated cases. It can be said that, for every 11 people, one has diabetes (IDF, 2015). The projection for 2040 is not encouraging since is estimated that 642 million individuals will live with diabetes, much of which will be underdiagnosed (IDF, 2015). The prevalence DM is increasing due to population growth and aging, greater urbanization, the progressive prevalence of obesity and sedentary life, as well as the greater survival of patients with DM (ADA, 1988).

According to the ADA (2016), the DM diagnosis can be confirmed by four different criteria, three criteria being used for glycaemia values, and one criterion is glycated hemoglobin:

1. Symptoms of hyperglycemia (polyuria, polyphagia and excessive weight loss) and plasma glucose ≥ 200 mg/dL; plasma glucose defined as that collected at any time of the day, without observing the interval since the last meal; or
2. Fasting plasma glucose of at least eight hours ≥126 mg/dL; or
3. Plasma glucose ≥200 mg/dL, two hours after the ingestion of 75 g of anhydrous glucose dissolved in water; or
4. Glycated hemoglobin (Hb A1c) ≥ 6.5%.

The criterion used should be repeated once if the value was below the expected value for hyperglycemia in the first assessment (ADA, 2016). Criterion 4 should be repeated at least once again at a time other than the first exam (ADA, 2016).

The chronic nature of DM, the severity of its complications, and all the necessary resources to manage, make it a very onerous disease (Cameron et al., 2001). In the U.S., it was estimated that the costs of health care for an individual with DM were two to three times higher than for an individual without the disease (Feldman et al., 1994). Intangible costs (e.g., pain, anxiety and loss of quality of life) also have a major impact on the lives of people with diabetes and their families, which is difficult to quantify (Feldman et al., 1994).

6.2 DIABETIC PERIPHERAL NEUROPATHY

Among the main complications of DM, diabetic neuropathy (DNP) is the most prevalent, reaching a frequency of 50% up to 100%, depending on the diagnostic test used, compared to 25%–30% of retinopathy and 20% of diabetic nephropathy (Shaw and Zimmet, 1999).

The definition of DPN for clinical practice is the presence of symptoms and/or signs of peripheral nerve dysfunction in diabetic individuals after exclusion of other causes (Tesfaye et al., 2010). Sensorimotor neuropathy is the most prevalent type of DPN because it is easier to diagnose, being considered the typical diabetic neuropathy (Testfay et al., 2010). Autonomic neuropathy is identified later because it relies on less clinically viable methods, becoming the second most prevalent form (Milech et al., 2016). Thus, clinical manifestations of DPN vary widely, ranging from specific, nonspecific sensory, somatic or autonomic symptoms to asymptomatic presentations (Tesfaye et al., 2010; Milech et al., 2016).

The pathophysiology of DPN is characterized by progressive loss of nerve fibers, which presents a systemic pattern with distal to proximal involvement, related to the diameter of the fibers. The thin fibers are preferentially affected in the initial stages of the disease, followed by the involvement of the large fibers, evidenced by reduction of the nerve conduction velocity or reduction of the vibratory sensitivity (Yagihashi et al., 2007). During the active process of degeneration of nerve fibers, symptoms appear as pain or paresthesia (positive symptoms). When the fibers were already lost, there is loss of sensitivity (negative symptoms). The risk of the negative symptoms is precisely the lack of complaint of the diabetic patient and, consequently, the greater risk of plantar ulcers (Yagihashi et al., 2007).

Several hypotheses have been proposed to elucidate the pathophysiological mechanisms that trigger DPN. Such mechanisms could be indirect or direct. Indirect mechanisms are associated with enzymatic glycation of structural proteins and vascular disease.

Enzymatic glycation refers to the process by which glucose chemically and irreversibly binds to the amino group of the proteins without the presence of enzymes activity. The degree of glycation is directly related to the degree of glycaemia (Cotran et al., 1996). The glycation of long-term proteins in the vessel walls is potentially pathogenic. In the large vessels, it can cause entrapment of the low-density lipoprotein in the intima, thus accelerating the atherogenesis processes. In the capillaries, the plasma proteins, albumin-like type, are fixed in the glycosylated basement membrane, being responsible, in part, for the greater thickness of the basal membrane, characteristic of diabetic microangiopathy. The latter, when installed in *vasa nervorum*, contributes to the establishment of diabetic neuropathy (Cotran et al., 1996). These vascular changes are also related to endoneural hypoxia and oxidative stress. Constant hyperglycemia may either decrease the synthesis of growth factors in neurons or Schwann cells, or interrupt its retrograde transport to the neuronal cell body (Stevens et al., 1995).

Direct damage of nervous fibers could occur by hyperactivity of the polyol pathway as glucose have free accesses to endoneural plasmatic site. Given the increase on glucose metabolism inside the nervous fiber, hyperglycemia leads to an increase in intracellular osmolarity with a consequent influx of water, which can result in osmotic cellular damage and axono-glial disruption (Fregonesi et al., 2004).

As a consequence of peripheral nervous degeneration, and loss of protective sensation of the feet, diabetic foot is one of the main complications of DPN.

This condition is considered a challenge for public health given its management and the consequences that can lead to amputations and disability (Boulton et al., 1983). Patients with DPN frequently evolve to a degree of denervation of their lower limbs which may culminate in anesthesia (Boulton et al., 1983). The absence of pain or sensitivity discourages the care of the feet, increasing the risk of ulcerative lesions (Boulton et al., 1983).

Even that sensory fibers degeneration leads to a loss of sensation, during the degenerative processes, up to 50% of all DPN patients may experience painful symptoms (Michigan University, 2005). Neuropathic pain is the main reason patients seek treatment (Vileikyte et al., 2003).

The diagnosis of painful DPN is clinical and by exclusion, based on complaints of distal localization pain, symmetrical distribution and nocturnal exacerbation, with descriptions of symptoms such as burning pain, electric shock sensation, stabbing pain, lancinating pain, tingling (paresthesia) and contact pain (allodynia), caused by daytime clothing or bed sheets at night (Vileikyte et al., 2003; Vinik et al., 2005; Poole et al., 2009; Tesfaye et al., 2011). Altered temperature perception can be described as "walking barefoot on cold stone" or "walking barefoot on warm sand" (Tesfaye et al., 2011). These symptoms start at the feet and progress to the entire lower limb symmetrically (Vileikyte et al., 2003; Vinik et al., 2005; Poole et al., 2009; Tesfaye et al., 2011). Neuropathic pain is associated with the presence of fatigue, depression, and low levels of health-related quality of life (Vileikyte et al., 2003; Vinik et al., 2005; Poole et al., 2009; Tesfaye et al., 2011). Nerve conduction tests are important in excluding other causes of pain, such as radiculopathies, for example. In addition, the sensory, motor and autonomic dysfunction also contributes to proprioceptive changes responsible for maintaining normal posture and locomotion. Such conditions increase the risk for falls in individuals with DPN (Maurer et al., 2005).

6.3 PREVENTION AND MANAGEMENT OF DIABETES AND DIABETIC NEUROPATHY

There is no prevention for T1DM. Its management is chronic and occurs through the continuous use of exogenous insulin to control glycemic levels and the complications related to hyperglycemia. As for T2DM, prevention is related mainly to lifestyle changes: adequate nutrition; maintenance of adequate body mass; regular physical exercise and no longer smoking or consuming alcohol (Ang et al., 2014; Knowler et al., 2002). Management of T2DM could be performed to use of antiglycemic drugs, insulin replacement and lifestyle changes. The adequate glycemic management, inhibits the appearance of complications of DPN, as well as its intensity and extension in T1DM and reduces its progression in T2DM (Ang et al., 2014; Knowler et al., 2002; Martin et al., 2014; No author listed 1995; DCCRG, 1995; Vincent et al., 2004; Javed et al., 2015). Some studies also suggest that good metabolic control may improve already established neuropathy (Ang et al., 2014; Knowler et al., 2002; Martin et al., 2014; DCCT Research Group, 1995; DCCRG, 1995; Vincent et al., 2004; Javed et al., 2015).

Regarding management of DPN after installed, first-choice treatment is pharmacologic (Javed et al., 2015). Some therapeutic options are based on the pathogenesis of DPN, such as the use of angiotensin converting enzyme inhibitors (ACE inhibitors) or antioxidant agents such as alpha-lipoic acid, benfotiamine and thioctic acid (Javed et al., 2015). However, the pharmacologic treatment of sensorimotor neuropathy is symptomatic, especially when the main complaint is neuropathic pain (Javed et al., 2015). The main therapeutic options for the paresthesia and pain in patients with DPN are drugs acting on the nervous system such as antidepressants (amitriptyline, imipramine, nortriptyline, duloxetine and venlafaxine) and anticonvulsant (pregabalin and gabapentin) (Javed et al., 2015). Given the drug interaction and adverse effects of the available pharmacologic treatments, adherence is low (Javed et al., 2015; Praet and van Loon, 2008). In addition, the treatment of DPN should be multiprofessional and the identification of effective and safe adjuvant interventions is still necessary (Tesfaye et al., 2013).

Physical exercise plays an important role in prevention and management of T2DM and DM related complications (Colberg et al., 2010). During physical exercise, several adaptive cardiovascular and metabolic physiological responses occur. Muscle contraction increases blood glucose uptake and this increase is greater in intensities exceeding 50% of maximal oxygen consumption (Suh et al., 2007). The magnitude of this increased glucose uptake is influenced by the intensity and duration of physical exercise. The energy for this increase in consumption comes from muscle stores of glycogen, triglycerides, free fatty acids and glucose from hepatic origin (Suh et al., 2007). Both aerobic and resistance training improve insulin action and can assist with the management of blood glucose levels, lipids, blood pressure, cardiovascular risk, mortality and quality of life.

However, exercise must be undertaken regularly for continued benefits (Colberg et al., 2010; Suh et al., 2007; Hayashino et al., 2012). In order to achieve all the benefits, metabolic (better glycemic control) or not, related to physical exercise, the ADA recommends that physical exercise should be performed for at least 150 minutes per week, with moderate intensity (50%–70% of maximal heart rate), and should be performed three days a week at intervals of no more than two days (Colberg et al., 2010). Data from meta-analysis investigating patients with T2DM who performed aerobic exercise compared to those who did not perform any intervention, showed that this type of exercise is able to reduce HbA1c effectively by approximately 0.5%, in addition to improving response to insulin, decreases lipids levels in the blood, reduces visceral adipose tissue and triglycerides, but without changes in cholesterol (Boulé et al., 2001).

Therefore, physical exercise is one of the recommendations that composes changes in lifestyle and, by collaborating with the metabolic control of diabetes, is a protection factor for T2DM onset and complications of any type of DM (references 33 to 41) Despite this, most people with T2DM and those with DPN are not active, mirroring the inertia of a lifetime of habits and motivational barriers such as lack of interest, lack of time and depression (ADA, 1988). In addition, physical disabilities and perceived discomfort when exercising, as pain or altered foot sensitivity, are challenges to adherence to physical activity (Praet and van Loon, 2008; Morrato et al., 2007).

6.4 WHOLE BODY VIBRATION AS AN ADJUVANT TREATMENT IN DIABETES AND DIABETIC NEUROPATHY

Among the alternatives aimed to increase overall physical activity, whole body vibration (WBV) has been shown to be a new effective option in healthy subjects and individuals with several health conditions (Cochrane, 2011). It is assumed that vibration activates muscle spindles and evokes muscle contractions induced by a complex spinal and supraspinal neurophysiological mechanism known as tonic vibration reflex, allowing muscular activity enhancement even in static positions (Zaidell et al., 2013). Furthermore, WBV requires significantly less time than conventional training and, therefore, reached a satisfactory compliance in previously inactive patients (Lam et al., 2012). Furthermore, it is supposed that vibration reduces pain both to gate control theory (Melzack and Wall, 1965) and diffuse central noxious inhibitory control (Coghill et al., 1994).

6.4.1 Effects of Whole Body Vibration in Diabetes

Evidence of positive effects of WBV is available from a systematic review including 70 participants with T2DM from two randomized controlled trials (Robinson et al., 2016). Patients investigated were of both gender with T2DM diagnoses without established complications and HbA1c >10%, or fasting blood glucose >250 ml/dl, whose ages ranged from adult to elderly (Behboudi et al., 2011; Sañudo et al., 2013; del Pozo-Cruz et al., 2013). The characteristics of each study is presented in Table 6.1.

The exposure used for patients with T2DM was the intermittent (vibration exposure with intervals at the same proportion), with peak acceleration between 1 and 2 g and total session duration ranging from 12 (8–16) to 14 (16–24) minutes. Individuals stood on the vibrating platform in a 100 to 110° squat position (considering total knee extension as 180°) (Behboudi et al., 2011; Sañudo et al., 2013; del Pozo-Cruz et al., 2013; del Pozo-Cruz et al., 2014). The vibratory stimulus was not isolated as patients performed exercises over the vibrating platform. The results of WBV intervention were investigated after eight and twelve weeks (Behboudi et al., 2011; Sañudo et al., 2013; del Pozo-Cruz et al., 2013; del Pozo-Cruz et al., 2014).

After the 12 week program of upper and lower limb exercises performed on the vibrating platform, participants in the intervention group exhibited significantly lower levels of HbA1c at the time of follow-up when compared to the control group, with a mean difference of −0.55% (95% CI: −0.15 to −0.76) (del Pozo-Cruz et al., 2014). The effect size for HbA1c improvement after WBV and exercise was close to the one found after aerobic or resistance training reported previously in two meta-analyses (Goldstein et al., 2004; Umpierre et al., 2013). Although the vibratory stimulation was not isolated from exercises in the proposed interventions, session duration was considerably lower in the WBV studies (8–24 minutes) than in the aerobic or resistance training studies (40–75 minutes) (Goldstein et al., 2004; Umpierre et al., 2013). This fact reinforces the role of WBV to potentiate the effects of exercise in shorter intervention time.

TABLE 6.1
Characteristic of the Included Studies

Author, Year	Follow-up	Participants	Intervention Group/ Comparison Group	Gender IG %/CG %	Age in Years Mean (sd) IG/CG	Description of Intervention	Description of Comparison	Outcomes	Results
Behboudi et al. (2011)	8-wk	DM2 diagnosis, males, <250 mg/dl 12-h FBG, non smoking or in regular exercise programs.	WBV + Aerobic exercise (AE): 10; /AE: 10; control (C): 10.	WBV: 100 (M) /AE: 100 (M) C:100 (M)	WBV: 49.20 (3.94); /AE: 53.10 (6.57); C: 52.30 (6.17)	30-min of increasing aerobic program, stood (110° squat positioning) on a platform (2 mm amplitude; 30 Hz; 1-min vibration and 1-min of rest).	AE: 30-min of increasing aerobic program; C: keep routine activities.	VO_{2max} (1 mile walk test); BMI; % BFM (caliper and Sirri formula); insulin, 12-h FBG, HbA1c (cubital blood).	After 4 and 8-wk intervention, VO_{2max} significantly increased only in AE group. BMI, % BFM, 12-h FBG, HbA1c and insulin did not change significantly in AE or WBV groups. 12-h FBG was significantly higher in C group than post intervention WBV and AE groups.
Del Pozo-Cruz et al. (2013)	12-wk	DM2 diagnosis by ADA criteria, HbA1c <10%, not receiving physical therapy.	WBV + exercises: 19/usual-care control group (C): 20.	WBV: 55 (M); 45 (F) /C: 50 (M); 50 (F)	WBV: 71.60 (8.54) /C: 66.80 (10.83)	Eight up and lower body extremities exercises with progressively 30 to 60-s duration (30-s interval between them), in a side-altering platform (1 to 2 g; 12 to 16 Hz; 4 mm peak to peak amplitude) in a 100° knees flexed squat position.	C: Keep nutritional and exercise habits.	TUG; Postural sway on the Wii Balance Board (WBB): AP and ML CoP excursion with eyes open (EO) and closed (EC), feet apart (FA) and together (FT).	After 12-wk, significant between-group differences in CoP excursions with EC (FA and FT) were found. Participants in the WBV group exhibited significantly lower CoP excursions with their EC after the intervention, while participants in the control group experienced a non-significant greater excursion with EO (ML). No significant difference in the TUG values post intervention.

(Continued)

TABLE 6.1 (CONTINUED)
Characteristic of the Included Studies

Author, Year	Follow-up	Participants	Intervention Group/ Comparison Group	Gender IG %/CG %	Age in Years Mean (sd) IG/CG	Description of Intervention	Description of Comparison	Outcomes	Results
Sañudo et al. (2013)	12-wk	The same participants of del Pozo-Cruz et al., 2013.	WBV group (WBV): 20 / usual-care control group (C): 20.	WBV:55 (M); 45 (F)/C: 50 (M); 50 (F)	WBV: 72 (8)/C: 67 (11)	WBV: description on study del Pozo-Cruz et al., 2013	C: Keep nutritional and exercise habits.	Body composition: waist circumference, waist-to-hip ratio, weight, height, % BFM, heart rate and blood flow, femoral artery diameter, maximum systolic velocity, maximum diastolic velocity, time averaged mean, pulsatility index and resistance index, mean velocity, and peak blood velocities.	After 12-wk of WBV intervention, weight, waist circumference, waist-to-hip ratio, %BFM, blood flow and maximum diastolic velocity, significantly improved and compared with C group. Mean velocity, maximum diastolic velocity and peak blood velocities showed significant differences within WBV analysis.

(Continued)

TABLE 6.1 (CONTINUED)
Characteristic of the Included Studies

Author, Year	Follow-up	Participants	Intervention Group/ Comparison Group	Gender IG %/CC %	Age in Years Mean (sd) IG/CG	Description of Intervention	Description of Comparison	Outcomes	Results
Del Pozo-Cruz et al. (2014)	12-wk	The same participants of del Pozo-Cruz et al., 2013.	WBV group (WBV): 19/ usual-care control group (C): 20.	WBV: 55 (M); 45 (F) /C:50 (M); 50 (F)	WBV: 71.60 (8.54)/C: 66.80 (10.83)	WBV: description on study del Pozo-Cruz et al., 2013.	C: Keep nutritional and exercise habits.	HbA1c; 12-h FBG; Cholesterol, triglycerides, atherogenic index, high density lipoprotein (HDL) and low-density lipoprotein (LDL); TUG; 6MWT distance; 30 s-STS test. Feasibility, adherence, compliance and safety.	HbA1, 12-h FBG, cholesterol, triglycerides and atherogenic index significantly decreased in WBV group compared to C group. No significant changes were detected for HDL, LDL and LDL/HDL as well as TUG. 6MWT distance and 30s-STS test significantly improved in WBV group compared to C group. There was no report of negative effects during treatment. Drops-out were regarding lack of time or interest and 76% of all participants completed the 12-wk program.

6MWT, six-minute walk test distance; 12-h FBG, 12 hours fasting blood glucose; 30s-STS: 30-second seat to stand.ADA, American Diabetes Association; AE, aerobic exercise; AP, anetro-posterior; % BFM, percentage of body fat mass; BMI, body mass index; C, control group; CG, comparison group; CoP, center of pressure; DM2, diabetes mellitus type 2; EO, eyes open; EC, eyes closed; FA, feet apart; FT, feet together; HbA1c, glycated hemoglobin; HDL, high density lipoprotein; IG, intervention group; LDL, low-density lipoprotein; ML, medio-lateral; SD, standard deviation; TUG, timed up and go test; VO_{2max}, maximal oxygen uptake; WBB: Wii Balance Board; WBV, whole body vibration; wk, week;

Body mass index did not decrease after the WBV interventions. Although WBV has gained popularity as a modality for weight loss, it does not have the ability to generate large energy expenditure to substitute conventional aerobic exercise (Umpierre et al., 2013). However, positive effects could be seen in some of the blood and physical markers of cardiovascular risk (cholesterol, triglycerides, atherogenic index, body weight, waist circumference and waist-to-hip ratio) that improved after WBV combined with exercises (del Pozo-Cruz et al., 2014) in individuals with T2DM. Improvements in physical capacity and lower limbs strength.

Despite the slight beneficial effect of WBV intervention on glycemic control, a paramount outcome for T2DM management, caution is required in extrapolating this result to practice. First, a significant reduction in glycemic values was found in comparison with no intervention and WBV was not investigated alone, but in addition to exercise. Similar caution must be taken regarding blood markers and functional capacity.

Adverse effects, such as hypoglycemia, discomfort, and musculoskeletal injuries, are highly reported in studies performing exercise interventions (Goldstein et al., 2004), however they were not reported in the studies assessing WBV in patients with T2DM· WBV performed close to the parameters presented in the primary studies and combined with low-level exercises seems to be a safe, feasible and less time-consuming intervention to help improve the glycemic control, cardiovascular risk markers, and functional capacity of individuals with T2DM in an exposure-dependent way compared to no intervention. However, given the methodological weaknesses of the primary studies and the heterogeneous protocols, confidence is limited on the decreasing effect of WBV on 12-h FBG. Further well-designed trials are still required to strengthen the current evidence and clarify whether the effect should be attributed to vibration, exercise or the combination of both.

6.4.2 Effects of Whole Body Vibration in Diabetic Neuropathy

Among the important outcomes for individuals with DPN, only glycemic profile (Lee et al., 2013), neuropathic pain (Hong et al., 2013) and balance were assessed in primary clinical studies (Hong et al., 2013; Andersen et al., 2004; Kordi Yoosefinejad et al., 2015). There is no information about effects of WBV in plantar sensitivity. WBV is associated with a slight improvement of the glycemic profile but quality of the evidence is low. Neuropathic pain and balance seem to improve in patients with DPN after WBV intervention, but evidence to support those outcomes is very low. Further studies are likely to change the estimated effect not supporting the current use of WBV in patients with DPN for neuropathic pain relief and improvement of balance and plantar tactile sensitivity.

Clinical studies assessed individuals of both genders with T1 and T2DM aged from adult to elderly with DPN. Most of studies are of poor methodological quality (uncontrolled or controlled pre-test/post-test) (Hong et al., 2013; Andersen et al., 2004; Kordi Yoosefinejad et al., 2015) and only one study is a randomized controlled trial (RCT) (Lee et al., 2013). Follow-up ranged from four to six weeks. WBV intervention was applied through intermittent vibratory protocols with a total exposure time shorter than 15 minutes. Frequencies ranged from 15 to 30 Hz, peak to peak amplitude ranged from 1 to 5 mm. Intervention was performed two to three times a week. All the studied protocols in patients with DPN were intermittent with at least 30 seconds of resting between vibration expositions (Lee et al., 2013; Hong et al., 2013; Kordi Yoosefinejad et al., 2015). Table 6.2 shows the main information about these studies.

Neuropathic pain is the most disabling symptom in patients with DPN, impairing quality of life (Lee et al., 2013). Usually, neuropathic pain is difficult to manage since patients often present low adherence to drug treatment (Tesfaye et al., 2013). In fact, the participants included in the studies investigating the effects of WBV in neuropathic pain were taking gabapentin, opiate analgesics and/or non-steroid anti-inflammatories to control their pain symptoms, but no participants reported satisfaction from these treatments (Hong et al., 2013). In primary clinical studies, WBV reduced both acute and long-term pain in patients with NDP. It is supposed that vibration reduces pain both to gate control theory (Melzack and Wall, 1965) and diffuse central noxious inhibitory control (Coghill et al., 1994). Unfortunately, it is not possible to affirm that WBV decreases neuropathic pain as quality of the evidence is very low and placebo effects cannot be completely excluded.

Previous studies with healthy individuals have shown that continuous vibration protocols acutely decreased the plantar tactile sensitivity (Schlee et al., 2012; Sonza et al., 2015). The studies assessing individuals with DPN used intermittent vibration protocols but none of the studies investigated the effects on plantar tactile sensitivity (references 56 to 59 and 62 to 65). Furthermore, there is no previous direct or indirect information about the impact of intermittent vibration protocol on this outcome. Since plantar tactile sensitivity is a predictor for foot ulceration and amputation in individuals with DPN (Dyck et al., 2013), further studies should assess this outcome to provide comprehensive information about the safety of this intervention.

There is strong evidence that WBV improves balance in frail populations, for example, elderly individuals (Lam et al., 2012; Orr, 2015). Given this indirect evidence and the benefits on balance provided by an RCT investigating WBV in patients with DPN, it is possible to consider that WBV should be used to improve balance in these patients. However, studies were required to understand whether improvement in balance is enough to prevent falls in this population.

6.4.3 Clinical Application

WBV can be used as an adjunctive intervention to motivate patients with T2DM to adhere to regular exercise. When performed in a 100 to 110° squat position together with low-level intensity exercises in intermittent exposure, with peak acceleration between 1 and 2 g and total session duration ranging from 12 (8–16) to 14 (16–24) minutes, is expected to slightly decrease glycemic levels, improve cardiovascular risk factors, improve functional capacity and balance. For individuals with DPN, WBV can be used to the same objectives but with attention to feet sensitivity. Results are expected after twelve weeks intervention but as evidence has moderate to low methodological quality, further studies can change the magnitude or effect directions.

TABLE 6.2
Characteristics of the Studies Investigating Individuals with Diabetic Neuropathy

Author, Year	Study Design	Participants	Age	Gender (masc/fem)	Intervention	Comparison	Outcomes	Follow-up	Conclusion
Kessler and Hong (2013)	Uncontrolled pretest post-test	DM 1 and 2, with DPN	56.12 (6.78)	6/2	**WBV:** Four bouts of three-min WBV (25 Hz; 5 mm); 30-s of rest between bouts; three times a week.	Not applicable	Pain intensity: NPD each week and VAS each pre and post session and duration (in hours). Balance: gait parameters.	Four weeks	A four-week WBV intervention reduced neuropathic pain, and improved gait parameters.
Lee et al. (2013)	RCT	DM 2, with DPN ≥ 65 years, either two or more falls during the previous 12 months or one fall plus a TUG test > 15 sec or recurrent unexplained falls.	**WBV + BE:** 76.31 (4.78) **BE:** 74.05 (5.42) **CG:** 75.77 (5.69)	**WBV + BE:** 9/10 **BE:** 7/11 **CG:** 8/10	**WBV + BE:** BE (as the comparison group) and WBV: three bouts of three-min with one-min of rest between bouts; three times a week. • first week: 15 Hz and 2 mm, • second and third weeks: 20 Hz and 1 mm, • fourth and fifth weeks: 25 Hz and 2 mm, • sixth week: 30 Hz and 3 mm	**CG:** without intervention **BE:** 60-min balance exercise twice a week, which progressive strength, balance, and functional mobility training, for six weeks.	Glycemic profile: HbA1C Balance: Postural stability (CoP sway and velocity moment at force plate). Dynamic stability (OLST, BBS, FRT and TUG).	Six weeks	A six-week balance exercise program involving WBV significantly improved HbA1c levels and balance, in comparison with CG and BE groups.

(Continued)

TABLE 6.2 (CONTINUED)
Characteristics of the Studies Investigating Individuals with Diabetic Neuropathy

Author, Year	Study Design	Participants	Age	Gender (masc/fem)	Intervention	Comparison	Outcomes	Follow-up	Conclusion
Kordi Yoosefinejad et al. (2015)	Controlled pretest-post-test	DM 1 or 2 with DPN; HbA1C < 8.5%; BMI between 25; age between 50 and 70 years	**WBV:** 57 (1.8) **CG:** 57 (1.5)	**WBV:** 6/4 **CG:** 6/4	**WBV:** synchronous plate, two times a week (30 Hz, 2 mm). Application time increased every two weeks: 30-s/45-s/one-min.	**CG:** without intervention	Balance: OLST, TUG, eight different positions to perform on the force plate.	6 weeks	A six-weeks WBV intervention significantly improved TUG time in comparison with CG.

BBS, Berg balance scale; BE, balance exercises; BMI, body mass index; CG, control group; CoP, center of pressure; DM, diabetes; DPN, diabetic peripheral neuropathy; FRT, Functional reach test. OLST, one leg stance test; RCT, randomized controlled trial; TUG, timed up and go; VAS, visual analogue scale; WBV, whole body vibration.

REFERENCES

American Diabetes Association (2016). Classification and diagnosis of diabetes. *Diabetes Care*, 39 (Supplement 1):S13–S22.

American Diabetes Association (1988). Consensus statement: Report and recommendations of the San Antonio conference on diabetic neuropathy. American Diabetes Association American Academy of Neurology. *Diabetes Care*, 11(7):592–597.

Andersen, H., et al. (2004). Muscle strength in type 2 diabetes. *Diabetes*, 53(6):1543–1548.

Ang, L., et al. (2014). Glucose control and diabetic neuropathy: Lessons from recent large clinical trials. *Current Diabetes Reports*, 14(9):528.

Balducci, S., et al. (2006). Exercise training can modify the natural history of diabetic peripheral neuropathy. *Journal of Diabetes and its Complications*, 20(4):216–223.

Behboudi, L., et al. (2011). Effects of aerobic exercise and whole body vibration on glycaemia control in type 2 diabetic males. *Asian Journal of Sports Medicine*, 2(2):83–90.

Boulé, N. G., et al. (2001). Effects of exercise on glycemic control and body mass in type 2 diabetes mellitus: A meta-analysis of controlled clinical trials. *JAMA*, 286(10):1218–1227.

Boulton, A. J., et al. (1983). Dynamic foot pressure and other studies as diagnostic and management aids in diabetic neuropathy. *Diabetes Care*, 6(1):26–33.

Cameron, N. E., et al. (2001). Vascular factors and metabolic interactions in the pathogenesis of diabetic neuropathy. *Diabetologia*, 44(11):1973–1988.

Castaneda, C., et al. (2002). A randomized controlled trial of resistance exercise training to improve glycemic control in older adults with type 2 diabetes. *Diabetes Care*, 25(12):2335–2341.

Center for Disease Control and Prevention (2011). National diabetes fact sheet: General information and national estimates on diabetes in the United States.

Cochrane, D. J. (2011). Vibration exercise: The potential benefits. *International Journal of Sports Medicine*, 32(2):75–99.

Coghill, R. C., et al. (1994). Distributed processing of pain and vibration by the human brain. *The Journal of Neuroscience: The Official Journal of the Society for Neuroscience*, 14(7):4095–4108.

Colberg, S. R., et al. (2010). American College of Sports Medicine, and American Diabetes Association: Exercise and type 2 diabetes. *Medicine & Science in Sports & Exercise*, 42(12):2282–2303.

Cotran, R. S., Kumar, V., and Robbins, S. L. (1996). *Robbins Patologia Estrutural e Funcional*, 5th edition. Rio de Janeiro: Guanabara Koogan.

DCCT Research Group (1995). Effect of intensive diabetes treatment on nerve conduction in the diabetes control and complications trial. *Annals of Neurology*, 38(6):869–880.

del Pozo-Cruz, J., et al. (2013). A primary care-based randomized controlled trial of 12-week whole-body vibration for balance improvement in type 2 diabetes mellitus. *Archives of Physical Medicine and Rehabilitation*, 94(11):2112–2118.

del Pozo-Cruz, B., et al. (2014). Effects of a 12-wk whole-body vibration based intervention to improve type 2 diabetes. *Maturitas*, 77(1):52–58.

Diretrizes da Sociedade Brasileira de Diabetes (2015-2016). Milech [et al.]; organização: Oliveira JEP, Vencio S. São Paulo: A.C. Farmacêutica, 2016. Disponível em www.diabetes.org.br/sbdonline/images/docs/DIRETRIZES-SBD-2015-2016.pdf. Last accessed Feb. 2016.

do Nascimento, P. S., et al. (2010). Treadmill training increases the size of A cells from the L5 dorsal root ganglia in diabetic rats. *Histology and Histopathology*, 25(6):719–732.

Dyck, P. J., et al. (2013). Assessing decreased sensation and increased sensory phenomena in diabetic polyneuropathies. *Diabetes*, 62(11):3677–3686.

Feldman, E. L., et al. (1994). A practical two-step quantitative clinical and electrophysiological assessment for the diagnosis and staging of diabetic neuropathy. *Diabetes Care*, 17(11):1281–1289.

Fregonesi, C., et al. (2004). Etiopatogenia da neuropatia diabética. *Arq Ciênc Saúde*, 8(2):147–150.

Goldstein, D. E., et al. (2004). Tests of glycemia in diabetes. *Diabetes Care*, 27(7):1761–1773.

Haskell, W. L., et al. (2009). Physical activity: Health outcomes and importance for public health policy. *Preventive Medicine*, 49(4):280–282.

Hayashino, Y., et al. (2012). Effects of supervised exercise on lipid profiles and blood pressure control in people with type 2 diabetes mellitus: A meta-analysis of randomized controlled trials. *Diabetes Research and Clinical Practice*, 98(3):349–360.

Hong, J. (2011). Whole body vibration therapy for diabetic peripheral neuropathic pain: A case report. *Health Science Journal*, 5(1):66–71.

Hong, J., Barnes, M., and Kessler, N. (2013). Case study: Use of vibration therapy in the treatment of diabetic peripheral small fiber neuropathy. *Journal of Bodywork and Movement Therapies*, 17(2):235–238.

International Diabetes Federation (IDF) (2015). *Diabetes Atlas*. 7th edition.

Javed, S., et al. (2015). Treatment of painful diabetic neuropathy. *Therapeutic Advances in Chronic Disease*, 6(1):15–28.

Johnson, P. K., et al. (2014). Effect of whole body vibration on skin blood flow and nitric oxide production. *Journal of Diabetes Science and Technology*, 8(4):889–894.

Kessler, N. J. and Hong, J. (2013). Whole body vibration therapy for painful diabetic peripheral neuropathy: A pilot study. *Journal of Bodywork and Movement Therapies*, 17(4):518–522.

Knowler, W. C., et al., and Diabetes Prevention Program Research Group (2002). Reduction in the incidence of type 2 diabetes with lifestyle intervention or metformin. *New England Journal of Medicine*, 346(6):393–403.

Kordi Yoosefinejad, A., et al. (2014). The effectiveness of a single session of Whole-Body Vibration in improving the balance and the strength in type 2 diabetic patients with mild to moderate degree of peripheral neuropathy: A pilot study. *Journal of Bodywork and Movement Therapies*, 18(1):82–86.

Kordi Yoosefinejad, A., et al. (2015). Short-term effects of the whole-body vibration on the balance and muscle strength of type 2 diabetic patients with peripheral neuropathy: A quasi-randomized-controlled trial study. *Journal of Diabetes and Metabolic Disorders*, 14(1):45.

Lam, F. M. H., et al. (2012). The effect of whole body vibration on balance, mobility and falls in older adults: A systematic review and meta-analysis. *Maturitas*, 72(3):206–213.

Lee, K., Lee, S., and Song, C. (2013). Whole-body vibration training improves balance, muscle strength and glycosylated hemoglobin in elderly patients with diabetic neuropathy. *The Tohoku Journal of Experimental Medicine*, 231(4):305–314.

Martin, C. L., Albers, J. W., Pop-Busui, R., and DCCT/EDIC Research Group (2014). Neuropathy and related findings in the diabetes control and complications trial/epidemiology of diabetes interventions and complications study. *Diabetes Care*, 37(1):31–38.

Maurer, M. S., Burcham, J., and Cheng, H. (2005). Diabetes mellitus is associated with an increased risk of falls in elderly residents of a long-term care facility. *The Journals of Gerontology. Series A, Biological Sciences and Medical Sciences*, 60(9):1157–1162.

Melzack, R. and Wall, P. D. (1965). Pain mechanisms: A new theory. *Science*, 150(3699): 971–979.

Michigan University (2005). The Michigan Diabetes Research and Training Center: Michigan Neuropathy Screening Instrument.

Moriyama, C. K., et al. (2008). A randomized, placebo-controlled trial of the effects of physical exercises and estrogen therapy on health-related quality of life in postmenopausal women. *Menopause*, 15(4):613–618.

Morrato, E. H., et al. (2007). Physical activity in U.S. adults with diabetes and at risk for developing diabetes, 2003. *Diabetes Care*, 30(2):203–209.

Orr, R. (2015). The effect of whole body vibration exposure on balance and functional mobility in older adults: A systematic review and meta-analysis. *Maturitas*, 80(4):342–358.

Poole, H. M., Murphy, P., and Nurmikko, T. J. (2009). Development and preliminary validation of the NePIQoL: A quality-of-life measure for neuropathic pain. *Journal of Pain and Symptom Management*, 37(2):233–245.

Praet, S. F. E. and van Loon, L. J. C. (2008). Exercise: The brittle cornerstone of type 2 diabetes treatment. *Diabetologia*, 51(3):398–401.

Robinson, C. C., et al. (2016). The effects of whole body vibration in patients with type 2 diabetes: A systematic review and meta-analysis of randomized controlled trials. *Brazilian Journal of Physical Therapy*, 20(1):4–14.

Sañudo, B., et al. (2013). Whole body vibration training improves leg blood flow and adiposity in patients with type 2 diabetes mellitus. *European Journal of Applied Physiology*, 113(9):2245–2252.

Schlee, G., Reckmann, D., and Milani, T. L. (2012). Whole body vibration training reduces plantar foot sensitivity but improves balance control of healthy subjects. *Neuroscience Letters*, 506(1):70–73.

Shaw, J. and Zimmet, P. (1999). The epidemiology of diabetic neuropathy. *Diabetes Reviews*, 7:245–252.

Snowling, N. J. and Hopkins, W. G. (2006). Effects of different modes of exercise training on glucose control and risk factors for complications in type 2 diabetic patients: A meta-analysis. *Diabetes Care*, 29(11):2518–2527.

Sonza, A., et al. (2015). Whole body vibration at different exposure frequencies: Infrared thermography and physiological effects. *The Scientific World Journal*, 2015:452657.

Stevens, M., Feldman, E., and Greene, D. (1995). The aetiology of diabetic neuropathy: The combined roles of metabolic and vascular defects. *Diabetic Medicine*, 12(7):566–579.

Suh, S.-H., Paik, I.-Y., and Jacobs, K. (2007). Regulation of blood glucose homeostasis during prolonged exercise. *Molecules and Cells*, 23(3):272–279.

The Diabetes Control and Complications Research Group (1995). The effect of intensive diabetes therapy on the development and progression of neuropathy. *Annals of Internal Medicine*, 122:561.

Tesfaye, S., et al., and Toronto Diabetic Neuropathy Expert Group (2010). Diabetic neuropathies: Update on definitions, diagnostic criteria, estimation of severity, and treatments. *Diabetes Care*, 33(10):2285–2293.

Tesfaye, S., et al., and Toronto Expert Panel on Diabetic Neuropathy (2011). Painful diabetic peripheral neuropathy: Consensus recommendations on diagnosis, assessment and management. *Diabetes/Metabolism Research and Reviews*, 27(7):629–638.

Tesfaye, S., Boulton, A. J., and Dickenson, A. H. (2013). Mechanisms and management of diabetic painful distal symmetrical polyneuropathy. *Diabetes Care*, 36(9):2456–2465.

Thomas, D. E., Elliott, E. J., and Naughton, G. A. (2006). Exercise for type 2 diabetes mellitus. *The Cochrane Database of Systematic Reviews*, (3):CD002968.

Umpierre, D., et al. (2013). Volume of supervised exercise training impacts glycaemic control in patients with type 2 diabetes: A systematic review with meta-regression analysis. *Diabetologia*, 56(2):242–251.

Vileikyte, L., et al. (2003). The development and validation of a neuropathy and foot ulcer specific Quality of Life Instrument. *Diabetes Care*, 26:2549–2555.

Vincent, A. M., et al. (2004). Oxidative stress in the pathogenesis of diabetic neuropathy. *Endocrine Reviews*, 25(4):612–628.

Vinik, E. J., et al. (2005). The development and validation of the norfolk QOL-DN, a new measure of patients' perception of the effects of diabetes and diabetic neuropathy. *Diabetes Technology & Therapeutics*, 7(3):497–508.

Wallberg-Henriksson, H., Rincon, J., and Zierath, J. R. (1998). Exercise in the management of non-insulin-dependent diabetes mellitus. *Sports Medicine (Auckland, N.Z.)*, 25(1):25–35.

Yagihashi, S., Yamagishi, S.-I., and Wada, R. (2007). Pathology and pathogenetic mechanisms of diabetic neuropathy: Correlation with clinical signs and symptoms. *Diabetes Research and Clinical Practice*, 77 Suppl 1(3): S184–S189.

Zaidell, L. N., et al. (2013). Experimental evidence of the tonic vibration reflex during whole-body vibration of the loaded and unloaded leg. *PloS One*, 8(12):e85247.

7 Effects of Whole Body Vibration in Patients with Chronic Obstructive Lung Disease

Dulciane N. Paiva, Patrícia E. M. Marinho, Litiele E. Wagner, Marciele S. Hopp, and Armèle Dornelas de Andrade

CONTENTS

7.1 INTRODUCTION

Chronic Obstructive Pulmonary Disease (COPD) is the fourth most frequent cause of mortality worldwide, without a tendency in decreasing or stabilizing its epidemiological growth, with the prospect of becoming the third cause of mortality in the world by 2020 (Vogelmeier et al., 2017; Mendes et al., 2011; de Sousa et al., 2011). Smoking is the main risk factor for the onset of COPD; in addition to cigarette smoke, occupational dust, chemical irritants, environmental pollution, low socioeconomic status, and severe respiratory infections in childhood constitute other factors. Some individual factors also contribute to COPD development such as bronchial hyper-responsiveness, malnutrition, prematurity, and recurrent respiratory infections. The genetic factor determined by alpha-1 antitrypsin deficiency occurs in about 3%–10% of the caucasian population, and may also be associated with the occurrence of COPD (Vogelmeier et al., 2017; de Sousa et al., 2011; Miravitles, 2004).

COPD is characterized as a non-reversible but treatable syndrome. Chronic airflow obstruction is usually progressive and associated with the abnormal inflammatory response of the lungs to noxious particles or toxic gases, causing hyperinflation due to pulmonary elastic tissue destruction, altering chest wall mobility and causing mechanical disadvantage to the inspiratory muscles. In this context, it also produces respiratory muscle weakness, leading inspiratory accessory muscles to also be recruited (Mendes et al., 2011).

COPD is considered a systemic disease. Parenchymatous and airway inflammation produce changes of variable intensity, in addition to dyspnea, coughing, wheezing, secretion production, and recurrent respiratory infections, leading to systemic consequences such as physical deconditioning, muscle weakness, weight loss, and malnutrition (Vogelmeier et al., 2017; Wagner, 2008).

When exposed to stress situations, COPD patients have functional limitations and are unable to meet the needs of increased ventilatory demand, which forces them to limit activities, resulting in chronic sedentarism and consequent reduction of muscle strength, muscle mass and their aerobic capacity, which can lead to social isolation, anxiety, depression, and dependence (Zanchet, 2005; Dourado, 2004).

A reduction of exercise tolerance occurs due to changes in respiratory mechanics, gas exchange, increased inspiratory neural drive, and development of pulmonary vascular insufficiency, and musculoskeletal dysfunction (O'Donnell et al., 2017). Hyperinflation causes mechanical disadvantage to the diaphragm, and changes in the ventilation/perfusion ratio lead to hypoxemia during exercise, and hypercapnia in some patients. The occurrence of pulmonary hypertension and right ventricular insufficiency may also be emphasized. Skeletal muscle dysfunction results from factors such as malnutrition, use of corticoids, primary skeletal myopathy, or other neuromuscular dysfunctions (Gea et al., 2015).

7.2 PULMONARY REHABILITATION

The treatment of COPD is pharmacological and scientific evidence in recent decades has shown the efficacy of a concomitant non-pharmacological treatment. Pulmonary Rehabilitation (PR) aims to reduce symptoms, improve quality of life (QoL), and increase physical and emotional participation in daily activities. Implementation of a PR program allows for improving exercise capacity, decreasing perception of dyspnea, anxiety, and depression, as well as reducing hospital admissions (Vogelmeier et al., 2017). The effects of a PR program aim to reduce fatigue (McCarthy et al., 2015), increase musculoskeletal function (Vogelmeier et al., 2017; Spruit et al., 2013) and QoL (Bernard et al., 1999; Benzo and Karpman, 2014), and may prolong the survival of these patients (Benzo and Karpman, 2014).

Physical activity is recommended for COPD patients, and it may increase exercise tolerance (Vogelmeier et al., 2017). On the other hand, aerobic conditioning (Clark et al., 2000) and resistance training (Panton et al., 2004) are associated with a high level of dyspnea perception, which may lead to reduced participation in physical activities (Normandin et al., 2002; Vogiatzis, 2011; Prefaut et al., 1995). Thus, implementation of peripheral muscle training for COPD patients should involve assessing the risks of exacerbation, acute dyspnea and hypoxemia during resistance training (Panton et al., 2004; Kongsgaard et al., 2004) or aerobic conditioning (Oliveira et al., 2010; O'Driscoll et al., 2011; Martin and Davenport, 2011); and, due to the need of implementing physical activity for COPD patients as well as their potential limitations, other exercise modalities should be considered. Directing treatment toward new therapeutic approaches that act in a systemic way may instill new hope for improving the QoL of these patients.

7.3 WHOLE BODY VIBRATION TRAINING IN CHRONIC OBSTRUCTIVE PULMONARY DISEASE

The treatment of chronic diseases increasingly focuses on prevention strategies and comprehensive and multidisciplinary care (McCarthy et al., 2015; Spruit et al., 2013). In this context, whole body vibration (WBV) appears as a complementary method to pulmonary rehabilitation of COPD patients who remain symptomatic or who perpetuate their (reduced) functional capacity, despite the continuity of medical treatment and conventional rehabilitation.

WBV is characterized by an external stimulus that produces an oscillation/vibration to an individual who is on an oscillating or vibrating platform (VP) (Figure 7.1), where the central mechanism is the tonic vibration reflex that causes involuntary muscular contractions through monosynaptic cell connections that especially act in the lower extremities (Ritzmann et al., 2010; Armstrong et al., 2008).

The idea of WBV training arose in the early 1970s in order to train Russian cosmonauts and avoid loss of muscle mass and bone minerals during space flights. In the 1990s it was associated with training professional athletes. During the last decade there has been a growing interest in the use of WBV for therapeutic purposes, and since then some meta-analyses and systematic reviews have investigated its effects, albeit obtaining inconsistent results that should be interpreted with caution, since study protocols use different parameters and are carried out with reduced sample sizes which compromise the validity and interpretation of

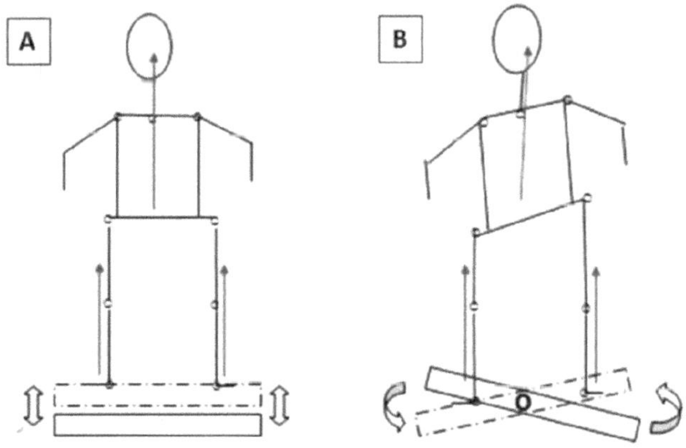

FIGURE 7.1 Vibrating platform (A) and oscillating platform (B). (Adapted with permission of Bernardo-Filho M. *Plataforma Oscilante/Vibratória: Fundamentos e a Prática Clínica*, 1st ed., Brazil, 2014, Editora Andreoli.)

the results (Conway et al., 2007; Nordlund and Thorstensson, 2007; Cardinale and Wakeling, 2005).

WBV is a therapeutic alternative for the treatment of COPD during rehabilitation, and although it has been applied in several clinical situations (such as in patients with neuromuscular diseases) (El-Shamy, 2014; Sañudo et al., 2013), in postmenopausal women (Lai et al., 2013), and in elderly individuals (Zhang et al., 2014), improving balance, potency, hormonal and postural control, bone mineral density, and muscle strength (Verschueren et al., 2004; Rittweger, 2010; Zha et al., 2012; Von Stengel et al., 2011; Merkert et al., 2011), its effects on COPD patients remain unclear.

The first study on evaluating the effects of WBV on cardiopulmonary parameters was developed in 1966 by Hood et al. (1966) in a study entitled *Cardiopulmonary Effects of Whole Body Vibration in Man*. However, the first clinical study developed in patients with chronic lung disease was published by Roth and colleagues in 2008 (Roth et al., 2008), where the effects of WBV on muscle strength and potency, as well as on bone mass and lung function in patients with cystic fibrosis were evaluated. After eleven months of home treatment, it was shown that WBV caused an average increase of 4.7% in muscle power and of 6.6% in jump speed without a significant drop in peripheral oxygen saturation (SpO_2) or blood pressure during training, in addition to a slight increase in forced expiratory volume in the first second of forced vital capacity (FEV_1) and trabecular density in the spine. Subsequently, Rietschel et al. 2008) developed a study involving ten stable cystic fibrosis patients aged between 24 and 47 years subjected to three months of WBV training (20–25 Hz), evidencing an increase in strength and muscular power regardless of FEV_1 and body mass index (BMI), in addition to reduced Sit-to-Stand Test time, without significant changes in FEV_1 and forced vital capacity (FVC).

In, 2012, Gloeckl et al. (2012) conducted a clinical trial to investigate the effects of WBV inserted in a multidisciplinary PR program over a three-week period, evaluating 82 patients with stage III-IV COPD. Patients were allocated to intervention groups (IG) who underwent three minutes of dynamic squatting exercise on an alternating VP (24 and 26 Hz) three times a week, and a control group (CG) who were submitted to the same squatting exercise time on the ground. PR effectiveness was assessed by increased distance covered in the Six-Minute Walk Test (6MWT) in the group submitted to WBV (64 ± 59 m) ($P < 0.01$) and in the CG (37 ± 52 m) ($P < 0.01$), with a difference of 27 m between groups [95% CI, 1–53] ($P < 0.046$). In this study, the 6MWT was performed according to the *American Thoracic Society* guidelines (ATS, 2002), where the baseline value used was the best of the two tests on the first and second day to exclude any learning effect (Hernandes et al., 2011). Moreover, the authors used the Sit-to-Stand Test, asking patients to sit and rise from a standard chair (43 cm height) as quickly as possible and with arms crossed over the chest, measuring the time patients required to perform the test (Bohannon, 2006). The Sit-to-Stand Test examines a body movement that is closely linked to the activities of daily living (ADL) and is considered an appropriate tool to determine the functional status of COPD patients (Canuto et al., 2010), and in the aforementioned study, patients submitted to WBV performed the test in less time.

It is important to note that the duration of WBV training in the initial studies of Roth et al. (2008) and Rietschel et al. (2008) were developed in three and

12 months, respectively, and that Gloeckl et al. (2012) showed promising results in a short period of three weeks of training. In detriment to the exposure time to WBV, Gloeckl et al. (2015) (three weeks) and Salhi et al. (2015) (12 weeks) showed no difference in muscle strength or exercise capacity after the WBV training period.

Pleguezuelos et al. (2013) developed a clinical trial in which 51 patients with stable COPD (stage IV with FEV_1 of 34.3%) underwent WBV in three weekly sessions for six weeks, and were subsequently assessed for functional capacity, while the control group consisted of patients who did not perform the intervention. The authors evaluated the isokinetic strength of the knee flexors/extensors and the exercise capacity assessed by the distance traveled on the 6MWT as the primary outcome, and they assessed inspiratory muscle strength through their maximal inspiratory pressure (MIP) and maximal expiratory pressure (MEP) components as secondary outcomes, not showing any significant differences between the groups regarding the maximal isokinetic force of concentric knee flexors/extensors. However, the authors emphasize that the sample submitted to WBV training showed a significant increase in the distance walked on the 6MWT ($P<0.001$), as well as a decrease in maximal oxygen desaturation in the 6MWT after the training period ($P=0.01$). It is also emphasized that these authors evidenced higher MIP and MEP at the end of the study in the group that used WBV for training.

Braz Júnior et al. (2015) conducted a randomized, crossover-controlled trial with 11 COPD patients (FEV_1: 14.63 ± 11.14% pred) subjected to WBV training (35 Hz) for 12 weeks (three supervised sessions per week), instructing the patients to remain in a semisquatting static position (knee flexion of 120°–130°) and feet 200 mm apart on the vibrating platform. The training was performed three times a week on alternate days covering four weeks of low intensity training (2 mm), and eight weeks high intensity training (4 mm). In this study, the training was ten minutes during the first four weeks, with 30 seconds of low intensity training interspersed with 60 seconds of rest. During the fifth and eighth weeks, the training was conducted for 15 minutes, and from nine to 12 weeks the training was 20 minutes, with 60 seconds of high intensity training and 30 seconds of rest in an anatomical position. Individuals in the control situation received no intervention. The authors showed that the distance walked on the 6MWT by subjects submitted to WBV was higher (64 m) than for those in the control group, and that quality of life improved significantly after the training period.

Salhi et al. (2015), in considering that WBV training improves muscle strength in healthy individuals and that resistance training (RT) is an important component of a PR program, investigated the effects of 12 weeks of WBV or RT training being implemented after 15 minutes of aerobic training with warm-up. The authors evaluated 62 COPD patients randomized into the WBV group or in the control group who underwent conventional RT. The authors evidenced an increase in the distance walked on the 6MWT, in the maximum capacity of exercise, in the strength of the quadriceps and in QoL after the training period; however, pointing out that only 30% of the patients reached the clinically significant minimum difference for the distance walked on the 6MWT.

Regarding the acute effect of WBV, a clinical study conducted by Furness and colleagues in 2013 (Furness et al., 2013), evaluated 17 COPD Functional Dependent patients (stages III-IV) regarding the profile of the subjective and acute responses obtained after a single session of WBV and a single session of simulated WBV (S-WBV). Dyspnea perception was evaluated through the Borg scale, heart rate (HR), and SpO_2. Study participants completed a single WBV session (25 Hz and 2 mm amplitude), and seven days later performed a single S-WBV session (25 Hz and 0 mm amplitude). The sessions consisted of five one-minute vibrations each in a knee flexion position (20°), interspersed with one (01) minute rest. Changes in the dyspnea perception remained stable from "very slight" to "slight" between WBV and S-WBV. It is important to emphasize that HR was higher in those submitted to WBV (11 ± 0.88) than those submitted to S-WBV (8 ± 0.70), without significant difference between the analyzed groups ($P=0.67$). No changes were observed in SpO_2 for any of the groups.

In 2014, Furness et al. (2014) performed a non-randomized crossover (Phase II) clinical trial which aimed to evaluate the long-term benefits of WBV training on the performance of ADLs and the gait of 15 COPD patients (stages II-III) by observing the acute effects on the perception of dyspnea, HR, SpO_2, acute exercise tolerance, and functional performance through the Timed-Up-and-Go (TUG) test and Five-Chair Stands Test. Participants were initially allocated to a group who underwent WBV intervention for six weeks, and after an interval of two weeks the same group underwent S-WBV intervention, also for six weeks. An alternating lateral vibration platform (25 Hz and amplitude of 2 mm) was used, and the group submitted to simulation was submitted to WBV of 25 Hz and amplitude of 0 mm. The authors did not observe a difference in dyspnea perception or SpO_2 in WBV and S-WBV ($P>0.05$). WBV increased HR up to 12 bpm ($P \leq 0.05$), while the group submitted to simulation showed no increase. Moreover, WBV improved TUG performance (13%), Five-Chair Stand Test (18%) and Gait Test ($P=0.01$) without having an effect on ADLs ($P>0.33$).

A study conducted by Greulich et al. (2014) aimed to investigate the effects of WBV on 40 COPD patients hospitalized for exacerbation. The sample was submitted to 30 minutes daily of respiratory physiotherapy focused on bronchial clearance, with a specific group receiving WBV treatment (26 Hz) for two to three minutes daily. Although the hospitalization time was the same for both groups, those submitted to WBV showed improved exercise capacity as well as a significant reduction of interleukin-8 levels upon hospital discharge. It is also important to note that such studies did not lead to adverse events related to WBV. According to Gloeckl et al., 2015), in order to obtain a training effect, the imposed load on the WBV training must be increased over time with use of different amplitudes and different additional WBV exposure loads, considering that there are reports of using WBV in regions closer to the thorax in order to increase expectoration; however, emphasizing, that there is no conclusive data on this practice.

Table 7.1 summarizes the main characteristics of the studies performed between the years 2012 to 2017, considering the great variability of WBV training protocols in COPD patients regarding the population type evaluated and the platform used, sample size, study design, training exposure time and interval between sessions, frequency and amplitude, body position, and adopted training protocol, as well as the evaluated outcomes.

TABLE 7.1
Characteristics of Studies Evaluating the Effects of Whole Body Vibration in Patients with Chronic Obstructive Pulmonary Disease

Author (Year)	Population	*n*	Study Outline	Outcomes Assessed	Training Time (Weeks)	Exposure Time and Interval between Sessions	Frequency (Hz) Amplitude (mm)	Body Position	Types of Training	Type of Vibrating Platform	Results
Spielmanns et al. (2017a, b)	COPD patients (stages II-IV)	28	Randomized clinical trial	Functional capacity (6MWT and Sit-to- Stand Test), pulmonary volumes by plethysmography and spirometry, CRQ and CAT	12 weeks	Once a week with 90-minute session (hiking, climbing, dancing, dumbbells, thera-band, and VP)	24–26 Hz 3 mm	Orthostasis/ Standing with knee flexion (10–20)	Squatting exercises (three sets of 20 repetitions) on a VP	Alternating lateral vibration (Galileo, Novotec Medical, Pforzheim, Germany)	No significant effects on exercise capacity
Gloeckl et al. (2017a)	COPD patients (FEV_1: 34±9% pred)	74	Randomized clinical trial	Neuromuscular performance, balance, postural muscle power (muscular work per time), maximal quadriceps muscle strength, Sit-to-Stand test, 6MWT	three weeks	four sets of two min, three times/week	24–26 Hz	Orthostasis/ standing with knee and hip flexion (90°–120°) at every squat	WBV group: squatting exercises on a VP. Control group: squatting on the ground. Four sets of squatting for two min, three times/week on non-consecutive days	Alternating lateral vibration (Galileo, Novotec Medical, Pforzheim, Germany)	Increased balance in most areas. Improvement of postural balance and muscle power.
Gloeckl et al. (2017b)	COPD patients (stages III–IV)	10	Randomized clinical trial	Cardiopulmonary stress testing and plethysmography	three weeks	One–two sessions of protocol exercise	26 Hz 5 mm	Orthostasis/ standing with knee and hip flexion (90°–100°)	Biking for ten min, squatting exercises (in random order) during one to three min on the ground or on a VP	Alternating lateral vibration (Galileo, Novotec Medical, Pforzheim, Germany)	Unchanged lung volumes. Good tolerance and the absence of exacerbation of symptoms

(Continued)

TABLE 7.1 (CONTINUED)
Characteristics of Studies Evaluating the Effects of Whole Body Vibration in Patients with Chronic Obstructive Pulmonary Disease

Author (Year)	Population	*n*	Study Outline	Outcomes Assessed	Training Time (Weeks)	Exposure Time and Interval between Sessions	Frequency (Hz) Amplitude (mm)	Body Position	Types of Training	Type of Vibrating Platform	Results
Spielmanns et al. (2017b)	COPD patients (stages I-III)	27	Randomized clinical trial	Functional capacity (6MWT and five repetitions: Sitting-Rising Test, Legg Press Strength, Berg Balance Scale, CAT, SGRQ)	12 weeks (two weekly sessions)	30 min	6–10 Hz 4–6 mm	Isometric squat position with knee flexion (150°)	Ten min walking or biking and stretching, 15 min of VP and five min of deceleration	Alternating lateral vibration (Galileo, Novotec Medical, Pforzheim, Germany)	Significant and clinically relevant improvement in exercise capacity compared to calisthenic exercises
Salhi et al. (2015)	COPD patients (FEV_1<50% pred or diffusion capacity (DL, CO) <50% pred)	61	Randomized clinical trial	Pulmonary function, body composition, exercise capacity, quadriceps strength, quality of life, 6MWT	12 weeks	Three times/ week for 30 sec to one min	27 Hz 2 mm	Squatting with knee flexion (120°–130°), deep squat with knee flexion (90°), wide-stance squat and lunge	15 minutes of resistance training, high squatting, deep squat and wide- stance squat and lunge	Vertical vibration (Fitvibe, Gymna, Belgium)	Increased distance in the 6MWT, maximum exercise capacity and quality of life

(Continued)

TABLE 7.1 (CONTINUED)
Characteristics of Studies Evaluating the Effects of Whole Body Vibration in Patients with Chronic Obstructive Pulmonary Disease

Author (Year)	Population	n	Study Outline	Outcomes Assessed	Training Time (Weeks)	Exposure Time and Interval between Sessions	Frequency (Hz) Amplitude (mm)	Body Position	Types of Training	Type of Vibrating Platform	Results
Braz Júnior et al. (2015)	Stable COPD patients	11	Pilot randomized crossover study	6MWT, SGRQ	12 weeks Washout period of similar duration to the training period	Three weekly sessions	35 Hz 2–4 mm	Semi-squat with knee flexion (120130)	First four weeks: 10 min (30 s of WBV/60 s of rest in standing position). Fifth to eighth weeks: 15 min. Ninth to 12th weeks: 20 min (60 sec high intensity/30 sec of rest). Stretching of upper and lower limbs before WBV	Vertical vibration (MY3, Power Plate, London, UK)	Increase in the distance walked in the 6MWT and improved domains of the SGRQ
Greulich et al. (2014)	COPD patients hospitalized due to exacerbation	40	Randomized clinical trial	Pulmonary function test, Functional capacity (6MWT and Sitting-Rising Test), SGRQ, CAT, laboratory analysis (C-reactive protein, leukocytes, alpha-1antitrypsin and interleukin-8)	8.63 6.16±days	Three×two min/ day.	20–35 Hz 4 mm	Orthostasis/ standi ng with flexed knees	Control group: (five min mobilization/ five min of passive mobilization, breathing exercises for ten min). WBV group: same pattern of exercises associated to VP training	Alternating lateral vibration (Galileo, NovotecMedical, Pforzheim, Germany)	WBV is safe, feasible and can present positive effects on functional capacity and quality of life in hospitalized COPD patients

(Continued)

TABLE 7.1 (CONTINUED)
Characteristics of Studies Evaluating the Effects of Whole Body Vibration in Patients with Chronic Obstructive Pulmonary Disease

Author (Year)	Population	*n*	Study Outline	Outcomes Assessed	Training Time (Weeks)	Exposure Time and Interval between Sessions	Frequency (Hz) Amplitude (mm)	Body Position	Types of Training	Type of Vibrating Platform	Results
Furness et al. (2014)	COPD patients (stages II-III)	15	Randomized clinical trial	Sitting-rising Test, Assessment of dyspnea by the BorgCR-10 Scale	Six weeks with two weeks of washout, totaling 14 weeks of intervention	Two sessions/ week	25 Hz 2 mm	WBV group with knee flexion at 53° and simulated WBV group with knee flexion at 44°	Standing position with knee flexion	Alternating Vibration (Amazing Super Health, Melbourne, AUS)	Improvement in the performance of activities of daily living and gait. Without exacerbation of COPD
Furness et al. (2013)	COPD patients (stages II-III)	14	Randomized clinical trial	Spirometry, evaluation of dyspnea by ScaleBorg CR-10, vital signs	One WBV session followed by a placebo WBV session after seven days	One min training /one min rest for five repetitions	25 Hz 2 mm	Orthostasis/ standing with knee flexion (20°)	Knee flexion (20°). During rest: VP in anatomical posture	Alternating Vibration (Amazing Super Health, Melbourne, AUS)	Effective in not causing exacerbation. No changes in the HR and SpO_2 during WBV training

(Continued)

TABLE 7.1 (CONTINUED)
Characteristics of Studies Evaluating the Effects of Whole Body Vibration in Patients with Chronic Obstructive Pulmonary Disease

Author (Year)	Population	n	Study Outline	Outcomes Assessed	Training Time (Weeks)	Exposure Time and Interval between Sessions	Frequency (Hz) Amplitude (mm)	Body Position	Types of Training	Type of Vibrating Platform	Results
Pleguezuelos et al.(2013)	Patients with severe COPD	51	Randomized clinical trial	6MWT, Isokinetic test, inspiratory and expiratory muscle strength	Six weeks	18 sessions in three sessions/week	35 Hz 2 mm	Orthostasis/standing with hip flexion (30°) and knee flexion (55°)	Ten minutes of warmup and after, upper limbs, lower limbs and back exercises, followed by WBV (six sets of four reps of 30 s/60 s rest) and ten minutes of stretching	Vertical vibration Gymnauniphy. Nv. Pasweg (6a 3740 Bilzen. Belgium)	Did not alter isokinetic muscle strength of knee flexors/extensors in concentric scheme. Increased in the distance walked in the 6MWT with reduction of the maximum oxygen desaturation
Furness et al. (2012)	COPD patients (stage II)	20	Nonrandomized clinical trial of crossover type	Dyspnoea (Scale Borg CR-10), vital signs, Sitting-Rising Test Timed Up-and-Go Test, Five-Chair Test, stride length, stride time, and stride velocity	14 weeks	12 sessions during six weeks, with two sessions a week with 48-hour intervals. Washout period of two weeks, followed by six weeks of intervention (control)	25 Hz 2 mm	Orthostasis/standing with knee flexion (20°)	WBV session of 5 60 s/60 s of rest	Alternating lateral Vibration (Amazing Super Health, Melbourne, AUS)	Questions the effectiveness of WBV in increasing or maintaining exercise tolerance and functional performance of the lower limbs

(Continued)

TABLE 7.1 (CONTINUED)
Characteristics of Studies Evaluating the Effects of Whole Body Vibration in Patients with Chronic Obstructive Pulmonary Disease

Author (Year)	Population	*n*	Study Outline	Outcomes Assessed	Training Time (Weeks)	Exposure Time and Interval between Sessions	Frequency (Hz) Amplitude (mm)	Body Position	Types of Training	Type of Vibrating Platform	Results
Gloeckl et al. (2012)	COPD patients (stages III-IV)	72	Randomized clinical trial	6MWT, Sitting-Rising Test, Pulmonary Function, CRQ	Three weeks	Three sessions/week with three sets of three min	24–26 Hz 6 mm	Orthostasis/standing with knee and hip flexion (90°100°)	Squatting exercises on a VP with control group performing squats on the ground	Alternating lateral vibration (Galileo, Novotec Medical, Pforzheim, Germany)	Both groups improved their performance on the 6MWT and the quality of life. VP presents itself as a viable modality for COPD patients that may potentiate the effects of rehabilitation program

6MWT, Six-Minute Walk Test; CAT, COPD Assessment Test; COPD, Chronic Obstructive Pulmonary Disease; CRQ, Chronic Respiratory Questionnaire; HR, Heart rate; SGRQ, Saint Georges Respiratory Questionnaire; SpO_2, Peripheral oxygen saturation; VP, Vibrating platform; WBV, Whole Body Vibration.

Regarding the systematic reviews and meta-analyses performed to evaluate the effects of WBV on musculoskeletal and respiratory system of COPD patients, we note that despite the small number of randomized clinical trials, preliminary evidence suggests that WBV training increases functional capacity, maximum knee extension strength, and quality of life (Gloeckl et al., 2015). However, evidence on the specific frequency, intensity, type, duration, and gravitational properties of the vibration platform and subsequent WBV have not yet been described, and such determinations may improve the training of peripheral muscles in patients with COPD, since these patients present evident peripheral muscular dysfunction.

It is important to note that some recommendations of the International Society of Musculoskeletal and Neural Interactions (ISMNI) on interventions with WBV include information on the vibration and brand of the device, direction, frequency, peak-to-peak displacement, influence of gravitational force, and accuracy and assessment of slippage and foot position on the vibration platform that should be considered in studies using WBV (NHMRC, 2009).

Yang et al. (2016) evaluated the effect of WBV on lung function, exercise capacity and QoL of COPD patients through a randomized clinical trial using the PEDro scale to evaluate the methodological quality of the selected studies, with the level of evidence evaluated through the GRADE approach. The authors included the Cochrane Central Register of Controlled Trials, PubMed, CINAHL, EMBASE, PEDro, AMED, PsycINFO, ClinicalTrials.gov and Current Controlled Trials databases, selecting four studies involving 206 participants. Quality was assessed as good for two studies, and no major benefits of WBV on lung function were found. Two studies demonstrated an increase in QoL and exercise capacity assessed through 6MWT, and did not show any adverse events. The authors concluded that WBV improves exercise capacity and QoL in patients with COPD; however, they attest that there is insufficient data to prove the effects of WBV on lung function. Gloeckl et al. (2017b) recently demonstrated an improvement in exercise capacity in patients with COPD undergoing dynamic squatting exercises with and without WBV. The authors evaluated ten patients with severe COPD (FEV_1 of 38 ± 8% pred) on two consecutive days. On the first day, the cardiopulmonary stress test was limited by symptom through a cycle ergometer. The next day, six sessions of squatting exercises were performed in random order for one, two, or three minutes with or without WBV, while the metabolic demands were measured simultaneously. The authors evidenced that the combination of squatting exercises with WBV induced a similar cardiopulmonary response in patients with severe COPD compared to squatting exercises without WBV, concluding that WBV can be a viable and safe modality of exercise even for patients with greater severity of the disease.

Spielmanns et al. (2017a,b) investigated whether WBV training can be beneficially applied within a three-month low-frequency exercise program in 28 COPD patients (stages II-IV). In that study, participants were randomized to performing squatting with and without WBV.

The training lasted 90 minutes once a week. The Sitting-Rising Test, 6MWT, COPD Assessment Test (CAT), and the Chronic Respiratory Disease Questionnaire (CRQ) were analyzed before and after the intervention. The authors showed that implementing WBV in the context of a low frequency exercise program did not improve the exercise capacity of COPD patients, observing improvement in CAT and CRQ in those subjects submitted to WBV.

A systematic review conducted by Cardim et al. (2016) evaluated randomized trials that assessed the effect of WBV treatment on exercise capacity, QoL, performance in activities of daily living, and lower limb muscle strength, including four articles with 185 subjects. All individuals in the groups submitted to WBV demonstrated improvement in distance covered in 6MWT compared to the control group, and only one article reported improvement in QoL and ADLs. The only article that assessed muscle strength found no difference between groups, and the quality of evidence for the functional exercise capacity outcome was considered moderate. The authors concluded that WBV seems to benefit COPD patients, increasing their capacity for functional exercise, without producing adverse effects with moderate evidence of quality and a strong degree of recommendation.

A recent randomized clinical trial developed by (Spielmann et al., 2017a,b) showed that WBV provides a significant clinical increase in exercise capacity compared to calisthenics in patients with moderate to severe COPD.

According to Gloeckl et al. (2015), there is currently no validated WBV training protocol to be used with COPD patients, and there is great variability as to the type of WBV (whether vertical or alternating), frequency (Hz), peak-to-peak displacement, exposure time to WBV, as well as the impact of the association of different types of exercise on maximal oxygen uptake and tolerance in COPD patients. However, it is indicated that frequencies greater than 20 Hz are used in platforms that promote lateral alternation, and frequencies lower than 35 Hz for platforms with vertical displacements when the therapeutic objective is to increase the neuromuscular frequency for activation of lower limb muscles. Static and dynamic squatting exercise is the most commonly used type of exercise, with no evidence of which is the most efficient and least costly to the patient. It is suggested that a combination of both types of exercises is more efficient when combined with high frequencies of vibration and the use of additional load. According to Ritzmann et al. (2013), the orthostasis/standing position with knee flexion at 60° seems to be more effective in activating the knee extensor muscles and plantar flexors. It should be noted that some platforms have bars that support upper extremities, but that the benefits of concomitant upper limb training are still not fully elucidated.

7.4 CONCLUSIONS

Progress in studies on the use of WBV combined with PR programs in patients with COPD is extremely relevant since it reaches the intricacies of public health. The cost effectiveness ratio also emerges as an important factor considering that the health system presents an economic overhead imposed by direct expenses with medications, examinations, consultations, and hospital services, as well as transportation expenses and rental/purchase of household equipment associated to COPD. In addition to the above, there are also indirect costs reflected by losses caused by reduction or interruption of productivity due to illness or early death (de Sousa et al., 2011; Vogelmeier et al., 2017).

WBV has been used in patients with stable or exacerbated COPD with different degrees of impairment, and there is currently no validated WBV training protocol for this patient population. There is great variability as to the type of WBV (whether vertical or alternating), frequency (Hz), peak-to-peak displacement, and duration of exposure to full-body vibration in all the evaluated studies, which makes data inconclusive on the short-term and long-term effects of exposure to WBV on maximal oxygen uptake during the association of different types of exercise on the vibrating platform.

Important considerations in the application of WBV in patients with COPD are clinical signs of exacerbation of chronic disease. An aspect present in most evaluated studies is the absence of serious adverse effects, such as injury or increase of respiratory symptoms, which were well tolerated by the majority of patients with chronic obstructive pulmonary disease.

There is indication for using frequencies greater than 20 Hz on platforms that promote lateral alternation and frequencies lower than 35 Hz for platforms that promote vertical displacements when the therapeutic objective is to increase the neuromuscular frequency for activation of the muscles of the lower limbs (Cochrane, 2011). Static and dynamic squatting exercise is the most commonly used type of exercise, with no evidence of which is the most efficient and least costly to the patient. It is speculated that a combination of both is more efficient (Ritzmann et al., 2013).

Some authors have pointed to an increase in balance and muscle power after WBV training in COPD patients, and postulate that neuromuscular adaptation is an important mechanism through which whole body vibration promotes increased exercise capacity in patients with chronic obstructive pulmonary disease (Glöeckl, 2017a,b).

REFERENCES

Armstrong, W. J., et al. (2008). The acute effect of whole-body vibration on the Hoffmann Reflex. *Journal of Strength and Conditioning Research*, 22(2):471–476.

American Thoracic Society (2002). ATS/ERS statement on respiratory muscle testing. *Am J Resp Crit Care Med.* 166(4):518-624.

Benzo, R. and Karpman, C. (2014). Gait speed as a measure of functional status in COPD patients. *International Journal of Chronic Obstructive Pulmonary Disease*, 9:1315-1320.

Bernard, S., et al. (1999). Aerobic and strength training in patients with chronic obstructive pulmonary disease. *American Journal of Respiratory and Critical Care Medicine*, 159(3):896–901.

Bernardo-Filho, M., Cabezuelo, M. J. P., and Paiva, D. N. (2014). *Plataforma Oscilante/ Vibratória: Fundamentos e a prática clínica. Paiva, Cabezuelo Bernardo Filho 9788560416332.*

Bohannon, R. W. (2006). Reference values for the five-repetition Sit-to-Stand Test: A descriptive meta analysis of data from elders. *Perceptual and Motor Skills*, 103(1):215–222.

Braz Júnior, D. S., et al. (2015). Whole-body vibration improves functional capacity and quality of life in patients with severe chronic obstructive pulmonary disease (COPD): A pilot study. *International Journal of Chronic Obstructive Pulmonary Disease*, 10:125–132.

Canuto, F. F., et al. 2010 Neurophysiological comparison between the Sit-to-Stand test with the 6-Minute Walk test in individuals with COPD. *Electromyography and Clinical Neurophysiology*, 50(1):47–53.

Cardim, A. B., et al. (2016). Does whole-body vibration improve the functional exercise capacity of subjects with COPD? A meta-analysis. *Respiratory Care*, 61(11):1552–1559.
Cardinale, M. and Wakeling, J. (2005). Whole body vibration exercise: Are vibrations good for you? *Commentary. *British Journal of Sports Medicine*, 39(9):585–589.
Clark, C. J., et al. (2000). Skeletal muscle strength and endurance in patients with mild COPD and the effects of weight training. *The European Respiratory Journal*, 15(1):92–97.
Cochrane, D. J. (2011). The potential neural mechanisms of acute indirect vibration. *Journal of Sports Science and Medicine*, 10(1):19–30.
Conway, G. E., Szalma, J. L., and Hancock, P. A. (2007). A quantitative meta-analytic examination of whole-body vibration effects on human performance. *Ergonomics*, 50(2):228–245.
de Sousa, C. A., et al. (2011). Doença pulmonar obstrutiva crônica e fatores associados em São Paulo, SP, 2008–2009. *Revista de Saúde Pública*, 45(5):887–896.
Dourado, V. Z. and Godoy, I. (2004). Recondicionamento muscular na DPOC: Principais intervenções e novas tendências. *Revista Brasileira de Medicina do Esporte*, 10(4): 331–334.
El-Shamy, S. M. (2014). Effect of whole-body vibration on muscle strength and balance in diplegic cerebral palsy. *American Journal of Physical Medicine and Rehabilitation*, 93(2):114–121.
Furness, T., et al. (2012). Efficacy of a whole-body vibration intervention to effect exercise tolerance and functional performance of the lower limbs of people with chronic obstructive pulmonary disease. *BMC Pulmonary Medicine*, 12(1):71.
Furness, T., et al. (2014). Benefits of whole-body vibration to people with COPD: A community-based efficacy trial. *BMC Pulmonary Medicine*, 14(1):38.
Furness, T., et al. (2013). Whole-body vibration as a mode of dyspnoea free physical activity: A community-based proof-of-concept trial. *BMC Research Notes*, 6(1):452.
Gea, J., Pascual, S. Casadevall, C., Orozco-Levi, M., Barreiro E. (2015). Muscle dysfunction in chronic obstructive pulmonary disease: update on causes and biological findings. *J Thorac Dis,* 7(10):E418-E438.
Gloeckl, R., et al. (2012). Effects of whole body vibration in patients with chronic obstructive pulmonary disease – A randomized controlled trial. *Respiratory Medicine*, 106(1): 75–83.
Gloeckl, R., Heinzelmann, I., and Kenn, K. (2015). Whole body vibration training in patients with COPD. *Chronic Respiratory Disease*, 12(3):212–221.
Gloeckl, R., et al. (2017a). What's the secret behind the benefits of whole-body vibration training in patients with COPD? A randomized, controlled trial. *Respiratory Medicine*, 126:17–24.
Gloeckl, R., et al. (2017b). Cardiopulmonary response during whole-body vibration training in patients with severe COPD. *ERJ Open Research*, 3(1).
Greulich, T., et al. (2014). Benefits of whole body vibration training in patients hospitalised for COPD exacerbations – A randomized clinical trial. *BMC Pulmonary Medicine*, 14(1):60.
Hernandes, N. A., et al. (2011). Reproducibility of 6-minute walking test in patients with COPD. *European Respiratory Journal*, 38(2):261–267.
Hood, W. B., et al. (1966). Cardiopulmonary effects of whole-body vibration in man. *Journal of Applied Physiology*, 21(6):1725–1731.
Kongsgaard, M., et al. (2004). Heavy resistance training increases muscle size, strength and physical function in elderly male COPD-patients–A pilot study. *Respiratory Medicine*, 98(10):1000–1007.
Lai, C.-L., et al. (2013). Effect of 6 months of whole body vibration on lumbar spine bone density in postmenopausal women: A randomized controlled trial. *Clinical Interventions in Aging*, 8:1603–1609.

Martin, A. D. and Davenport, P. W. (2011). Extrinsic threshold PEEP reduces post-exercise dyspnea in COPD patients: A placebo controlled, double-blind cross-over study. *Cardiopulmonary Physical Therapy Journal*, 22(3):5–10.

McCarthy, B., et al. (2015). Pulmonary rehabilitation for chronic obstructive pulmonary disease. *Cochrane Database of Systematic Reviews*, (2):CD003793.

Mendes, F. A. R., et al. (2011). Analysis of cardiovascular system responses to forced vital capacity in COPD. *Revista Brasileira de Fisioterapia*, 15(2):102–108.

Merkert, J., et al. (2011). Combined whole body vibration and balance training using Vibrosphere. *Zeitschrift für Gerontologie und Geriatrie*, 44(4):256 261.

Miravitlles, M. (2004). Avaliação econômica da doença pulmonar obstrutiva crônica e de suas agudizações: Aplicação na América Latina. *Jornal Brasileiro de Pneumologia*, 30(3):274–285.

National Health and Medical Research Council of Australia (2009). NHMRC additional levels of evidence and grades for recommendations for developers of guidelines. www.nhmrc.gov.au.

Nordlund, M. M. and Thorstensson, A. (2007). Strength training effects of whole-body vibration? *Scandinavian Journal of Medicine and Science in Sports*. 17(1):12-17.

Normandin, E. A., et al. (2002). An evaluation of two approaches to exercise conditioning in pulmonary rehabilitation. *Chest*, 121(4):1085–1091.

O'Donnell, D. E., et al. (2017). Advances in the evaluation of respiratory pathophysiology during exercise in chronic lung diseases. *Frontiers in Physiology*, 8:82.

O'Driscoll, B. R., et al. (2011). A crossover study of short burst oxygen therapy (SBOT) for the relief of exercise induced breathlessness in severe COPD. *BMC Pulmonary Medicine*, 11:23.

Oliveira, C. C., et al. (2010). Influence of respiratory pressure support on hemodynamics and exercise tolerance in patients with COPD. *European Journal of Applied Physiology*, 109(4):681–689.

Panton, L. B., et al. (2004). The effects of resistance training on functional outcomes in patients with chronic obstructive pulmonary disease. *European Journal of Applied Physiology*, 91(4):443–449.

Pleguezuelos, E., et al. (2013). Effects of whole body vibration training in patients with severe chronic obstructive pulmonary disease. *Respirology*, 18(6):1028–1034.

Préfaut, C., Varray, A., and Vallet, G. (1995). Pathophysiological basis of exercise training in patients with chronic obstructive lung disease. *European Respiratory Review*, 5(25):27–32.

Rietschel, E., et al. (2008). Whole body vibration: A new therapeutic approach to improve muscle function in cystic fibrosis? *International Journal of Rehabilitation Research*, 31(3):253–256.

Rittweger, J. (2010). Vibration as an exercise modality: How it may work, and what its potential might be. *European Journal of Applied Physiology*, 108(5):877–904.

Ritzmann, R., Gollhofer, A., and Kramer, A. (2013). The influence of vibration type, frequency, body position and additional load on the neuromuscular activity during whole body vibration. *European Journal of Applied Physiology*, 113(1):1–11.

Ritzmann, R., et al. (2010). EMG activity during whole body vibration: Motion artifacts or stretch reflexes? *European Journal of Applied Physiology*, 110(1):143–151.

Roth, J., et al. (2008). Whole body vibration in cystic fibrosis–A pilot study. *Journal of Musculoskeletal and Neuronal Interactions*, 8(2):179–187.

Salhi, B., et al. (2015). Effects of whole body vibration in patients with COPD. *COPD: Journal of Chronic Obstructive Pulmonary Disease*, 12(5):525–532.

Sañudo, B., et al. (2013). Changes in body balance and functional performance following whole-body vibration training in patients with fibromyalgia syndrome: A randomized controlled trial. *Journal of Rehabilitation Medicine*, 45(7):678– 684.

Spielmanns, M., et al. (2017a). Low-volume whole-body vibration training improves exercise capacity in subjects with mild to severe COPD. *Respiratory Care*, 62(3):315–323.

Spielmanns, M., et al. (2017b). Whole-body vibration training during a low frequency outpatient exercise training program in chronic obstructive pulmonary disease patients: A randomized, controlled trial. *Journal of Clinical Medicine Research*, 9(5):396–402.

Spruit, M. A., et al. (2013). An official American thoracic society/European respiratory society statement: Key concepts and advances in pulmonary rehabilitation. *American Journal of Respiratory and Critical Care Medicine*, 188(8):e13–e64.

Verschueren, S. M., et al. (2003). Effect of 6-month whole body vibration training on hip density, muscle strength, and postural control in postmenopausal women: A randomized controlled pilot study. *Journal of Bone and Mineral Research*, 19(3):352–359.

Vogelmeier, C. F., et al. (2017). Global strategy for the diagnosis, management, and prevention of chronic obstructive lung disease 2017 report. GOLD executive summary. *American Journal of Respiratory and Critical Care Medicine*, 195(5):557–582.

Vogiatzis, I. (2011). Strategies of muscle training in very severe COPD patients. *European Respiratory Journal*, 38(4):971–975.

Von Stengel, S., et al. (2011). Effects of whole-body vibration training on different devices on bone mineral sensity. *Medicine & Science in Sports & Exercise*, 43(6):1071–1079.

Wagner, P. D. (2008). Possible mechanisms underlying the development of cachexia in COPD. *European Respiratory Journal*, 31(3):492–501.

Yang, X., et al. (2016). Effects of whole body vibration on pulmonary function, functional exercise capacity and quality of life in people with chronic obstructive pulmonary disease: A systematic review. *Clinical Rehabilitation*, 30(5):419–431.

Zanchet, R. C., Viegas, C. A. A., and Lima, T. (2005). A eficácia da reabilitação pulmonar na capacidade de exercício, força da musculatura inspiratória e qualidade de vida de portadores de doença pulmonar obstrutiva crônica. *Jornal Brasileiro de Pneumologia*, 31(2):118–124.

Zha, D.-S., et al. (2012). Does whole-body vibration with alternative tilting increase bone mineral density and change bone metabolism in senior people? *Aging Clinical and Experimental Research*, 24(1):28–36.

Zhang, L., et al. (2014). Effect of whole-body vibration exercise on mobility, balance ability and general health status in frail elderly patients: A pilot randomized controlled trial. *Clinical Rehabilitation*, 28(1):59–68.

8 Whole Body Vibration, Cognition, and the Brain

Eddy A. van der Zee, Marelle Heesterbeek, Oliver Tucha, Anselm B. M. Fuermaier, and Marieke J. G. van Heuvelen

CONTENTS

8.1 INTRODUCTION

8.1.1 History

Scientific interest into the impact of vibrating stimuli goes back to the beginning of the 20th century when Hamilton (1918) described an unusual disease in limestone cutters. After the introduction of air-hammers (main vibrations of 75 Hz during 8–10 hours per shift), the majority of the cutters developed vibration-induced white fingers with altered sensory perceptions (Hamilton, 1918). Many vibration studies that were published in the 20th century focused on the detrimental effects of mechanical vibrations in the work environment, for example when operating tools (e.g. sledgehammer, form machines) or while riding in a vehicle (e.g. truck, helicopter, tank). The latter vibrations affect the whole body and for such vibrations the term whole body vibration (WBV) is introduced. The work-related "bad vibrations" often consists of prolonged, (random) vibrations in multiple directions, with lower frequencies (1–25 Hz), and variable magnitudes. Reviews of the literature on work-related vibrations show that exposure to such levels of vibrations mainly leads to increased health risks of the musculoskeletal, as well as the peripheral nervous system (Seidel and Heide, 1986; Hulshof van Zanten, 1987).

In 1987, Nazarov and Spinak started to use WBV as a training modality for athletes. During WBV the whole body of an individual is exposed to vibrations that are mechanically transferred from a vibrating device such as a platform. In contrast to the work-related random "bad vibrations", interventions with controlled vibrations were used. The vibrations are generally mild with small amplitudes (1–2 mm according to manufacturer's settings, but in practice much lower) and higher frequencies (10–60 Hz) compared to the work-related vibrations. The scientific interest in WBV as a training modality in the fields of sports and fitness as well as for rehabilitation and medical therapies has increased ever since. More and more authors reported positive effects of WBV and suggested it to be a safe and effective way to train the musculoskeletal system and improve physical performance. Reported effects of WBV, for example, include increased muscle strength (Ebid et al., 2012) and reduced knee osteoarthritis symptoms (Salmon et al., 2012). In addition, WBV improves physiological and health-related components of physical fitness such as higher bone density (Prioreschi et al., 2012) and lower blood pressure (Figueroa et al., 2012). In the elderly, WBV improves mobility, balance, general health status (Zhang et al., 2014; Turbanski et al., 2005), as well as body composition, insulin resistance and glucose regulation (Bellia et al., 2014).

While the world started to embrace WBV for training of skeletal muscles and physical functions, interest in the detrimental and/or beneficial impact of WBV on cognitive function and the brain studied in both humans and animals started to appear gradually. Below, after briefly introducing some basic aspects of WBV, we will summarize findings of WBV on cognition and the brain in humans and animals.

8.1.2 Mechanical Principles

During WBV a vibration source (platform) transfers mechanical vibrations to the body. In most vibration devices the applied mechanical oscillations are periodic with a sinusoidal shape. The intensity of WBV can be controlled by adjusting the amplitude (A) or acceleration (a, peak or root mean square [rms]), frequency (f) and time of exposure (t). The peak-to-peak amplitude can be calculated as the difference between the minimal and maximal position and represents the displacement of the vibrating source (in case of vertical vibrations). Frequency is the number of vibrations per unit time. For commercially available devices applied in sport and rehabilitation, two different types of vibration transmission can be distinguished (Figure 8.1, left panel). On the one hand, synchronous vertical vibration transmission, where the whole platform, and thus the body, oscillates linearly up and down. On the other hand, side-alternating vibration transmission, where the platform rotates around the fulcrum causing reciprocating displacement of the left and right side of the body.

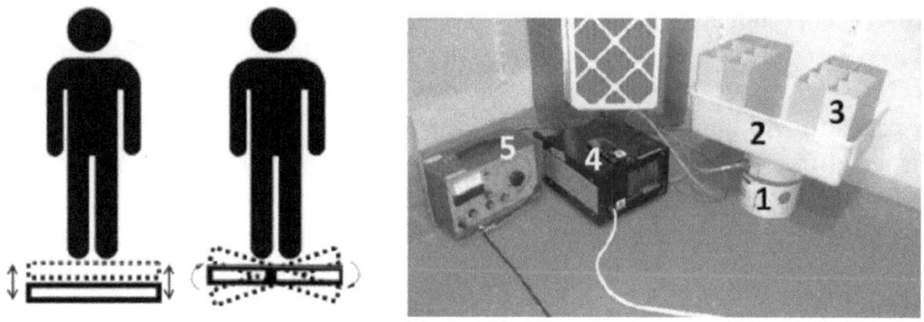

FIGURE 8.1 Left panel: Standing on a platform with synchronous vertical vibrations (left) versus side-alternating vibrations (right). Right panel: Set-up of the mouse platform: a cage (2), (length: 44,5 cm, width: 28 cm; height: 16 cm) is connected to a vibrator (1). Mice are placed in separate boxes (3) to avoid social interactions (e.g. fights between males); 4 = amplifier; 5 = oscillator.

8.1.3 Active Versus Passive

WBV can be applied in an active as well as a passive manner. During active WBV, exercises are performed while standing (often squatted) on or interacting with the vibrating source. For passive WBV no active contribution is required. Often, when passive WBV is applied the person is standing upright or sitting on the vibration source. Passive WBV is also referred to as passive exercise, since the body is moving without active performance. Passive WBV may be a suitable form of exercise for those who are not able to perform physical exercise.

8.1.4 Aim of This Chapter

The aim of this chapter is to give an overview of current knowledge regarding the impact of WBV on cognition and the brain in animals and humans including putative underlying working mechanisms. Cognition is defined as mental processes of perception, information processing, learning and thinking. Important aspects are memory and executive functions like attention, planning, inhibition or working memory. WBV-induced cognitive effects can be established at three time points: (1) acute effects, which are measured during the exposure to WBV; (2) short-term effects, which are measured right after a single session of WBV; and (3) chronic or long-term effects, which are measured after a series of WBV sessions (e.g. three times a week for two months) and are suggested to last for some period of time. As the term cognition is often a less suitable term for animals we will instead refer to learning and memory performance in the findings from animal studies.

8.2 FINDINGS FROM ANIMAL STUDIES

Few studies examined the impact of WBV on cognition and the brain in animals. In one study, rats were chronically exposed to four hours of WBV (30 Hz, 0.5 g setting) per day for four, six or eight weeks. These rats showed memory impairment and signs of brain damage (Yan et al., 2015). As WBV sessions used in training are of a much shorter duration, we used sessions ranging from five to 30 minutes in our studies. To test the effect of passive WBV on cognition, a scale model of a human vibration platform was built in collaboration with the University of Aberdeen, Scotland, by which different frequencies and g-forces could be applied to mice in a sinusoidal way. The used mouse platform (Figure 8.1; right panel) provides vertical vibrations, although the vibrations are transferred to the cage through the center point only. The peak-to-peak displacement at the center is 40, 29, and 14 μm for the left-right, front-back and up-down directions, respectively. In the corners of the cage these values are higher: 60, 75, and 54 μm, respectively. Mice, housed individually in their home cages, were placed in the separate boxes (see Figure 8.1) in the cage which was directly connected to the vibrating platform. Both young and old mice (approximately three and >18 months of age, respectively) received WBV (30 Hz) or pseudo-WBV (mice were placed in the cage, but in the absence of the actual vibrations). Of note, on days that a behavioral test was performed mice received the WBV session *after* the behavioral test to avoid an acute effect of WBV on the performance. During WBV mice quietly explored the box, sometimes display rearing (against the wall of the box) or grooming, or lying down. As such, this intervention cannot easily be compared to either active or passive WBV, and for that reason we solely refer to WBV in our mouse studies.

Initial studies in Aberdeen on motor performance were done in NMRI mice, an outbred strain widely used in general biology as well as in pharmacology and toxicology. Three frequencies (20, 30, and 45 Hz) and two g-forces (0.2 and 1.9 g settings according to the apparatus) were compared for the capacity to improve motor performance, using a rotarod (motor learning and motor coordination), hanging wire (muscle strength) and balance beam test (sensory motor abilities). WBV sessions of 10 minutes were used, five days a week and for a period of five weeks. As WBV is known to improve neuromuscular performance (Cardinale and Wakeling, 2005), we hypothesized that it would improve motor performance in mice as well. Indeed, the results revealed that motor performance was significantly improved, and most notably so at 30 Hz and the 1.9 g setting, in WBV-treated mice as compared to pseudo-WBV mice (unpublished observations). Thereafter, in Groningen (the Netherlands) we set out to test brain functioning (e.g. cognition by way of learning and memory performance) in two other mouse strains: the C57Bl/6J (inbred) strain (hereafter referred to as B6), and the ICR(CD1) (outbred) strain (hereafter referred to as CD1). The use of different mouse strains is relevant as they represent different "personalities" as seen in the human population. It has been postulated that personality type might have a moderating role on performance in relation to WBV (Ljungberg, 2008 for review). All procedures concerning animal care and treatment were in accordance with the regulations of, and approved by, the ethical committee for the use of experimental animals of the University of Groningen.

8.2.1 Learning and Memory Performance

Learning and memory performance was studied in a Y-maze reference test. In this test, one arm of the Y maze was baited with a small food crumb. The mice entered the Y maze through the start arm, and then had to choose one of the two arms. A visit to the baited arm was considered a correct choice, and each day mice were trained six times. The young WBV mice reached 75% or more correct choices (the criterion of having mastered the task) at day four, and the pseudo-WBV at day six. There was a statistically significant better performance of the WBV mice at day three, four and five (all $P<0.05$). In total, young WBV mice had 21% more correct choices as compared to pseudo-WBV mice during the learning phase. Next, a reversal test was performed, by which the food reward was relocated to the other arm of the Y maze. In contrast to the learning phase, there were no differences between the WBV and pseudo-WBV mice at any day of the reversal learning phase. The data of the old mice revealed a similar picture. Old WBV mice reached 75% or more correct choices at day four, and the pseudo-WBV mice at day five. There was a statistically significant better performance of the WBV mice at day one and day two. In total, old WBV mice had 12% more correct choices as compared to the pseudo-WBV mice during the learning phase. As was seen in young mice, WBV did not result in any improvement during the reversal phase. The cognitive improvements during the learning phase were not due to the sound produced by the vibrating plate. In a separate experiment, the WBV sessions were performed in the room where the mice were housed (both the WBV and the pseudo-WBV animals). Also, in this setting the WBV mice outperformed the pseudo-WBV animals during the learning phase.

Besides B6 mice, we also used CD1 mice. CD1 mice have a preference for a striatal-driven over a hippocampal-driven strategy to solve spatial tasks, most likely because their hippocampus is weaker. This means that the CD1 mouse tends to rely on the use of landmarks (and acquired habits) to orient themselves in space, rather than using a ("mental") spatial map known to depend on hippocampal functioning. The B6 mice have a preference for a hippocampal strategy. The preferred strategy can be tested in a cross maze training protocol, by which the mice are challenged to use their preferred strategy at the testing day. In 11 B6 mice tested in our lab, 36% preferred the striatal strategy (and hence 64% the hippocampal strategy; Hagewoud et al., 2010). In contrast, in 20 CD1 mice tested in the same cross maze 87% preferred the striatal strategy (and hence 13% the hippocampal strategy; a significant strain difference [$P=0.0447$; Fisher Exact test]). WBV was not able to shift the striatal preference of CD1 mice ($n=15$) to a hippocampal strategy, although the preference slightly changed from 13% to 25% hippocampal ($P=0.6045$ Fisher Exact test).

In contrast to the B6 mice, in CD1 mice WBV did not result in a better performance in the Y-maze training (n=eight per group for young mice and $n=11$ per group for aged mice). This could be due to the poor (strain-specific) functioning of their hippocampus, leaving it insensitive to WBV, if WBV indeed acts on the hippocampus. In line with having a poorly functioning hippocampus, the CD1 mice struggled to deal with the reversal phase in the Y-maze task (as this reversal test requires additional and specific activation of hippocampal activity; Havekes et al., 2007). They needed on average 18 more trials to master this task than did the B6 mice, while there was no difference between the strains in the initial training phase.

We decided to use a different learning task in the CD1 mice: the novel object recognition task (see Figure 8.2, panel B). In this task mice have to discriminate a novel object from a familiar object (called the novel object recognition (NOR) task;

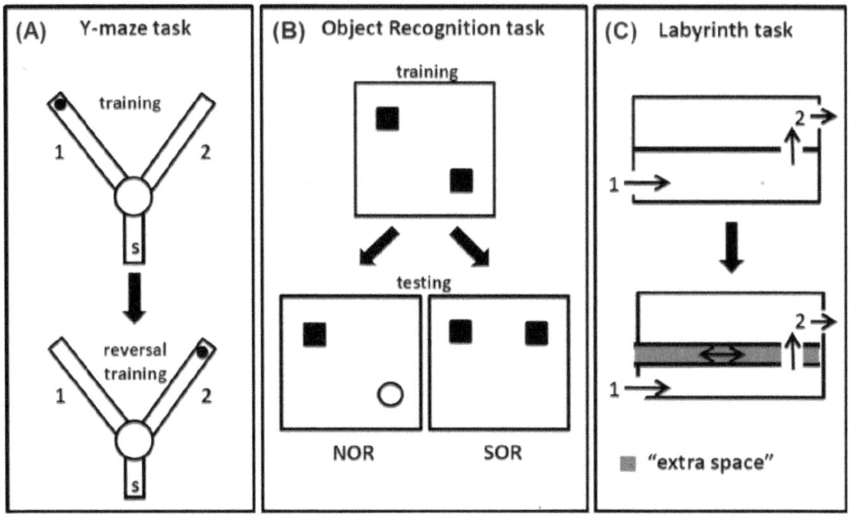

FIGURE 8.2 Schematic representation of the Y-maze task (panel A: s = start arm); black dot represent the food reward in the baited arm (in arm 1 in this example) which was reversed during reversal training (now in arm 2); the Object Recognition task (panel B); after training with two identical objects in the arena (■ familiar object; ○ novel object) a NOR or SOR was performed, or a NOR followed by a SOR; the Labyrinth task (panel C); the arrows indicate the trained route from the entrance (1) to the exit (2), initially without and thereafter with an extra space (grey area) created to be explored (double-headed arrow) by the mouse.

this task is hippocampus-independent), or to discriminate a replaced familiar object from a familiar object that is still in the same position (called the spatial object recognition (SOR) task; this task is hippocampus-dependent). We started with WBV sessions of ten minutes, similar as done for B6 mice. However, it turned out that, besides a significant improvement in balance beam performance ($P<0.01$ already reached after three weeks of WBV in young mice, and $P<0.05$ reached after six weeks in aged mice; n = ten per group), both NOR and SOR performance did not significantly improve by WBV despite a trend towards improvement in the NOR. We hypothesized that for the CD1 mice, a different duration of the WBV sessions might be necessary. We therefore compared the effects of five versus 30 minutes WBV sessions instead of ten minutes sessions. The results showed that the five minutes WBV protocol ($n = 12$ per group) was the best for CD1 mice, with a statistically significant improvement in the NOR and the balance beam test, but not in the SOR test (Keijser et al., 2017). The 30 min WBV protocol (n = ten per group) did not reveal significant improvements for any of the test, indicating that the duration of 30 minutes sessions was too long (Keijser et al., 2017).

Taken together, the results showed that WBV can improve learning and memory performance in mice, depending on the task, the mouse strain and the duration of the WBV session. Future research is needed to understand the exact relationship between WBV and the underlying brain mechanisms critically involved in mastering the tasks mentioned above.

8.2.2 Arousal-Reducing Effects

Besides the improvement in cognitive performance observed in the WBV mice it seemed, based on overt behavior, that these mice were slightly less aroused as were the pseudo-WBV individuals. We performed a series of experiments in young mice (three months of age; all males) to determine the potential arousing-reducing effects of WBV. Firstly, we determined the amount of locomotor activity measured in the home cage directly after a ten minutes WBV session in B6 mice (n=eight per group). WBV mice were 19% less active (and hence less aroused) in the first five minutes when they were returned to their home cage than pseudo-WBV mice, although this difference did not reach statistical significance. Secondly, we examined how WBV mice responded to an unexpected change in the environment, as it is known that aroused mice pay less attention to changes in their environment (Benus et al., 1991). We used the balance beam test, commonly used to measure sensory motor abilities. Mice (B6; n=eight per group) had to walk over a thin one-meter long squared wooden strip (diameter 5 mm; WBV mice outperform pseudo-WBV mice; Keijser et al., 2017) to reach their home cage. After four weeks of WBV, at 66 and 33 cm distance from the home cage a mark (a small rubber band) was placed around the wooden strip. The time it took for the mice to cross this novel environmental element was measured. At the first mark (at 66 cm), WBV mice needed 4.5 ± 2.2 (average ± s.e.m.) seconds whereas pseudo-WBV mice needed 11.8 ± 6.3 seconds to cross the mark. This is consistent with less-aroused mice being more daring to deal with an environmental change. At the second mark (at 33 cm), WBV mice needed 2.5 ± 0.9 seconds and pseudo-WBV mice 1.4 ± 1.0 seconds to cross the mark. Although not statistically significant, the WBV mice were more daring to cross the unexpected change in the environment when first encountered; at the second encounter this effect was no longer present. Thirdly, we tested CD1 mice (n=eight per group) in a simple labyrinth (see Figure 8.2, panel C). The mice were trained to walk through this labyrinth within ten seconds. When they reached this criterion, the labyrinth was changed in such a way that an additional space was created in the middle. In the next test trial, it was measured how much time the mice spent in this unexpected, new area in the labyrinth. WBV mice explored the new space for 32 ± 11 seconds, whereas the pseudo-WBV mice spent only 12 ± 4 seconds in the new area, which was significantly different ($F_{1,12}$=3,89, $P<0.05$; one-way ANOVA). Apparently, WBV made the mice more daring to explore the novel environment. Finally, we tested mice (B6; n=six per group) in a standard elevated plus maze with two open and two closed arms. Although the WBV mice did not spend significantly more time in either the open or closed arms of the elevated plus maze (reduced activity in the open arms is considered to be a reflection of enhanced anxiety), they showed significantly less activity in the maze than the pseudo-WBV mice (P=0.012; two-sample t-test). The enhanced activity of the pseudo-WBV mice was primarily caused by more often repeatedly entering the dark arms, most likely reflecting a higher level of arousal than expressed by the WBV mice (15.0 ± 2.1 versus 8.5 ± 1.3; P=0.009; two-sample t-test). Taken together, although the differences were often too subtle to reach statistical significance (also due to relative low numbers of animals per group), these results suggest that WBV reduces behavioral arousal. On the other hand, it also makes clear that WBV by itself does not induce anxiety or stress in mice.

8.3 FINDINGS FROM HUMAN STUDIES

8.3.1 Acute Effects

Several human studies examined acute effects, measured during WBV vs. control, on cognitive function. These studies were all performed from the perspective of "bad" vibrations in the work environment and are summarized in Table 8.1. Frequencies, magnitudes, and sometimes also direction of the vibrations were chosen in such a way that they were representative for vibrations during a truck/tractor drive or helicopter flight. Most studies used fixed frequencies, while some others used random frequencies from a frequency spectrum. Applied frequencies ranged from 1 to 25 Hz. Magnitudes are generally expressed in accelerations and varied from 0.53 to 3.5 ms^{-2} rms. If we calculate amplitudes based on frequency and acceleration data (under the assumption of pure sinusoidal vibrations), the peak-to-peak amplitudes generally vary between 0.07 and 25 mm. However, these amplitudes should be interpreted with caution, since in practice pure sinusoidal vibrations do not exist. A variety of cognitive measures were used with mean reaction time and number of errors or correct responses as outcome variables. Most studies revealed detrimental acute effects on cognitive performance or did not succeed to find effects (see Table 8.1). Two studies found some positive acute effects of WBV. In a study of Ishimatsu et al. (2016) lower reaction times during WBV vs. control were found on a sustained attention go no-go task. However, these lower reaction times went together with more errors suggesting a speed-accuracy trade-off. Zamanian et al. (2014) found improved performance on a divided attention (choice reaction time) task but not on a selective attention task, which holds true for three different vibration magnitudes without a specific magnitude effect. One other study (Ljungberg et al., 2004) examined a dose-response relation in which the magnitudes of the vibrations were varied (1.0, 1.6 and 2.5 ms^{-2} rms), but found no differences on a short-term memory task. This is to a large extent in agreement with the findings of Sherwood and Griffin (1990).

To summarize: acute effects were examined from work environment perspective. Mixed effects were found for cognition with some more evidence towards a detrimental effect, but without indications of a dose-response relationship.

8.3.2 Short-Term Effects

Ljungberg and Neely (2007b) were the first who examined short-term effects of WBV. They used the same set-up as described in Ljungberg and Neely (2007a, see Table 8.1). After 44 minutes of exposure, the subjects ($n=54$, male) went to another room and immediately performed a short-term "memory and search" attention task. After vibration the subjects performed the task faster (vibration only vs. control $d=0.17$; $P<0.05$), but with less accuracy ($d=-0.40$; $P<0.05$) and more errors ($d=-0.29$; $P<0.01$) indicating a speed-accuracy trade-off.

TABLE 8.1

Studies Regarding the Acute Effects of WBV Versus Control Summarized

Study	Population[a]	Design	Vibrations[b]	Outcome Measure	Results WBV vs Control[c]
Ishimatsu et al. (2016)	$N=19$ M/F 9/10 22.8 ± 4.4 yrs Healthy Elite runners	Cross-over Balanced WBV vs Control	1.0 ms^{-2} rms (z) 17 Hz 5.3 min/condition Sitting	Sustained Attention to Response Task (go no-go)	# Errors higher no-go trials ($d=0.72$*) # Errors higher go trials ($d=-0.41$ ns) Reaction time lower ($d=0.75$*) (speed-accuracy trade-off)
Zamanian et al. (2014)	$N=25$ M 20–30 yrs Students	Cross-over Partly balanced 3 WBV magnitude conditions vs Control	Sinusoidal/random waves (x,y and z) 0.53, 0.81, 1.12 ms^{-2} 3–7 Hz 3 min/condition × test Sitting	Selective Attention Divided Attention	Selective: Reaction time no effect ($d=0.02$ to 0.11) # Correct diverse (resp. $d=0.14$ ns; d=−0.39***; $d=0.14$ ns) Divided: Reaction time lower (resp. $d=0.25$ **;$d=0.29$***; $d=0.27$***) # Correct higher (resp. $d=0.76$***; $d=0.31$**; $d=0.54$***)
Newell and Mansfield (2008)	$N=21$ M/F 11/12 25.3 ± 5.2 yrs Healthy Students/staff	Cross-over 4 sitting positions × WBV (y/n)	1.4 ms^{-2} (x) 1.1 ms^{-2} (z) 1–20 Hz 3 min/condition Sitting	Visual Motor Choice Reaction Time	Reaction time higher*** # Errors higher**

(Continued)

TABLE 8.1 (CONTINUED)
Studies Regarding the Acute Effects of WBV Versus Control Summarized

Study	Population[a]	Design	Vibrations[b]	Outcome Measure	Results WBV vs Control[c]
Ljungberg and Neely (2007a)	Experiment 1 $N=24$ M 25 ± 2.4 yrs Healthy Students	Cross-over Balanced Noise (y/n) × WBV (y/n) (in four days)	Sinusoidal 1.1 ms^{-2} 2 Hz (x) 3.15 Hz (y) and 4 Hz (z) 44 min/day Sitting	Short-term memory test (not analyzed) Grammatical reasoning task	Reasoning task: Reaction time no effect (d=0.03) # Errors higher (d=−0.29 ns)
Sherwood and Griffin (1992)	$N=44$ M 18–35 years Fit	Two independent groups (WBV vs control)/cross-over	Sinusoidal 2.0 ms^{-2} (z) 16 Hz Sitting	Learning Memory Task	Independent groups: # Correct learning less (d=−0.47*) Cross-over: # Correct no effect
Sherwood and Griffin (1990) (abstract only)	$N=16$	Cross-over Control and 3 WBV magnitude conditions	Sinusoidal 1.0, 1.6 and 2.5 ms^{-2} rms 16 Hz	Short-term memory task (memory scanning)	Detrimental effect reaction time*** and # attentional lapses** # Errors higher only in 1.0 condition
Sandover and Champion (1984)	$N=7$ (Exp 1) $N=6$ (Exp 2) $N=7$ (Exp 3) M + F 19–55 yrs Fit	Cross-over 2 noise conditions × WBV (y/n) per experiment	Exp 1 1.18 ms^{-2} rms Exp 2 0.83 ms^{-2} rms Exp 3 0.59 ms^{-2} rms 2–25 Hz 12 min/session Sitting	Arithmetic task	# Errors higher in Exp 1** (no effect in Exp 2 and Exp 3) Time/question no effects

(Continued)

TABLE 8.1 (CONTINUED)
Studies Regarding the Acute Effects of WBV Versus Control Summarized

Study	Population[a]	Design	Vibrations[b]	Outcome Measure	Results WBV vs Control[c]
Harris and Shoenberger (1980) (abstract only)	$N=12$	Cross-over 2 noise conditions × WBV (y/n)	0.36 rms G_z sum-of-sines 30 min/condition	Complex counting task	Adverse effect
Rao and Ashley (1974)	$N=5$ M 19–35 yrs Students/staff	Cross-over Control-WBV-Control	0.23 G rms 2.5 Hz (peak) 20 min/condition Sitting	Choice Reaction Time	Reaction time higher ($d=-0.53$)*
Harris and Shoenberger (1970) (abstract only)	Highly trained	Cross-over 2 noise conditions × WBV (y/n)	0.25 G (z) 5 Hz 19 min/condition	Reaction time (two conditions appearance red light disappearance green light)	Adverse effect

ns = non-significant ($P>0.05$); d = Cohens d effect size (+ favor WBV, – favor Control).

[a] M = male; F = female

[b] x = front-back; y = left-right; z = vertical

[c] $P<0.05$; $^{**}P<0.01$ $^{***}P<0.001$

From a sports perspective and using commercially available vibration platforms, Amonette et al. (2015) compared cognitive performance after active WBV with performance after exercise only. Twelve young adults (eight male, mean age 28 years) completed a 25-minutes neuropsychological assessment after three different conditions: static squats only (five sets of two minutes), static squats with vertical vibrations (30 Hz, 4 mm peak-to-peak) and static squats with side-alternating vibrations (30 Hz, 4 mm peak-to-peak). Composite scores revealed slightly better scores after side-alternating WBV for verbal memory ($d=0.37$, P not reported) and inhibition ($d=0.43$), but not for vertical WBV (respectively $d=0.23$ and $d=-0.13$). For visual memory and reaction time (vertical and side-alternating WBV) differences between the sessions were small and inconsistent ($-0.20<d<0.13$).

After we found promising cognitive effects of passive WBV in mice in our Groningen research group, we initiated multiple studies to investigate whether these effects are present in humans as well. A pilot study in 12 healthy young adults (eight male, mean age 22.8 years; see Figure 8.3A for the platform used) was performed to identify the optimal passive WBV settings for short-term cognitive enhancement. We used a commercially available platform with vertical vibrations on which a chair was mounted. Executive functioning (attention and inhibition) was assessed immediately after each frequency (20, 30, 40, 50 and 60 Hz) × amplitude (2 and 4 mm as reported by manufacturer, but assessed on the chair between 0.19 and 0.74 mm) condition with a duration of two minutes per condition. Passive WBV (30 Hz, 0.5 mm) appeared the only condition that significantly improved attention and inhibition (Regterschot et al., 2014). These settings were used in a consecutive study among 113 healthy young adults (21 male, mean age 20.5 years). Again, we found positive effects of two minutes of passive WBV (30 Hz, 0.5 mm) on response inhibition and attention, but only when the tests were performed immediately after the passive WBV sessions ($d=0.13$; $P<0.05$; Regterschot et al., 2014). In other studies using the same device and similar settings we found improved attention/inhibition in 83 healthy young adults (40 male, mean age 22.5 yrs; $d=0.44$; $P<0.001$; Fuermaier et al., 2014a), in 17 adults with Attention Deficit Hyperactivity Disorder (ADHD) (eight male, mean age 24.2 years; $d=0.64$; $P<0.01$; Fuermaier et al., 2014a) and in 55 healthy schoolchildren (27 male, eight–13 years; three minutes WBV/condition; $d=0.10$; $P<0.01$; den Heijer et al., 2015).

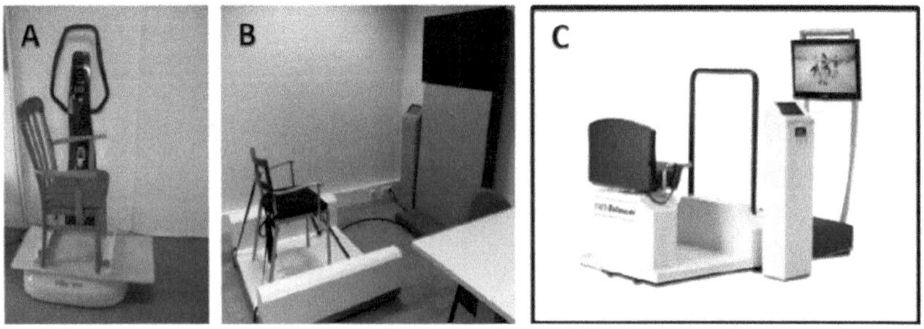

FIGURE 8.3 Platforms used for our human pilot studies (see text).

Summarized, side-alternating active WBV and passive WBV vertical vibration with limited durations (two to ten minutes) appeared to improve attention/inhibition measured immediately after the WBV, but effect sizes were generally small, except for young adults with ADHD who achieved larger improvements.

8.3.3 Chronic Effects

Following the short-term effect studies we performed two preliminary studies regarding the chronic effects of passive WBV. In a case study, an adult with ADHD (male, age 25 years) followed a daily regimen of 15 minutes WBV (vertical, 30 Hz, about 0.5 mm; for used platform see Figure 8.3A), three times a day and for ten weeks. We found positive effects on divided attention, vigilance, flexibility, inhibition and verbal fluency, but working memory, distractibility and reaction times remained unchanged (Fuermaier et al., 2014b). In a randomized-controlled pilot study with 34 healthy middle-aged to old adults (15 male, mean age 66 years with range 42–99 years), the experimental group underwent passive WBV (vertical, 30 Hz, about 0.5 mm; for used platform see Figure 8.3B) for four four-minute sessions per week over five weeks. The control group received attention-matched sham WBV following the same time schedule and using vibrations with the same amplitude but with a frequency of 1 Hz. In this setting accelerations are really low (0.001 g rms vs. 0.641 g rms for 30 Hz calculated from pure sinusoidal) and the vibrations are hardly experienced. After this five-week period we found trends towards improved performance for the experimental vs. control group for an attention/inhibition task ($d=0.35$; $P=0.051$) and a visual working memory task ($d=0.54$; $P=0.068$). Statistically controlled for age, both effects became significant ($P<0.05$). For several memory tasks and a tracking task we did not find any evidence for beneficial or detrimental effects of passive WBV ($-0.02<d<0.15$; $P>0.05$). Although these preliminary studies suggest beneficial chronic effects of passive WBV in at least some cognitive domains, more research is necessary to confirm this. Important research questions relate to the cognitive domain specificity of WBV as well as the population specificity. For example, older populations or clinical populations with cognitive impairment are more likely to benefit from WBV, as it is assumed that there is only little room for exercise related improvements of cognition in early adulthood (Hillman et al., 2008), but more in periods of cognitive decline. Passive WBV is especially attractive for people who are not able to be physically active anymore. Currently we are executing a study in people with dementia (Heesterbeek et al. 2018). In this study we investigate chronic effects of WBV alone and WBV integrated in a multi-sensory environment with simultaneous visual, auditory and passive gross-motor stimulation (see Figure 8.3C for the used platform). This integration of WBV will enhance the attractiveness of WBV and facilitate implementation in (clinical) practice.

8.4 POSSIBLE UNDERLYING MECHANISMS

At present, it is not known by which underlying mechanism(s) WBV exerts its positive effects on cognition and the brain. However, based on known physiological mechanisms in humans in combination with the results of the animals studies some likely candidates come to surface: skin mechanoreceptors, neurotransmitter systems and other brain factors involved in various aspects of brain functioning. Next, we will briefly discuss these candidates.

8.4.1 Skin Mechanoreceptors

The skin contains many specialized mechanoreceptors that subserve "touch" sensations contributing to proprioception and motor control. Mechanoreceptors located in various layers of the skin are excited by indentation of the skin by their preferred stimulus (for example vibrations, stretching of the skin or brushing). This is followed by transferring the information to the brain via the spinal cord reaching the thalamus. From there, the information is conveyed to the sensory areas of the neocortex and other areas in the basal forebrain, cerebellum and brainstem. The four types present in the glabrous skin are the Meissner and Pacinian corpuscles, Merkel cell-neurite complexes and Ruffini endings. Most likely all these types of cutaneous mechanoreceptors respond to WBV in their own way. Meissner and Pacinian corpuscles are fast-adapting types, mainly responding at the start and the end of the skin indentation. Both respond strongly to vibratory stimuli, but the Pacinian corpuscles respond predominantly to vibration frequencies exceeding 400 Hz (but also detect vibrations starting from 150 Hz). In contrast, Meissner corpuscles are sensitive to much lower frequencies, especially those of 20–40 Hz (Roudat et al., 2012 and references therein). The Merkel cell-neurite complexes and Ruffini endings are slow-adapting types, responding for a longer duration to a continuous skin indentation. These mechanoreceptors mainly respond to stretching of the skin or brushing.

For now, it seems that the Meissner corpuscles are the main candidates involved in activating the brain in response to 30 Hz WBV, but it should be noted that the platforms used often also generate (less intense) multiples of such a setting, by which these vibrations may also activate Pacinian corpuscles. It is unclear to what extent the transfer of WBV stimuli through skin mechanoreceptors is directly comparable between humans and animals (e.g. mice). Mice also have, next to the glabrous skin of the paws, extensive numbers of hairy skin mechanoreceptors (Roudant et al., 2012 for review) as well as whiskers which may add to the detection of touch including vibrations. The use of mice that lack specific types of mechanoreceptors (having a touch deficit) may shed more light on this issue.

8.4.2 Neurotransmitters

Improved cognitive performance could be induced by enhanced functioning of underlying neurotransmitter systems. It is known that the cholinergic system responds to behaviorally salient stimuli from the environment (Acquas et al., 1996), often in tandem with the dopaminergic system (Doig et al., 2014). Our results in mice (B6; n = eight per group) revealed that both systems are significantly affected after five weeks of WBV, as reflected by the significantly increased expression of the enzymes responsible for the production of the neurotransmitters. The expression was determined by way of immunohistochemistry applied to brain sections of WBV and pseudo-WBV brains, using highly selective and sensitive antibodies against the enzymes. Immunoreactivity for choline-acetyltransferase, the enzyme responsible for the production of acetylcholine, was measured in various brain regions. Statistically significant increases of ChAT immunoreactivity were found in the medial septum, hippocampus, neocortex and basolateral amygdale as compared to pseudo-WBV. On average, the increase in immunoreactivity, when statistically significant, ranged between 17% and 25% (with p-values ranging from 0.048 to 0.0009; Heesterbeek et al., 2017). These findings suggest increased cholinergic responsiveness due to passive WBV and are comparable to effects on the cholinergic system described for physical exercise (Van der Zee et al., 2012). Of interest, female NMRI mice that underwent the WBV protocol were less sensitive to a cholinergic antagonist reducing cognitive performance (Timmer et al., 2006), which seems to be in line with an increased capacity of the cholinergic system. In the striatum, we measured the expression of the enzyme tyrosine hydroxylase, critically involved in the production of dopamine. WBV enhanced the immunoreactivity by 48% (n=eight per group; $P<0.01$; two-sample t-test).

8.4.3 Other Factors

Next to changes in the cholinergic and dopaminergic systems, several WBV-specific findings indicate that the sensations caused by WBV are processed in various regions of the brain. Immunocytochemically examinations of the brains showed that mice in the passive WBV group had an increase in Glucose transporter 1 (Glut1) immunoreactivity throughout the brain, but most prominently in the dentate gyrus of the hippocampus (Lahr et al., 2009). Additionally, strongly significant increases in c-fos protein expression, a brain marker for neuronal activity, were found in brain areas involved in sensorimotor and learning/memory functions. These measures were done two hours after the last WBV session (the time point c-fos protein expression is highest upon a stimulus, followed by a gradual degradation of the protein), compared to pseudo-WBV. The strongest changes were found in the striatum, hippocampus (but much less so in the CD1 mouse than the B6 mouse), motor cortex and parts of the cerebellum. It must be noted that c-fos expression revealed a specific pattern in the brain, with several brain regions unaffected by WBV such as the biological clock. Moreover, as expected c-fos expression was back to low baseline values one day after the last WBV session, showing that the increase was a direct response to the WBV session (Van der Zee et al., 2012). The poorer c-fos response in the hippocampus of the CD1 mouse is in line with the behavioral observation that WBV is not able to improve hippocampus-dependent learning and memory performance in this strain.

Next to the above mentioned factors, an important role of the stretch reflex, the equilibrium system (balance) and increased blood flow to the brain in the cognitive improvements after WBV should not be underestimated. The effects of WBV on neuromuscular performance are supported by neurological processes such as changes in motor neuron excitability, synchronization and motor unit recruitment thresholds, involving the motor cortex and brain areas involved in motor control (e.g. the striatum and cerebellum). It is currently unknown whether the equilibrium system is critically involved in WBV and the WBV-induced changes in cognition and brain activity. Maintaining one's balance requires a constant feedback to the brain and WBV-induced corrections requires enhanced brain activity. Finally, heart rate can go up during WBV, which results in increased blood flow supporting the brain with oxygen and glucose, critical components for proper brain functioning.

To summarize, passive WBV can improve cognitive performance in men and mice, although the optimal conditions are not yet known. The response of different brain regions to (passive) WBV are very likely to contribute if improvements in cognitive performance are observed. Of note, in mice not all brain regions respond to WBV. For example, the biological clock (suprachiasmatic nucleus) located in the hypothalamus which dictates the temporal, circadian (~24 hours) rhythms in sleep/wake activity did not show any change in c-fos or in its output neuropeptide system vasopressin. In line with these brain findings WBV was not able to improve poor sleep/wake patterns notably seen in older mice.

Interestingly, a single session of WBV (10, 20, 40, 50, 80, 100, 150, and 200 Hz) applied to neural cell cultures for 30 minutes resulted in an increase in neurite outgrowth (except in the case of 150 and 200 Hz, with 40 Hz being the most effective frequency), indicative of enhanced vitality of the nerve cells (Koike et al., 2004). Part of the underlying mechanism of the increased neurite outgrowth was found to be the activation of a transcription factor (CREB) and the activation of a kinase (p38 MAPK) (Koike et al., 2004 and 2015). These observations may suggest that vibrations may also influence the brain directly (mechanically), without a role of all above mentioned factors.

8.5 CLINICAL IMPLICATIONS

We realize that effect sizes in our human studies were mainly small, and that WBV by itself might not have the capacity to replace established treatment to stimulate cognitive function. However, that does not mean it has no clinical value, as more longitudinal clinical studies are needed to find out the potential value of WBV for the following aspects: (1) which populations (children vs. adults vs. elderly; healthy people vs. psychiatric vs. neurological conditions), (2) which functions can be improved, and (3) by which treatment regimen. WBV could be a safe and cheap additional intervention in schools, homes for the elderly or nursing homes, which complement current and already established treatment methods. Our research team is now working on interventions in which WBV is included as an additional way to stimulate the brain in an attempt to improve cognition, for example in patients with dementia as mentioned in Section 3.3 of this chapter.

The animal studies indicated that our WBV protocol is not unpleasant and has the tendency to reduce behavioral arousal. Participants of the human WBV intervention using our device were asked to rate the experience on a scale from one (very unpleasant) to ten (very pleasant). The vast majority of the participants experienced the treatment as "neutral" to "slightly pleasant", irrespective of the age of the participants. This subjective rating was confirmed by a study in which the facial expression was monitored and analyzed of 40 individuals (20.5 ± 1.15 years of age; 30 females and ten males) during a four-minute WBV session (unpublished observations). Facial expressions were compared to the situation without receiving WBV at the start of the intervention. The results revealed no significant differences for the seven emotions distinguished: anger, disgust, fear, happiness, neutral, sadness and surprise. Taken together, these observations demonstrate that WBV interventions are perceived as a rather neutral experience suitable for clinical implementation. Altogether, WBV has the potential to be a safe, feasible and clinically relevant additional treatment to affect cognitive function of several populations, but more research is needed to confirm this.

8.6 CONCLUSION

In this chapter an overview of the current knowledge regarding the impact of WBV on cognition and the brain in animals and humans is presented. Results from animal studies show that, depending on the task, the mouse strain and the duration of the WBV session(s), WBV can improve learning and memory performance in mice. Also, mice tend to show reduced behavioral arousal after WBV. In human studies, short-term effects of both side-alternating active and vertical passive WBV with limited durations (two to ten minutes) are found. It is shown that in schoolchildren, young adults and young adults with ADHD, WBV can improve attention/inhibition when measured immediately after a single WBV session, with largest improvements found in young adults with ADHD. Furthermore, preliminary studies looking into the chronic effects of WBV show trends towards improved attention/inhibition and visual working memory after five weeks of WBV.

Multiple possible mechanisms that may underlie these effects of WBV on the brain are suggested. First of all, skin mechanoreceptors, especially the Meissner corpuscles, have been reported to be most sensitive to low frequencies (20–40 Hz) and may be the main candidates involved in brain activation following 30 Hz WBV. Second, improved cognitive performance could be induced by enhanced functioning of underlying neurotransmitter systems such as the cholinergic and dopaminergic system. Third, a variety of other factors such as changes in motor neuron excitability, synchronization and motor unit recruitment thresholds, as well as increased blood flow following in case of increased heart rate during WBV may play an important role in cognitive improvements due to WBV. Fourth, there are also some implications that WBV may influence the brain directly (mechanically), without a role of earlier mentioned factors.

Taken together (passive) WBV can improve cognitive functioning and is safe and feasible to apply in a clinical setting, but more (longitudinal) research is needed to (1) confirm the effectiveness of WBV for different populations, (2) examine which functions can be improved by WBV, (3) establish an optimal treatment regimen and (4) to examine which mechanisms underlie cognitive improvements found after WBV.

8.7 ACKNOWLEDGMENTS

We thank Dr. Gernot Riedel and Dr. Serena Deira (University of Aberdeen, Scotland) for supporting us with the WBV equipment for mice, and the undergraduate students Bettie Atsma, Gosse Beeksma, Vibeke Bruinenberg, Edwin Dasselaar, Nathaly EspitiaPinzón, Amerens Gaikema, Michael Jentsch, Charlotte de Jong, Mandy van der Klij, Maarten Lahr, Dafne Piersma, Ruben Regterschot, Peter Roemers,

Edwin Rutgers, Louise Taatgen, Carolien de Vries, as well as the technicians Wanda Douwenga, Jan Keijser, Folkert Postema and Kunja Slopsema for their valuable contribution to this chapter.

REFERENCES

Acquas, E., C. Wilson, and H. C. Fibiger. (1996). Conditioned and unconditioned stimuli increase frontal cortical and hippocampal acetylcholine release: Effects of novelty, habituation, and fear. *Journal of Neuroscience* 16:3089–3096.

Amonette, W. E., et al. (2015). Neurocognitive responses to a single session of static squats with whole body vibration. *Journal of Strength and Conditioning Research* 29(1): 96–100. doi:10.1519/JSC.0b013e31829b26ce.

Bellia, A., et al. (2014). Effects of whole body vibration plus diet on insulin-resistance in middle-aged obese subjects. *International Journal of Sports Medicine* 35:511–516.

Benus, R. F., et al. (1991). Heritable variation for aggression as a reflection of individual coping strategies. *Experientia* 47:1008–1019.

Cardinale, M. and J. Wakeling. (2005). Whole body vibration exercise: Are vibrations good for you? *British Journal of Sports Medicine* 39:585–589.

den Heijer, A. E., et al. (2015). Acute effects of whole body vibration on inhibition in healthy children. *Plos One* 10(11):e0140665. doi:10.1371/journal.pone.0140665.

Doig, N. M., et al. (2014). Cortical and thalamic excitation mediate the multiphasic responses of striatal cholinergic interneurons to motivationally salient stimuli. *Journal of Neuroscience* 34:3101–3117.

Ebid, A. A., et al. (2012). Effect of whole body vibration on leg muscle strength after healed burns: A randomized controlled trial. *Burns* 38:1019–1026. doi:10.1016/j.burns.(2012).02.006.

Figueroa, A., et al. (2012). Whole-body vibration training reduces arterial stiffness, blood pressure and sympathovagal balance in young overweight/obese women. *Hypertension Research* 35:667–672.

Fuermaier, A. B., et al. (2014a). Good vibrations--effects of whole body vibration on attention in healthy individuals and individuals with ADHD. *PloS One* 9(2):e90747.

Fuermaier, A. B., et al. (2014b). Whole-body vibration improves cognitive functions of an adult with ADHD. *Attention Deficit Hyperactivity Disorder* 6(3):211–20. doi:10.1007/s12402-014-0149-7.

Hagewoud, R., et al. (2010). Coping with sleep deprivation: Shifts in regional brain activity and learning strategy. *Sleep* 33(11):1465–1473.

Hamilton, A. (1918). Effect of the air hammer on the hands of stone-cutters. *Bulletin 236: U.S. Bureau of Labor Statistics* vol. 19:53–61.

Harris, C.S. and R.W. Shoenberger. (1980). Combined effects of broadband noise and complex waveform vibration on cognitive performance. *Aviation, Space, and Environmental Medicine* 51(1):1–5.

Harris, C.S. and R.W. Shoenberger. (1970). Combined effects of noise and vibration on psychomotor performance. *Aerospace Medical Research Lab Wright-Patterson Afb Ohio* http://oai.dtic.mil/oai/oai?verb=getRecord&metadataPrefix=html&identifier=AD0710595.

Havekes, R., M. Timmer, and E.A. Van der Zee (2007). Regional differences in hippocampal PKA immunoreactivity after training and reversal training in a spatial Y-maze task. *Hippocampus* 17(5):338–348.

Heesterbeek, M., E.A. van der Zee, and M.J.G. van Heuvelen (2018). Passive exercise to improve quality of life, activities of daily living, care burden and cognitive functioning in institutionalized older adults with dementia – a randomized controlled trial study protocol. *BioMed Central Geriatrics* 18:182.

Heesterbeek, M., et al. (2017). Whole body vibration enhances choline acetyltransferase-immunoreactivity in cortex and amygdala. *Journal of Neurology and Translational Neuroscience* 5(2):1079.

Hillman, C.H., K.I. Erickson, and A.F. Kramer. (2008). Be smart, exercise your heart: Exercise effects on brain and cognition. *Nature Reviews Neuroscience* 9:58–65.

Hulshof, C. and B.V. van Zanten. (1987). Whole-body vibration and low-back pain. *International Archive of Occupational and Environmental Health* 59(3):205–220.

Ishimatsu, K., et al. (2016). Action slips during whole-body vibration. *Applied Ergonomics* 55:241–247.

Keijser, J. N., et al. (2017). Whole body vibration improves attention and motor performance in mice depending on the duration of the whole-body vibration session. *African Journal of Traditional, Complementary and Alternative Medicines* 14:128–134.

Koike, Y., et al. (2004). Low-frequency vibratory sound induces neurite outgrowth in PC12M3 cells in which nerve growth factor-induced neurite outgrowth is impaired. *Tissue Culture Research Communications* 23:81–90.

Koike, Y., et al. (2015). Low-frequency, whole body vibration induced neurite outgrowth by PC12M3 cells with impaired nerve growth factor-Induced neurite outgrowth. *Journal of Novel Physiotherapies* 5:1 http://dx.doi.org/10.4172/2165-7025.1000249.

Lahr, M. M. H., et al. (2009). Whole body stimulation functions as a cognitive enhancer in young and old mice. In 8th *Dutch Endo-Neuro-Psycho Meeting Abstract* (Vol. 170).

Ljungberg, J. K. (2008). Combined exposures of noise and whole-body vibration and the effects on psychological responses, a review. *J Low Frequency Noise, Vibration and Active Control* 27(4):267–279.

Ljungberg, J. K., G. Neely, and R. Lundström. (2004). Cognitive performance and subjective experience during combined exposures to whole-body vibration and noise. *International Archive of Occupational and Environmental Health* 77(3):217–221.

Ljungberg, J. K. and G. Neely. (2007a). Stress, subjective experience and cognitive performance during exposure to noise and vibration. *Journal of Environmental Psychology* 27:44–54.

Ljungberg, J. K. and G. Neely. (2007b). Cognitive after-effects of vibration and noise exposure and the role of subjective noise sensitivity. *Journal of Occupational Health* 49(2):111–116.

Muller, S., et al. (2002). The effects of proprioceptive stimulation on cognitive processes in patients after traumatic brain injury. *Arch Physical Med Rehab* 83(1):115–121.

Nazarov, V. and G. Spivak. (1987). Development of athlete's strength abilities by means of biomechanical stimulation method. *Theory and Practice of Physical Culture* 12:37–39.

Newell, G. S. and N. J. Mansfield. (2008). Evaluation of reaction time performance and subjective workload during whole-body vibration exposure while seated in upright and twisted postures with and without armrests. *International Journal of Industrial Ergonomics* 38:499–508.

Prioreschi, A., et al. (2012). Whole body vibration increases hip bone mineral density in road cyclists. *International Journal of Sports Medicine* 33(8):593–599. doi:10.1055/s-0032-1301886.

Rao, B. K. N. and C. Ashley. (1974). Effect of whole body low frequency random vertical vibration on a vigilance task. *Journal of Sound and Vibration* 33(2):119–125.

Regterschot, G. R., et al. (2014). Whole body vibration improves cognition in healthy young adults. *PLoS One* 9(6):e100506. doi:10.1371/journal.pone.0100506.

Roudant, Y., et al. (2012). Touch sense – Functional organization and molecular determinants of mechanosensitive receptors. *Channels* 6:234–245.

Salmon, J. R., et al. (2012). Does acute whole-body vibration training improve the physical performance of people with knee osteoarthritis? *Journal of Strength and Conditioning Research* 26(11):2983–2989. doi:10.1519/JSC.0b013e318242a4be.

Sandover, J. and D. F. Champion. (1984). Some effects of a combined noise and vibration environment on a mental arithmetic task. *Journal of Sound and Vibration* 95(2):203–212.

Seidel, H. and R. Heide. (1986). Long-term effects of whole-body vibration: A critical survey of the literature. *International Archive of Occupational and Environmental Health* 58(1):1–26.

Sherwood, N. and M. J. Griffin. (1990). Effects of whole-body vibration on short-term memory. *Aviation, Space, and Environmental Medicine* 61(12):1092–1097.

Sherwood, N., and M. J. Griffin. (1992). Evidence of impaired learning during whole-body vibration. *Journal of Sound and Vibration* 152(2):219–225.

Timmer, M., E. A. Van der Zee, and G. Riedel. (2006). Whole body vibration and behaviour: Investigation of the role of various neurotransmitter systems. *Federation of European Neuroscience Societies Abstracts* 3(089.31).

Turbanski, S., et al. (2005). Effects of random whole-body vibration on postural control in Parkinson's disease. *Research in Sports Medicine* 13(3):243–256.

Van der Zee, E.A., et al. (2012). Whole body vibration and spatial learning: c-Fos and ChAT as neuronal correlates of cognitive improvements. *8th International Conference Methods Techn Behav Res.*

Yan, J.-G., et al. (2015). Neural systemic impairment from whole-body vibration. *Journal of Neuroscience Research* 93:736–744.

Zamanian, Z., et al. (2014). Short-term exposure with vibration and its effect on attention. *Journal of Environmental Health, Science and Engineering* 12(1):135. doi:10.1186/s40201-014-0135-1.

Zhang, L, et al. (2014). Effect of whole-body vibration exercise on mobility, balance ability and general health status in frail elderly patients: A pilot randomized controlled trial. *Clinical Rehabilitation* 28(1):59–68.

9 Effects of Whole Body Vibration in Adult Individuals with Metabolic Syndrome

D. da Cunha de Sá-Caputo, M. Fritsch Neves, and Mario Bernardo-Filho

CONTENTS

9.1 METABOLIC SYNDROME

Metabolic syndrome (MetS) is a complex escalating global public health problem and involves clinical challenges. Various approaches must be considered, and multi-professional actions are necessary for the management of the surplus energy intake, increasing obesity, and sedentary life habits (Oda, 2012; Kaur, 2014).

According to the International Diabetes Federation (IDF), it is estimated that about 20%–25% of adults worldwide have MetS. Furthermore, this population is twice as likely to die and three times as likely to have a myocardial infarction or stroke compared with people without MetS. In addition, people with MetS have a fivefold greater risk of developing type 2 *diabetes mellitus* (T2DM). Additionally, 230 million people in the world have diabetes, one of the most common chronic diseases and the fourth or fifth leading cause of death in the developed countries. The clustering of cardiovascular disease (CVD) risk factors that characterizes MetS has been considered as a possible driver for the CVD epidemic (Alberti et al., 2009; Alberti and Zimmet, 2005; Olijhoek et al., 2004). Moreover, MetS would be the first risk factor for atherothrombotic complications. Therefore, its presence or absence should be an indicator of long-term risk. On the other hand, the short-term (five to ten years) risk is better calculated using the classical algorithms (Niiranen, 2017), due to the inclusion of age, sex, total cholesterol or low-density lipoprotein (LDL), and smoking (Grundy, 2006).

The relevance of the approaches about MetS would be also related to the World Health Organization (WHO) International Classification of Disease (ICD-9) code (277.7) WHO (1999) which permits healthcare reimbursement. This shows that the term "metabolic syndrome" is institutionalized and a part of the medical vocabulary.

At the same time, different criteria to describe the association including parameters such as visceral obesity, high blood pressure (hypertension), dyslipidemia, high blood glucose (hyperglycemia), and insulin resistance (IR) have been used (Eddy et al., 2008). MetS has its origins in 1920 when Kylin, a Swedish physician, demonstrated the association of hypertension, hyperglycemia, and gout (Kylin, 1923). Later in 1947, Vague described that visceral obesity was commonly associated with the metabolic abnormalities found in CVD and T2DM (Vague, 1947). Following this, in 1965, a study was presented at the Annual Meeting of the European Association for the Study of Diabetes by Avogaro and Crepaldi which again described a syndrome which comprised hypertension, hyperglycemia, and obesity. Vague proposed that abdominal obesity may predispose to diabetes and CVD. In 1981, metabolically obese individuals were diagnosed as normal-weight individuals who might be characterized by hyperinsulinemia and possibly increased fat cell size (Ruderman et al., 1981). In 1982, it was reported that upper body obesity offered an important prognostic marker for glucose intolerance, hyperinsulinemia, and hypertriglyceridemia in women (Kissebah et al., 1982). In, 1987, a novel classification of obesity was proposed (visceral fat obesity vs. subcutaneous fat obesity) (Fujioka et al., 1987) and it was also demonstrated that essential hypertension is an insulin-resistant state (Ferrannini, 1987). The terms "syndrome X" or IR syndrome or MetS have been proposed to describe the process where individuals display a cluster of IR and compensatory hyperinsulinemia, high plasma triglycerides, low high-density lipoprotein (HDL) cholesterol levels, and hypertension with significant increased risk of CVD and the World Health Organization (WHO) defined the criteria for this syndrome. In 1999 The European Group for the Study of IR suggested a modified version of MetS to be used for nondiabetic subjects (Balkau and Charles, 1999). The definition of MetS was proposed in the Third Report of the NCEP by the Expert Panel on the Detection, Evaluation, and Treatment of High Blood Cholesterol in Adults (NCEP, 2001). In 2003, the American Association of Clinical Endocrinologists modified the definition to refocus on IR as the primary cause of MetS and thus it excluded diabetic subjects from MetS and returned to the name Study of IR (Einhorn et al., 2003). In 2004, the inclusion of high-sensitivity C-reactive protein (hs-CRP) was proposed as a component of MetS as hs-CRP is strongly related to obesity and IR, and was also established as an elevated risk factor of CVD (Ridker et al., 2004). After that, the International Diabetes Federation (IDF) suggested a new definition of MetS, where the abdominal obesity was a necessary component (Alberti et al., 2006). Different International Scientific Associations have also discussed the criteria to be used in the characterization of the MetS (Grundy et al., 2005; Kahn et al., 2005; Reaven, 2006; Klein et al., 2007; Alberti et al., 2009). The visceral adipose tissue volume and abdominal subcutaneous adipose tissue volume were compared considering the relationship with cardiometabolic risk factors, inflammatory markers, and markers of endothelial dysfunction, and oxidative stress through the Framingham Heart Study (Fox et al., 2007; Pou et al., 2007). It was suggested in 2009 that the MetS definition requiring obesity is dangerous as normal-weight individuals have a high mortality risk and are more prevalent than overweight or obese subjects. The WHO Expert Consultation reported that MetS is more an educational and comprehensive concept that focuses attention on the complex multifactorial health problems, being a pre-morbid condition rather than a clinical diagnosis. Thus, this approach could have limited practical utility as a diagnostic or management tool and therefore there is a narrow utility in epidemiological studies in which different criteria of MetS are compared (Simmons et al., 2010).

Putting together the previous considerations, several authors (Oda, 2012; Fox et al., 2007; Pou et al., 2007) have pointed out, that there are some limitations related to the concept of MetS, as it is shown in Figure 9.1.

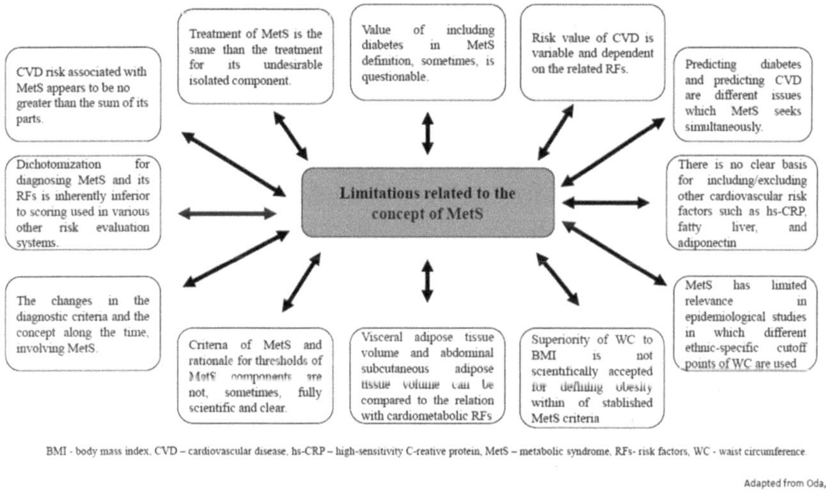

FIGURE 9.1 Limitations related to the metabolic syndrome concept.

9.2 MECHANISMS OF DEVELOPMENT OF THE METABOLIC SYNDROME

Some authors also consider that MetS would be the systemic presence of adipose tissue disease (Oda, 2012; Oda, 2008) characterized by an increased aggregation of activated macrophages into adipose tissue from bone marrow induced by chronic energy overload and with an increased number of crown-like structures in adipose tissue (Cinti et al., 2005). Nevertheless, MetS would be associated with other complex and interlinked pathophysiological mechanisms. Increased waist circumference (WC), hypertension, increased serum triglyceride levels, decreased serum HDL cholesterol levels, and impaired fasting glucose are well established and widely accepted parameters of MetS (Oda, 2012; Gundy et al., 2005; Alberti et al., 2009).

Although, there is important understanding of the syndrome (NCEP, 2001; Einhorn et al., 2003; Ridker et al., 2004; Alberti et al., 2006), the underlying cause of MetS continues to challenge the experts, most individuals with MetS have IR and abdominal obesity that are considered relevant factors (Olijhoek et al., 2004).

In addition, an IDF publication (2006) pointed out that (a) genetics, (b) physical inactivity, (c) ageing, (d) proinflammatory state, and (e) hormonal changes may also have a causal effect, but the role of these may vary depending on ethnic group. Furthermore, IR occurs when cells (liver, skeletal muscle, and adipose/fat tissue) become less sensitive and eventually resistant to insulin that facilitates the glucose absorption. Glucose, which remains in the blood, leads the necessity of more and more insulin (hyperinsulinemia). When the pancreas does not produce enough insulin, there is hyperglycemia characterizing the diagnosis of T2DM. Before the presence of T2DM occurs a build-up of triglycerides which further impairs insulin sensitivity. The glucose transporter type-4 (GLUT4) is a protein that plays a relevant role in the tissue and plasma glucose balance. Moreover, it is involved in the facilitated diffusion of glucose in insulin-sensitive tissues. Changes in the expression and/or translocation of the glucose transporter, especially in adipose and skeletal muscle cells, has been directly associated with IR (Klip, 2009; Klip et al., 2009; Caponi et al., 2013).

Following the IDF publication (2006), abdominal obesity is associated with IR and the MetS. Obesity contributes to hypertension, high serum cholesterol, low HDL cholesterol, and hyperglycemia, and is independently associated with higher CVD risk. The risk of strong health consequences in the form of T2DM, coronary heart disease, and a range of other clinical conditions, such as some types of cancer, has been shown to rise with an increase in body mass index (BMI), but the waist circumference (WC) would be more indicative of MetS than BMI.

Oda (2012) pointed out that high energy fast-food environment, sedentary life style, and other obesogenic socioeconomic environment factors have contributed to an obesity pandemic in the developed world. Moreover, genetic predisposition to obesity and proinflammatory reactions would be also associated with the MetS. As the adipose tissue secretes humoral substances, such as tumor necrosis factor-a (TNF-a), leptin, adiponectin, resistin, visfatin, monocyte chemoattractant protein-1, retinol binding protein-4, and adipocyte-type fatty acid binding protein, obesity has been considered as an endocrine and inflammatory disorder related with IR rather than anthropometric fatness.

9.3 COMORBIDITIES INVOLVING METABOLIC SYNDROME

Besides the complexity of the interconnected physiological, biochemical, clinical, and metabolic factors of the MetS (Kaur, 2014), chronic pain is also evident (Ites et al., 2011). The individuals with MetS can also show peripheral small fiber neuropathy, characterized by damage to the primary afferent nociceptors due to chronic hyperglycemia (Veves et al., 2009). In consequence, this damage causes peripheral sensitization, leading to central neuron hyperexcitability and spontaneous nerve impulse generation, presenting as chronic pain (Veves et al., 2009). In addition, authors have reported that obese people have more pain than normal-weight individuals (Hitt et al., 2007). Duruöz et al. (2012) have reported that low back pain is mostly common in individuals with MetS. In general, the mechanisms underlying the association between the metabolic disturbances and pain are not still well established; although some factors, such as the inflammation and the mechanical overload seem relevant (Barbe and Barr, 2006; Courties et al., 2015).

According to WHO (2016), more than 50% of the deaths and the disability related to CVD can be eliminated worldwide, with a combination of national efforts and individual actions to reduce major risk factors such as hypertension, high cholesterol, obesity, and smoking. Then, the morbidity and the mortality rate may be reduced by simple lifestyle interventions and easily modifiable behavior changes. In an undesirable approach, WHO estimates that 25% of healthier life years will be lost to CVD globally by 2020, if no action is taken to improve cardiovascular health and current trends. Putting together all the considerations, MetS would be related to a cluster of risk factors for CVD and T2DM, which occur together more often than alone (Fox et al., 2007). In, 2011, heart disease and stroke costs in the U.S. were about $320 billion on health care and lost productivity (Centers for Disease Control and Prevention, 2016). For that reason, WHO has stated that countries might adopt policies, guidelines, and programs to promote population-wide interventions, such as encouraging the practice of exercise and other actions (Fox et al., 2007; WHO, 1999).

The absence of physical activity has been presented as one of the main factors for chronic pain. Furthermore, authors (Beavers et al., 2013; Hwang and Kim, 2015) have suggested that the MetS is strongly associated with a poor physical activity and decreased cardiorespiratory endurance, flexibility, and muscular strength. Indeed, the presence of MetS was also related to a decrease in flexibility of the lower limbs (Chang et al., 2015; Sá-Caputo et al., 2014). Therefore, physical reconditioning has been proposed in clinical practice as a goal to be reached in the treatment of individuals with MetS. Despite this, it is important to consider that not all the forms of exercise are effective and safe. Although aerobic exercise (Kaur, 2014; Dixit et al., 2014 ; Blair et al., 2011) and resistance training have been related to the decreased CVD risk factors, obesity or MetS severity (Farias et al., 2013; WHO, 2014; Sossa et al., 2013), the presence of pain or even poor physical activity, most individuals do not exercise (Beavers et al., 2013; Callaghan et al., 2016).

Metabolic impairments due to the excessive fat accumulation in the obese are associated with increased risk for T2DM, CVD, disability, and mortality (Cade, 2008; Alley and Chang, 2007). The local adiposity, as verified in MetS individuals, disturbs the insulin receptors function within the muscle and is associated with insulin sensitivity (Stehno-Bittel, 2008; Wei et al., 2008). This complex and interlinked process is visualized in Figure 9.2, where it is observed that poor physical activity can lead to MetS.

The glycated hemoglobin values, post-intervention, in individuals with T2DM decreases in the physically active (Thomas et al., 2006). Moreover, Chang et al. (2015) evaluated the body composition, physical fitness, muscle endurance, flexibility, and cardiorespiratory endurance of community-dwelling elders, and it was verified that the presence of MetS is associated with decrease in flexibility independent

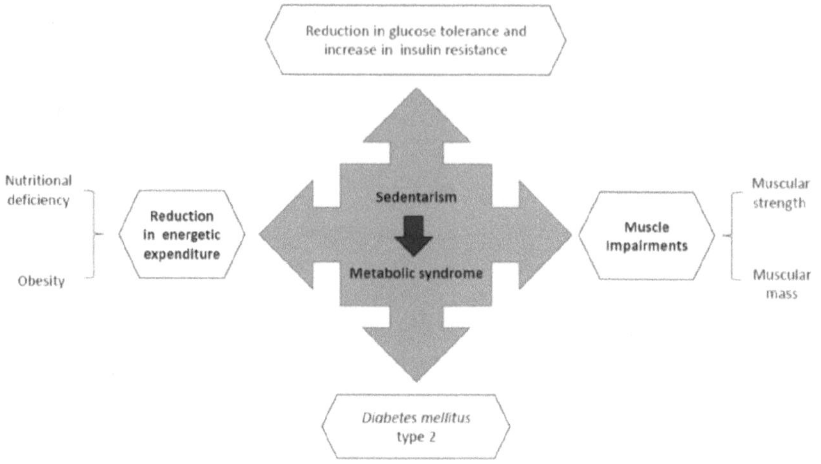

FIGURE 9.2 The relationship between the sedentarism and the development of metabolic syndrome and their associated clinical conditions.

on age, gender, and BMI. With a regular exercise practice, there is an improvement of several metabolic parameters, such as reduced visceral fat in adults (Ohkawara et al., 2007) without any change in body mass (Thomas et al., 2006), which suggests an increase in muscle mass due the exercise training.

In consequence, an excellent strategy for the management of individuals with MetS is regular exercise.

9.4 METABOLIC SYNDROME AND EXERCISE

In MetS, obesity, hypertension, and T2DM are associated with IR and with a decrease of the GLUT4 expression (Garvey et al., 1998). Aerobic exercise training is an effective and proper intervention for improving insulin sensitivity, because it increases the glucose transport in the insulin-sensitive tissues, especially in the skeletal muscle (Ploug et al., 1990). This benefit results from increases in both GLUT4 gene expression (Friedman et al., 1990) and the translocation of vesicles containing GLUT4 from the cytoplasm to the cell surface (Ren et al., 1994; Neufer et al., 1992), among other associated and important factors (Jensen and Richter, 2012; Klip, 2009). Positive effect in the insulin sensitivity and in the pancreatic β-cell function were shown after the practice of exercise (Tjonna et al., 2008). Moreover, the association of this practice with a low-calorie diet and salt restriction can reduce the blood pressure and the prevalence of MetS (Anderssen et al., 2007). Then, it is verified that the most appropriate intervention for the management of MetS, involves changes of the lifestyle, such as regular aerobic exercise and healthy food. The problem is there are few investigations about either the isolated effects of practice of aerobic exercise in individuals with MetS or the benefits of this practice in relation to various severe risk factors (Kemmler et al., 2009). The impact of impaired vascular function on CVD has been investigated for several decades. In this regard, major factors for the development and progression of CVD are associated with attenuated arterial function including structural and functional changes in the arterial wall, such as endothelial dysfunction and arterial stiffness. In addition, studies have reported reduced muscular strength and reduced muscle mass with vascular aging and CVD (Figueroa et al., 2014b; Iemitsu et al., 2008; Otsuki et al., 2008a).

The importance of physical activity in the maintenance or the improvement of general health is widely accepted, and exercise training has been known as a desirable nonpharmacological therapeutic modality to restore the impaired cardiovascular function. Whole body vibration (WBV) exercise can be considered as aerobic exercise, and a suitable and safe nonpharmacological intervention (Otsuki et al., 2008b; Rittweger, 2010).

9.5 GENERAL APPROACHES OF WHOLE BODY VIBRATION EXERCISE

WBV, as an exercise modality, has emerged as a feasible and useful strategy for improving overall health (Rittweger, 2010; Cochrane, 2011). In fact, studies have reported that WBV exercise improves body composition, muscular strength, and cardiovascular health concurrently. Furthermore, it is suitable for diseased populations who cannot perform traditional resistance or aerobic exercise training and the elderly (Rittweger, 2010; Cochrane, 2011; Shanb et al., 2017).

WBV exercise was initially developed in the middle of the last century in the former Soviet Union. The aim was the training of cosmonauts in order to prevent loss of bone mineral and muscle mass during space flights. Afterwards, reports on athletic training came (Nazarov and Spivak, 1987). It is widely used for personal training purposes, with numerous claims from manufacturers, encompassing improved muscular strength, bone mass, proprioception, balance, flexibility, quality of life, decrease of pain and falls risk, and beneficial metabolic effects, such as augmentation in lipolysis, glucose tolerance, growth hormone (GH), and testosterone secretion (Verschueren et al., 2004; Bautmans et al., 2005; Bruyere et al 2005; Gusi et al., 2006; Rehn et al., 2007; Cardinale et al., 2010; Slatkovska et al., 2010; Sá-Caputo et al., 2014). The individuals in the various trials involving WBV exercise vary in age, metabolic profile, physical activity levels or types of sports (Cochrane, 2011; Sierra-Guzmán et al., 2017). Some authors (Lorenzen et al., 2009; Rauch et al., 2010) observed the necessity for standardization of WBV exercise protocols, especially regarding the type of mechanical vibratory stimulus and the type of the oscillating/vibratory platform (OVP) (i.e. vertical and/or lateral plate displacement, or side alternating oscillation around a fixed center axis), duration of exercise bouts, interspersed time, number of repetitions, frequency, amplitude, peak-to-peak displacement and peak acceleration of the platform movement, and the position of the individual on the base of the OVP. In general, these parameters vary widely among several studies, resulting in manufacturers claiming effects observed in trials conducted with different OVP and specifications of the platform movement.

Bogaerts et al. (2009) have observed improvement in the maximum rate of *oxygen consumption* (VO_2max) in elderly inactive women after 12 months of WBV exercise, while Vissers et al. (2010) have found no such improvement after six months in overweight and obese middle-aged subjects. During acute WBV exercise, no or little effect on heart rate or oxygen uptake during WBV exercise have been observed (Rittweger et al., 2001; Hazell et al., 2008). It is suggested that there is a lack of understanding and agreement regarding the cardiovascular response elicited by WBV exercise and this stimulated studies with individuals with MetS.

The mechanisms proposed for the increase of the strength associated with WBV exercise include the tonic vibration reflex (TVR) (Cardinale and Lim, 2003), increased hormone secretion (Bosco et al., 2000; Cardinale et al., 2010), muscle hypertrophy (Delecluse et al., 2003), and stimulation of proprioceptive pathways. It is suggested that the TVR happens when the mechanical vibration causes a reflex muscle contraction by exciting the muscle spindles in the exposed muscle (Cardinale and Lim, 2003). Human studies have shown that the mechanical vibration induces muscle hypertrophy in healthy individuals as well as in populations with clinical disorders (Bogaerts et al., 2007; Machado et al., 2010; Milanese et al., 2012; Roelants et al., 2004), which cannot be overlooked as a possibility for strength improvement. Probably, WBV exercise would improve the efficiency of the proprioceptive feedback loop. In consequence, a more efficient proprioceptive feedback loop, caused by WBV exercise, could lead to increases in force production (Rittweger, 2010; Rosenberger et al., 2017).

Some mechanisms proposed for the improvement in the body composition (increase in muscle mass and decrease fat mass) with WBV exercise. It was described that additional gravitational load in individuals exposed to this type of exercise elicits an anabolic hormonal response (Bosco et al., 2000). Acute sessions of WBV exercise have acutely increased the serum levels of testosterone and human GH, and decreased cortisol levels (Bosco et al., 2000). The metabolic power increased with vibration training and the energy requirements during vibration training was similar to those of moderate intensity walking (Rittweger et al., 2002). Di Loreto et al. (2004) have reported that an acute bout of WBV exercise transiently reduced plasma glucose (no statistical difference), possibly by increasing glucose uptake and utilization by contracting muscle. Indeed, these considerations suggest WBV exercise as potential intervention for the treatment of obesity and sarcopenia. This would be possible by inducing predominance of anabolic hormones with increased energy expenditure that may lead to an increase in lean mass and reduction in fat mass. This is very important since the progression of arterial stiffness in older individuals is related to muscle tissue loss (Figueroa et al., 2016). It has been speculated that when the body is exposed to vibration, it evokes rhythmic muscle contractions (Da Silva et al., 2007; Kerschan-Schindl et al., 2001), which may induce changes in peripheral arteries. WBV exercise would increase mean velocity of the popliteal artery, which indicates vasodilation of the small vessels in the exposed muscles. This increase in blood flow may be due to vibration-induced reduction in blood viscosity (Kerschan-Schindl et al., 2001). The underlying mechanism for the significant increase in blood flow following vibration may be due to pulsatile endothelial stress resulting in increased endothelial nitric oxide (NOx), synthase activity, and NOx concentration (Maloney-Hinds et al., 2008; Rubin et al., 2003).

Although several biological actions of the WBV exercise have been reported, these effects in specific populations with MetS have not been clearly approached yet.

9.6 IMPROVEMENTS OF WHOLE BODY VIBRATION EXERCISE IN INDIVIDUALS WITH METABOLIC SYNDROME

There is a strong association of MetS and the risk of CVD and T2DM, and the use of different non-pharmacological interventions is recommended in this setting. However, there are a limited number of publications concerning WBV exercise and MetS. In a case report, Sá-Caputo et al. (2014) have shown, that WBV exercise using a protocol with low-frequencies (5–14 Hz) improved the anterior trunk flexibility in one individual with MetS. Carvalho-Lima et al. (2017) described improvements in parameters related to the quality of life (physical, psychological, environmental, social relations domains, and overall score) in individuals with MetS, using the WHO QOL-BREF questionnaire with a similar WBV exercise protocol previously published (Sá-Caputo et al., 2014).

As it has been pointed out, the effects of WBV exercise directly related with individuals with MetS have not been widely described yet. Nevertheless, the consequences of the WBV exercise in individuals with clinical conditions related to MetS are found in the current literature. Concerning obesity, some studies have reported the influence of the WBV exercise in overweight and obese people. Alvarez-Alvarado et al. (2017) investigated the effects of WBV exercise on aortic and leg arterial stiffness and observed reductions in arterial stiffness, central blood pressure (BP), and wave reflection in young obese women (Alvarez-Alvarado et al., 2017). Wong et al. (2016a) described the combined and independent effects of WBV exercise and L-citrulline supplementation on aortic hemodynamic and plasma NOx in obese post-menopausal (PM) women. The authors reported a decrease in brachial and aortic pressures as well as in augmented pressure (AP), and increased NOx levels. Augmentation index (AIx) and heart rate-corrected augmentation index (AIx@75) decreased in the WBV exercise + L-citrulline and WBV exercise + placebo groups, but not in the L-citrulline group. The improvement in AIx@75 in the WBV exercise + L-citrulline group was significant compared with the L-citrulline group L-citrulline supplementation and WBV exercise alone and combined decreased BP. The combined intervention reduced AIx@75 (Wong et al., 2016a). Wong et al. (2016b) studied the effect of an eight-week WBV exercise regimen on heart rate variability (HRV) and BP in obese PM women. There were group time interactions for brachial systolic blood pressure (SBP), diastolic blood pressure (DBP), LnHF- natural logarithm of low frequency (LnLF)/LnHF- natural logarithm of high frequency (LnHF), and normalized low frequency (nLF)/normalized high frequency (nHF) that decreased after WBV exercise, compared with no changes after control. There was an increase in nHF and decrease in nLF in the WBV exercise group compared with baseline, but the changes were not different than those in the control group (CG). No significant changes were observed in LnTP, LnLF, LnHF, or heart rate (HR) after eight weeks in either group (Wong et al., 2016a). Dipla et al. (2016) evaluated whether involuntary mechanical oscillations, induced by WBV elicit greater hemodynamic responses and altered neural control of BP in obese versus lean premenopausal women. During WBV exercise, obese women exhibited an augmented systolic BP response compared with lean women that was correlated with body fat percentage. The exaggerated BP rise was driven mainly by the greater increase in cardiac output index in obese versus lean women, associated with a greater stroke volume index in obese women. Involuntary contractions did not elicit a differential magnitude of responses in HR, HRV indices and systemic vascular resistance in obese versus lean women. However, there were greater spontaneous baroreceptor sensitivity (sBRS) responses in obese women. Figueroa et al. (2015a) investigated the effects of WBV exercise on ankle SBP and its associations with changes in pulse wave velocity (PWV) and aortic SBP in obese PM women. The baseline ankle SBP was higher in the WBV-high group compared with the WBV-normal group. WBV-high reduced ankle SBP compared with WBV-normal and control. Both WBV groups decreased brachial SBP, aortic SBP, leg pulse wave velocity (legPWV), and brachial-ankle pulse wave velocity (baPWV) compared with the CG. Reductions in legPWV were correlated with decreases in ankle SBP, brachial SBP, aortic SBP, and baPWV. Figueroa et al. (2015b) also studied the independent and combined effects of whole body vibration exercise (WBVE) and L-citrulline supplementation on PWV and muscle function in PM women. WBVE + L-citrulline decreased carotid-femoral pulse wave velocity (cfPWV) compared to both groups. All interventions decreased the femoral-ankle pulse wave velocity (faPWV) similarly. Leg movements (LM) index increased after WBVE + L-citrulline compared with L-citrulline. Both WBVE interventions increased leg strength compared to L-citrulline while decreased body fat percentage. Reductions in cfPWV were correlated with increases in LM index (Zaki et al., 2014). Zaki (2014) evaluated the impact of two exercise programs, WBV and resistance training on BMD and anthropometry in PM women. The BMD at the greater trochanter, at ward's triangle, and at lumbar spine were higher after physical training, using both WBV and resistive training. Both exercises were effective in BMI and waist-to-hip ratio. Simple and multiple regression analyses showed associations between physical activity duration and BMD at all sites (Figueroa et al., 2015a). Figueroa et al. (2014a) evaluated the effects of WBV on aortic hemodynamic and leg muscle strength in PM women. It was observed that the WBV significantly decreased aortic SBP, DBP, PP, AP, AIx and tension time index (TTI), while increased muscle strength compared with no changes after control. Figueroa et al., 2014b examined the impact of the WBVE on arterial stiffness, PWV, BP, and leg muscle function in PM women. There was a group-by-time interaction for arterial stiffness, BP, and strength as brachial-ankle PWV, leg PWV, systolic BP, DBP, and mean arterial pressure decreased and as strength increased after WBVE compared with no change after control. HR decreased, after WBVE, but there was no interaction. Leg lean mass and carotid-femoral PWV were not affected by WBV exercise training or control. Vissers et al. (2009) assessed the effect of WBV on oxygen uptake and carbon dioxide production in overweight and obese women. The ventilation of oxygen and carbon dioxide were consistently, significantly higher in the exercises with vibration compared with the exercises without vibration. Borg's scale scores only showed a significant difference between calf raises with and without vibration. Giunta et al. (2012) evaluated the acute effects of WBV alone or in combination with squatting plus external load (WBV + S) on serum GH levels and blood lactate in obese women. The WBV and WBV + S determined a GH increase. Lactate increased after both conditions. The lactate response was higher after WBV + S than after WBV. Baseline GH and GH peak values positively correlated to baseline lactate and lactate peak concentrations in both conditions. Figueroa et al. (2012) evaluated the WBVE on arterial function, autonomic function, and muscle strength in young overweight and obese women. There were decreases in baPWV, AIx, bSBP, aSBP), low-frequency power, and sympathovagal balance after WBVE compared with CG. Increases in high-frequency power and leg extension one-repetition maximum (1RM) occurred after WBVE compared with CG. Six weeks of WBVE decreased systemic arterial stiffness and aSBP via improvements in wave reflection and sympathovagal balance in young overweight/obese normotensive women. Milanese et al. (2013) explored the short-term effect of WBVE on anthropometry, body composition, and muscular strength in obese women. All changes in the two groups of WBVE, with or without radiofrequency (RF) were similar and these groups were combined in a single WBV intervention group. Individuals submitted to WBVE had significantly lower BMI, total body and trunk fat, sum of skinfolds, and body circumferences. Lower limb strength tests were increased in the WBV group.

In Table 9.1, additional details of the protocols used to evaluate the WBVE effects in populations with obesity are presented.

Concerning T2DM, improvements in these individuals are shown with use of WBVE. Baum et al. (2007) compared the WBVE with the strength training and a CG (flexibility training) on glycemic control in T2DM patients. After 12 weeks of training with three training sessions per week, fasting glucose concentrations remind unchanged. The area under curve and maximal glucose concentration of oral glucose tolerance test (OGTT) were reduced in the vibration and strength training group. Glycated hemoglobin (HbA1c) values tended to decrease below baseline date in the vibration training group while it increased in the two other intervention groups. Behboudi et al. (2011) investigated how aerobic exercise (AE) and WBVE affect glycaemia control in T2DM males. After eight weeks of exercise, no significant differences in any of the variables between AE and WBVE were found. A significant decrease in fasting glucose was observed in the exercise groups (AE and WBVE) compared with CG. Sañudo et al. (2013) verified the effect of a 12 week WBVE program on leg blood flow and body composition in people with type T2DM. Significant increments in mean velocity (Vmed), diastolic velocity (DV), weight, WC, waist-hip ratio (WHR), and body fat were observed after WBVE compared with CG. Within-group analysis showed significant differences in weight, WC, WHR, body fat, Vmed, peak blood velocity (PBV), and DV in the WBVE group. Del Pozo-Cruz et al. (2013) determined whether 12 week WBVE improved balance in individuals with T2DM. Significant between-group differences in the center of pressure (COP) excursions with eyes closed were found. Individuals in the WBVE group exhibited lower COP excursions with their eyes closed after the intervention. Yoosefinejad et al. (2014) investigated the effects of one session of WBVE on strength and balance of diabetic patients. Tibialis anterior muscle strength increase and decrease in Timed Up and Go Test (TUGT) parameters of T2DM individuals showed differences post-exercise in comparison to baseline. Del Pozo-Cruz et al. (2013) tested a 12 weeks WBVE on glycemic control, lipid-related cardiovascular risk factors and functional capacity among type T2DM individuals. There was a reduction in HbA1c and fasting blood glucose when compared to the CG. Lipid-related cardiovascular risk factors (cholesterol, triglycerides and atherogenic index) were also reduced. The balance was

TABLE 9.1
Additional Information about the Protocols used in Studies Involving the Whole Body Vibration Exercises in Obese Women

Study	Type of OVP and Biomechanical Parameters	Protocol
Alvarez-Alvarado et al. (2017)	Vertical OVP, 30-35 Hz and amplitude low-high.	Squats at 90° KA, semi-squats at 120° KA, wide-squat at 90° KA and calf raises. DE (60 beats per min) at a rate of three seconds for both eccentric and concentric contractions.
Wong et al. (2016a)	Vertical OVP, 25-40 Hz and amplitude 1–2 mm.	SDS with 90° KA, semi-squats with 120° KA, wide-stance semi-squats, and calf raises. DE with movements starting from an upright position into the assigned degree of KA (squats) and calf raises. The movements were at a rate of three seconds eccentric/two seconds concentric phases.
Wong et al. (2016b)	Vertical OVP, 25–40 Hz, amplitude low-high, 4.3 and 21.3 g.	WBVE consisted of four static and four dynamic leg exercises (normal, high, and wide-stance squats and calf raises). The movements were at a rate of three seconds eccentric/two seconds concentric phases.
Dipla et al. (2016)	Side alternating OVP, 25 Hz and PPD 6 mm.	Four minutes in steady-state conditions (seated baseline), six minutes WBV and four minutes during a seated recovery. Likewise, the "control" protocol involved a four-minute baseline, six minutes on the OVP (obtaining the same body position, without vibration) and four-minute recovery. After 30 seconds of standing on OVP, the participant assumed 18° KA, feet placed equidistant from the rotation axis.
Figueroa et al. (2015a)	Vertical OVP, 25–40 Hz and amplitude 1–2 mm.	SDS with 90° and 120° KA, wide-stand semi-squats, and calf raises. Duration of the sets and interset recovery time were increased (30–60 seconds) and decreased (60–30 seconds), respectively.
Figueroa et al. (2015a)	Vertical OVP, 25–40 Hz and amplitude 1–2 mm.	DS leg exercises. Exercises were at 90° and 120° KA. Duration of the exercise set (30–60 seconds), number of sets (one–five), total duration of training session (11–60 minutes), and decreasing the duration of rest periods (60–30 seconds).
Zaki (2014)	Side alternating OVP, 16 Hz.	Standing on the OVP performing stretching exercises. During the first session, the WBV group did three sets of one minute vibration with 16 Hz of vibration stimulus, separated by one minute resting periods. The training load increased, increasing by one set every session until zero sets of WBV with one minute rest sets.

(Continued)

TABLE 9.1 (CONTINUED)
Additional Information about the Protocols used in Studies Involving the Whole Body Vibration Exercises in Obese Women

Study	Type of OVP and Biomechanical Parameters	Protocol
Figueroa et al. (2014b)	Vertical OVP, 25–35 Hz and amplitude 1 mm.	DS for semi-squats and lunges with a 120° KA, squats with a 90° KA and calf raises. DE were at a rate of two seconds for the concentric and three seconds for the eccentric phase. The duration (30–45 seconds) and number of the sets (one–two) increased, while rest period 60 seconds.
Figueroa et al. (2014a)	Vertical OVP, 25–40 Hz and amplitude 1–2 mm.	Standing on OVP, DSS at a KA of 90° and 120°, lunges, and calf raises. DE were at a rate of two seconds for the concentric and three seconds for the eccentric phase. The duration (30–60 seconds) and number (one–six sets) of the sets per exercise increased, while the rest periods decreased (60–30 seconds)
Vissers et al. (2009)	Vertical OVP, 35 Hz and amplitude 4 mm.	Six exercises in random order: standing, dynamic squatting between 5° and 60° KA, dynamic calf raises, standing plus vibration, dynamic squatting plus vibration, and dynamic calf raises plus vibration. All exercises were at three seconds up and three seconds down.
Giunta et al. (2012)	Vertical OVP, 30 Hz and 2.85 xg.	Only WBV in static squat position. WBV+S, the subjects performed dynamic squats on the OVP. In only WBV, the stimulus was repeated for ten consecutive series with 50-second rest. In WBV+S, subjects performed ten consecutive series (separated by 50-second rest). The duration of WBV and WBV+S was 19 minutes and 30 seconds.
Figueroa et al. (2012)	Vertical OVP, 25–30 Hz and amplitude 1–2 mm.	DS semi squats with 120° KA, wide-stand semi-squat and calf raises. The duration of the sets and rest periods was increased (30–60 seconds) and decreased (60–30 seconds), respectively.
Milanese et al. (2013)	Vertical OVP, 40–60 Hz and amplitude 2–5 mm.	Subjects performed 20 sequential unloaded static leg and arm exercises. Each session lasted 19 minute (14 minute vibration training, five minute rest) and duration of each exercise was 30–60 seconds.

DE – dynamic exercises, DS – dynamic and static exercise, KA – knee angle, OVP – oscillating/vibratory platform, PPD – peak-to-peak displacement, SDS – static and dynamic squats, WBV – whole body vibration, WBV+S – WBV combined with squatting plus external load.

improved. Yoosefinejad et al. (2015) studied the effects of WBVE on balance and strength of diabetic individuals. Increased strength of tibialis anterior and quadriceps muscles. TUGT time decreased significantly in the WBV group.

In Table 9.2, additional details of the protocols used to evaluate the WBVE effects in populations with T2DM are presented.

Concerning the blood flow, which can present dysfunction in MetS individuals, some studies have described the improving effects of WBVE. Menéndez et al. (2016) analyzed the acute effects of isolated and simultaneous application of WBV and electromyostimulation (ES) on popliteal artery blood velocity (BV) in individuals with spinal cord injury. Simultaneous WBV+ES seems to produce a greater increase in mean blood velocity (MBV) and PBV of the popliteal artery than the

TABLE 9.2
Additional Information about the Protocols used in Studies Involving the Whole Body Vibration Exercises in Type 2 *Diabetes Mellitus* Individuals

Study	Type of OVP and Biomechanical Parameters	Protocol
Baum et al. (2007)	Vertical OVP, 30–35 Hz and amplitude 2 mm.	Eight different positions and exercises, 12–20 minutes, three sessions/week – 12 weeks.
Behboudi et al. (2011)	Side alternating OVP, 30 Hz and amplitude 2 mm.	Stand up and 110° semi-squat positioning, 16–24 minutes, three sessions/week – eight weeks.
Sañudo et al. (2013)	Side alternating OVP, 12–16 Hz and PPD 4 mm.	Eight different positions and exercises, 12–20 minutes, three sessions/week –12 weeks.
Del Pozo-Cruz et al. (2013)	Side alternating OVP, 12–16 Hz and PPD 4 mm.	Eight different positions and exercises, 12–20 minutes, three sessions/week – 12 weeks.
Yoosefinejad et al. (2014)	Vertical OVP, 30 Hz and amplitude 2 mm.	Stood on the plate with 30° KA, five× 30 seconds/ one minute rest between the bouts. Single session.
Del Pozo-Cruz et al. (2013)	Side alternating OVP, 12–16 Hz and PPD 4 mm.	Eight different positions and exercises, 12–20 minutes, three sessions/week –12 weeks.
Yoosefinejad et al. (2015)	Vertical OVP, 30 Hz and amplitude 2 mm.	Individuals stood barefooted on the OVP with 30° KA and contracted the muscles of the lower limbs during WBVE. Time increased every two weeks from 30 seconds initially to 45 seconds for third and fourth wks and to one minute for two last weeks, to sessions/week – six weeks.

KA – knee angle, OVP – oscillating/vibratory platform, PPD – peak-to-peak displacement, WBVE – whole body vibration exercise.

isolated (WBV or ES) or consecutive application of both stimuli. An increase was also found only with WBV. Johnson et al. (2014) determined the effects of low-frequency, low-amplitude WBV on whole blood NOx concentrations and on the skin blood flow (SBF) in individuals with symptoms of distal symmetric polyneuropathy. WBVE significantly increased SBF compared to the sham condition. Sañudo et al. (2013) observed the effect of a 12 week WBVE on leg blood flow and body composition in people with T2DM. Significant increases in the blood flow, Vmed, and DV were observed after WBV compared with the control group. Within-group analysis showed significant differences in Vmed (17%), PBV (12.6%), and DV (0.7%) due to the WBVE. Lohman et al. (2012) determined the effects of WBV (active and passive [no vibration]) and/or moist heat on SBF and ST in elderly, non-diabetic individuals following short-term exposure. Mean SBF increased (450%) after no WBV (passive) combined with moist heat and persisted for ten minutes after the intervention (379%). Significant increments in ST were also observed. Herrero et al. (2011a) investigated the effects of WBVE on BFV and muscular activity after vibration protocols in Friedreich's ataxia (FA) individuals. PBV increased after one, two, and three minutes of WBV with 30 Hz, as well as the rate of perceived exertion. Electromyography activity (EMG) of the muscles *vastus lateralis* (VL) and *vastus medialis* (VM) was increased (Herrero et al., 2011b). Herrero et al. (2011a) evaluated the effects of WBV on muscular activity and BFV after vibration treatments in patients with SCI. PBV increased after one, two, and three minutes of WBV with 30 Hz (with 20 Hz only after two and three minutes; no changes with 10 Hz). EMG increased with all frequencies. Lythgo et al. (2009) verified the effect of WBV on leg blood flow. Four-fold increase (33%) in MBV with 10–30 Hz and a two-fold increase (27%) in PBV with 20–30 Hz. Compared with the standing condition, squatting alone produced increases in mean and PBV. Lohman et al. (2007) determined the effects of short-duration, high-intensity, isometric weight bearing vibration exercise and vibration only on SBF. MBV significantly increased at both post-intervention time intervals with 30 Hz.

In Table 9.3, additional details of the protocols used to evaluate the effects of the WBVE in clinical conditions involving blood flow dysfunction are presented.

Concerning growth hormone (GH), there is substantial evidence that it is involved in the pathophysiology of obesity and MetS. The visceral obesity can be associated with GH deficiency and modalities of treatment to improve the GH secretion are desired (Lewitt, 2017). Studies with WBVE have described effects of WBVE on GH concentration. Di Giminiani et al. (2014) investigated the acute residual hormonal and neuromuscular responses (NM) after a single session of mechanical vibration with different acceleration loads. GH increased significantly over time only in the high vibration group (HVG). Giunta et al. (2013) compared GH, blood lactate (LA), and cortisol responses to different protocols involving WBVE. GH secretion significantly increased in all conditions immediately after the exercise session compared to other time points. A significantly larger increase was identified following squatting + external load + vibration (SEV) as compared to the other conditions. Elmantaser et al. (2012) compared the effects of two regimens of

TABLE 9.3
Additional Information about the Protocols used in Studies Involving the Whole Body Vibration Exercises in Clinical Conditions Involving Blood Flow Dysfunction

Study	Type of OVP and Biomechanical Parameters	Protocol
Menéndez et al. (2016)	Not informed the type of OVP, 10 Hz, PPD 5 mm.	Seated on wheelchairs with the feet on the OVP, ten × one minute/ one minute. Single session.
Johnson et al. (2014)	Side alternating OVP, 26 Hz, amplitude 2 mm.	Squat with knees flexed 30 to 40°, ten × 30 seconds/ one minute. Single session.
Sañudo et al. (2013)	Side alternating OVP, 12–16 Hz, PPD 4 mm.	Eight different dynamic and static exercises, eight × 30–60 seconds/30 seconds, three sessions/week – 12 weeks.
Lohman et al. (2012)	Vertical OVP, 50 Hz, PPD 5–6 mm.	Subject stands with one foot on the OVP for ten minutes, ten bouts × 60 seconds/two seconds. Single session.
Herrero et al. (2011a)	Side alternating OVP, 10–30 Hz, PPD 5 mm.	Subject was laid down and fixed to a tilt table with straps (60° KA – 0° the full knee extension). Tilt table was placed at 45° and the subject was for ten minutes before the WBVE, three minute continuous (constant) or fragmented (three bouts 60 second exposures, separated by 60 second rests), eight sessions in three weeks.
Herrero et al. (2011b)	Side alternating OVP, 10–30 Hz, PPD 5 mm.	Subject was laid down and fixed to a tilt table with straps (60° KA – 0° the full knee extension). Tilt table was placed at 45° and the subject was kept in that position for a period of ten minutes before the WBVE, three minute continuous (constant) or fragmented (three bouts 60 seconds, with 60 seconds rest), eight sessions in three weeks.
Lythgo et al. (2009)	Side alternating OVP, 5–30 Hz, amplitude 2.5–4.5 mm.	Squat position (50° KA), 14 × one minute and two × one minute bouts where no vibration was applied. Single session.
Lohman et al. (2007)	Vertical OVP, 30 Hz, amplitude 5–6 mm.	Isometric hold with 25°, 80° or at 100° KA, three × 60 seconds (180 seconds). Single session.

KA – knee angle, OVP – oscillating/vibratory platform, PPD – peak-to-peak displacement, WBVE – whole body vibration exercise.

WBVE on endocrine status, muscle function and markers of bone turnover. Side-alternating OVP at 22 Hz was associated with an immediate increase in serum GH after WBVE. None of the changes observed in the vertical OVP group reached statistical significance. Giunta et al. (2012) evaluated the acute effects of WBV alone on GH responses in severely obese females. WBV and WBV + S determined a significant GH increase, GH peaks occurring immediately after both exercise sessions. No significant differences were observed between GH peaks and GH net incremental area under the curve (nAUC) after both conditions, the whole pattern of GH responsiveness being comparable among all the subjects. Sartorio et al. (2011) compared the effects of maximal voluntary leg press exercise at equivalent force output and/or repeated vibration bouts on serum GH, LA, and muscle soreness. GH responses were significantly higher after tests with maximal voluntary contractions (MVC) and WBVE and MVC than after test with WBVE alone, with no difference between tests with MVC and combination of WBVE and MVC. Sartorio et al. (2010) compared GH responses to repeated bout of three different GH-releasing stimuli. Baseline GH levels increased significantly after the first bout of all the tested stimuli; it was significantly lower after WBVE than after MVC or MVC-WBVE, no differences being detected in-between. The second bout resulted in significantly lower GH increases than those elicited in the first bout in the three tests. All responses after the second bout of MVC and MVC-WBVE were significantly higher than those observed after WBVE alone. Cardinale et al. (2010) analyzed the effects of a single session of WBVE on anabolic hormones in aged individuals. No significant differences were identified in GH levels. Fricke et al. (2009) investigated the effect of side alternating WBV on hormone secretion and metabolism in healthy adults. WBVE increased GH levels in men and decreased GH levels in women. Kvorning et al. (2006) studied whether WBV combined with conventional resistance training (CRT) induces a higher increase in neuromuscular responses (NR) and hormonal measures compared with CRT or WBVE, respectively. GH increased in all groups, squatting + vibration showing higher responses than squatting alone and only WBV. Goto and Takamatsu (2005) examined the effects of WBV on the hormone and lipolytic responses. In the WBVE trial, GH concentration increased slightly after the session in both trials, but no significant difference was found at any point between trials. Moreover, no significant difference was observed in GH peak values between trials. Di Loreto et al. (2004) studied if the concentrations of glucose and some hormones were affected by WBVE, and reported that WBVE did not change GH concentration. Bosco et al. (2000) evaluated the acute responses of blood hormone concentrations and NM performance after WBVE and noted a significant increase on plasma GH concentration.

In Table 9.4, additional details of the protocols used to evaluate the effects of the WBVE on GH concentration are presented.

TABLE 9.4
Additional Information about the Protocols used in Studies Involving the Whole Body Vibration Exercises on Growth Hormone Concentration

Study	Type of OVP and Biomechanical	Protocol
Di Giminiani et al. (2014)	Vertical OVP, 20–40 Hz, PPD 0.2–0.9 mm.	Push-up position on the vibration device while flexing the elbow at 90°, series of 20 trials × ten seconds of WBVE with a ten-second pause between each trial and a four-minute pause after the first ten trials. Single session.
Giunta et al. (2013)	Vertical OVP, 35 Hz, PPD 4 mm.	In S condition the individuals stood and performed DS without vibration; in S+V the subjects performed DS in the same position of S, but with vibration; in SE condition, the subjects performed DS with external load; in SEV, the individuals performed the DS with S and external load. In S, 12 DS (three seconds down and three seconds up) for 72 seconds total without vibration. The stimulus was repeated for ten series with a 50-second rest. In S+V, 12 DS with vibration. In SE condition, the same protocol with an additional external load. In SEV, same protocol of SE standing with vibration. A randomized cross-over design consisting of four different training protocols, administered in separate days, with a washout period of one week between protocols.
Elmantaser et al. (2012)	Side alternating OVP/Vertical OVP, 18–22 Hz/32–37 Hz, PPD 4 mm/0.085 mm.	To stand still on the OVP platforms without shoes and, in the case of GP, with slight flexion of the knees, three times per week, ten minutes, 18 Hz (four weeks) and three times per week, 20 minutes, 22 Hz (four weeks), eight weeks.
Giunta et al. (2012)	Vertical OVP, 30 Hz, PPD 4 mm.	Individuals exposed to WBV alone or in combination with squatting plus external load (WBV+S). Treatments done in random order and separated by at least three days. In the WBV, the individuals remained on an OVP for 72 seconds. The stimulus repeated for ten series with a 50-second rest. The duration of WBV and WBV+S was 19 minutes and 30 seconds.
Sartorio et al. (2011)	Vertical OVP, 35 Hz, PPD 4 mm.	The individuals in test A, initially seated on a leg press machine and WBVE was delivered while individual was on an OVP with the 110° KA; In test B, the subjects were initially in the same position of test A, and then followed in the same static position of test A, with 110° KA but without WBVE; in test C, initially like in B, and then received the WBV, like in test A. In test A; 30 seconds in the leg press and 30-seconds WBVE - 15 series. In test B; like in A and performed three five-second MVC, separated by five-second rest. Then followed a 30-second in the same static position of test A - 15 series. In test C, individuals initially performed the three five-second MVC like in test B, then received the 30-second WBVE, like in test A. These two periods repeated without resting in-between for a total of 15 series. Single session.

(Continued)

TABLE 9.4 (CONTINUED)
Additional Information about the Protocols used in Studies Involving the Whole Body Vibration Exercises on Growth Hormone Concentration

Study	Type of OVP and Biomechanical	Protocol
Sartorio et al. (2010)	Vertical OVP, 35 Hz, PPD 5 mm.	Stood in static position with the 110° KA. In MVC, individuals performed three × five-second MVC, separated by five-second resting periods in between. Then, 30 second of rest. These two MVC rest cycles were repeated 15 times (total of 15 minutes). In MVC-WBVE, individuals performed the three × five-second MVC like in the MVC condition, and then received the 30-second WBVE bout, as in the WBVE condition. These two MVC-WBVE cycles were repeated 15 times (total of 15 minutes). Single session.
Cardinale et al. (2010)	Vertical OVP, 30 Hz, PPD 4 mm.	Slight knee flexion. five one-minute sessions separated by one-minute rest. Single session.
Fricke et al. (2009)	Side alternating OVP, 26 Hz, amplitude 1 mm.	Barefoot, in a half-squat position. Two series of five × WBVE, 60 second each with 60 second break between WBVE episodes. The two series were interrupted by a break of six minute sitting on a chair. Single session.
Kvorning et al. (2006)	Side alternating OVP, 20–25 Hz, PPD 4 mm.	Barefooted and the foot position and squatting technique with 90° KA was identical between groups. S group: six sets × eight repetitions of WL squat performed on the floor with two minute rest between sets. S + V group: like in S, but on the vibrating platform. V group: like in S without WL performed on the OVP, nine weeks.
Goto and Takamatsu (2005)	Side alternating OVP, 26 Hz, amplitude 2.5 mm.	A static squat position on the OVP with 120° KA, with shoes. Ten × 60 second with one minute rest between sets. Single session.
Di Loreto et al. (2004)	Vertical OVP, 30 Hz, PPD 4 mm.	Stood with the 70° KA. Ten × one minute with one minute rest between each treatment and with five minute rest after the five series (total 25 minute). Single session.
Bosco et al. (2000)	Vertical OVP, 26 Hz, PPD 4 mm.	100° KA. Ten × one minute with one minute rest between the vibration sets (a rest period lasting six minutes after five vibration sets). Single session.

DS – dynamic and static, EL – external load, KA – knee angle, MVC – maximal voluntary contractions, OVP – oscillating/vibratory platform, PPD – peak-to-peak displacement, S – squatting alone, SEV – squatting + external load + vibration, S + V – squatting + vibration, WBVE – whole body vibration exercise.

9.7 CONCLUSION

Considering the several aspects involved in MetS and the cardiovascular risk increase due to the cluster of the symptoms, various approaches and clinical interventions are necessary, including the nonpharmacological ones, such as WBV exercise.

In general, individuals with MetS are sedentary and have difficulty in performing regular exercise, and this regular practice is very important in their management.

Thus, the use of the WBV exercise as a modality of exercise in these individuals could be a safe and feasible strategy.

Research directly involving individuals with MetS is scarcer yet. However, studies involving the use of WBV exercise and the comorbidities associated with MetS, such as T2DM, and obesity, have been described. The results of these investigations have shown the strong relationship of the WBV exercise with the improvements in muscle strength, flexibility, balance, functionality, body composition, blood flow, and concentration of biochemical parameters. However, more research is needed to evaluate the effects of the WBV exercise in subjects with MetS in order to establish better biomechanical parameters, WBV exposition and rest time, and the individuals position during the WBV exercise.

REFERENCES

Alberti, K. G. and Zimmet, P. (2005). The metabolic syndrome–A new worldwide definition. *The Lancet*. 366:1059–1062.

Alberti, K. G., Zimmet, P., and Shaw, J. (2006). Metabolic syndrome: A new world-wide definition. A consensus statement from the International Diabetes Federation. *Diabetic Medicine*. 23:460–480.

Alberti, K. G., et al. (2009) Harmonizing the metabolic syndrome: A joint interim statement of the International Diabetes Federation Task Force on Epidemiology and Prevention; National Heart, Lung, and Blood Institute; American Heart Association; World Heart Federation; International Atherosclerosis Society; and International Association for the Study of Obesity. *Circulation*. 16:1640–1645.

Alley, D. E. and Chang, V. W. (2007). The changing relationship of obesity and disability, 1988–2004. *JAMA*. 298:2020–2027.

Alvarez-Alvarado, S., et al. (2017). Benefits of whole-body vibration training on arterial function and muscle strength in young overweight/obese women. *Hypertension Research*. 40:487–492.

Anderssen, S., et al. (2007). Combined diet and exercise intervention reverses the metabolic syndrome in middle-aged males: Results from the Oslo Diet and Exercise Study. *Scandanavian Journal Medicine and Science Sports*. 17:687–695.

Avogaro, P. and Crepaldi, G. (1965). Essential hyperlipidemia, obesity and diabetes. *Diabetologia*. 1:137.

Balkau, B. and Charles, M. A. (1999). Comment on the provisional report from the WHO consultation. European group for the study of insulin resistance (EGIR). *Diabetic Medicine*. 16:442–443.

Barbe, M. F. Barr, A. E. (2006). Inflammation and the pathophysiology of work-related musculoskeletal disorders. *Brain Behavior, and Immunity*. 20:423–429.

Baum, K., Votteler, T., and Schiab, J. (2007). Efficiency of vibration exercise for glycemic control in type 2 diabetes patients. *International Journal of Medical Sciences*. 31:159–163.

Bautmans, I., et al. (2005). The feasibility of whole body vibration in institutionalised elderly persons and its influence on muscle performance, balance and mobility: A randomised controlled trial [ISRCTN62535013]. *BMC Geriatrics*. 22:5–17.

Beavers, K. M., (2013). Health ABC Study. The role of metabolic syndrome, adiposity, and inflammation in physical performance in the Health ABC Study. *Journals of Gerontology Series A Biological Sciences and Medical Sciences*. 68:617–623.

Behboudi, L., et al. (2011). Effects of aerobic exercise and whole body vibration on glycaemia control in type 2 diabetic males. *Asian Journal of Sports Medicine*. 2:83–90.

Blair, S. N., Cheng, Y., and Holder, J. S. (2011). Is physical activity or physical fitness more important in defining health benefits? *Medicine & Science in Sports & Exercise.* 33:379–99.

Bogaerts, A., et al. (2007). Impact of whole-body vibration training versus fitness training on muscle strength and muscle mass in older men: A 1-year randomized controlled trial. *Journals of Gerontology Series A Biological Sciences and Medical Sciences.* 62:630–635.

Bogaerts, A., et al. (2009). Effects of whole body vibration training on cardiorespiratory fitness and muscle strength in older individuals (a 1-year randomised controlled trial). *Age Ageing.* 38:448–454.

Bosco, C., et al. (2000). Hormonal responses to whole-body vibration in men. *European Journal of Applied Physiology.* 81:449–454.

Bruyere, O., et al. (2005). Controlled whole body vibration to decrease fall risk and improve health-related quality of life of nursing home residents. *Archives of Physical Medicine and Rehabilitation.* 86:303–307.

Cade, W. T. (2008). Diabetes-related microvascular and macrovascular diseases in the physical therapy setting. *Physical Therapy.* 88:1322–1335.

Callaghan, B. C., et al. (2016). Association between metabolic syndrome components and polyneuropathy in an obese population. *JAMA Neurology.* 73:1468–1476.

Caponi, P. W., et al. (2013). Aerobic exercise training induces metabolic benefits in rats with metabolic syndrome independent of dietary changes. *Clinics.* 68:1010–1017.

Cardinale, M. and Lim, J. (2003). Electromyography activity of vastus lateralis muscle during whole-body vibrations of different frequencies. *Journal of Strength and Conditioning Research.* 17:621–624.

Cardinale, M., et al. (2010). Hormonal responses to a single session of wholebody vibration exercise in older individuals. *British Journal of Sports Medicine.* 44:284–288.

Carvalho-Lima, R.P., et al. (2017). Quality of life of patients with metabolic syndrome is improved after whole body vibration exercises. *African Journal of Traditional Complementary and Alternative Medicine.* 14:59–65.

Centers for Disease Control and Prevention (2016), http://millionhearts.hhs.gov/[accessed 18.02.16].

Chang, K.V., et al. (2015). Reduced flexibility associated with metabolic syndrome in community-dwelling elders. *PLoS One.* 10: 0117167. doi:10.1371/journal.pone.0117167. eCollection2015.

Cinti, S., et al. (2005). Adipocyte death defines macrophage localization and function in adipose tissue of obese mice and humans. *Journal of Lipid Research.* 46:2347–2355.

Cochrane, D. J. (2011). Vibration exercise: The potential benefits. *International Journal of Sports Medicine.* 32:75–99.

Courties, A., et al. (2015). Metabolic stress-induced joint inflammation and osteoarthritis. *Osteoarthritis Cartilage.* 23:1955–1965.

Da Silva, M.E., et al. (2007). Influence of vibration training on energy expenditure in active men. *Journal of Strength and Conditioning Research.* 21:470–475.

Del Pozo-Cruz, J., et al. (2013). A primary care-based randomized controlled trial of 12-week whole-body vibration for balance improvement in type 2 diabetes mellitus. *Archives of Physical Medicine and Rehabilitation.* 94:2112–2118.

Delecluse, C., Roelants, M., and Verschueren, S. (2003). Strength increase after whole-body vibration compared with resistance training. *Medicine & Science in Sports & Exercise.* 35:1033–1041.

Di Giminiani, R., et al. (2014). Hormonal and neuromuscular responses to mechanical vibration applied to upper extremity muscles. *PLoS One.* 9(11):e111521. doi:10.1371/journal.pone.0111521.

Di Loreto, C., et al. (2004). Effects of whole-body vibration exercise on the endocrine system of healthy men. *Journal of Endocrinological Investigation.* 27:323–327.

Dipla, K., et al. (2016). Exaggerated haemodynamic and neural responses to involuntary contractions induced by whole-body vibration in normotensive obese versus lean women. *Experimental Physiology.* 101:717–730.

Dixit, S., Maiya, A., and Shastry, B. (2014). Effect of aerobic exercise on quality of life in population with diabetic peripheral neuropathy in type 2 diabetes: A single blind, randomized controlled trial. *Quality of Life Research.* 23:1629–1640.

Duruöz, M. T., et al. (2012). Evaluation of MetSin patients with chronic low back pain. *Rheumatology International.* 32:663–667.

Eddy, D. M., Schlessinger, L., and Heikes, K. (2008). The metabolic syndrome and cardiovascular risk: Implications for clinical practice. *International Journal of Obesity.* 32:5–10.

Einhorn, D., et al. (2003). American College of Endocrinology position statement on the insulin resistance syndrome. *Endocrine Practice.* 9:237–252.

Elmantaser, M. et al. (2012). A comparison of the effect of two types of vibration exercise on the endocrine and musculoskeletal system. *Journal of Musculoskeletal and Neuronal Interactions.* 12:144–154.

Farias, D. L., et al. (2013). Elderly women with metabolic syndrome present higher cardiovascular risk and lower relative muscle strength. *Einstein.* 11:174–179.

Ferrannini, E., et al. (1987). Insulin resistance in essential hypertension. *The New England Journal of Medicine.* 317:350–355.

Figueroa, A., et al. (2012). Whole-body vibration training reduces arterial stiffness, blood pressure and sympathovagal balance in young overweight/obese women. *Hypertension Research.* 35:667–672.

Figueroa, A., et al. (2014a). Effects of whole-body vibration exercise training on aortic wave reflection and muscle strength in postmenopausal women with prehypertension and hypertension. *Journal of Human Hypertension.* 28:118–122.

Figueroa, A, et al. (2014b). Whole-body vibration exercise training reduces arterial stiffness in postmenopausal women with prehypertension and hypertension. *Menopause.* 21:131–136.

Figueroa, A., et al. (2015a). Impact of L-citrulline supplementation and whole-body vibration training on arterial stiffness and leg muscle function in obese postmenopausal women with high blood pressure. *Experimental Gerontology.* 63:35–40.

Figueroa, A., Kalfon, R., and Wong, A. (2015b). Whole-body vibration training decreases ankle systolic blood pressure and leg arterial stiffness in obese postmenopausal women with high blood pressure. *Menopause.* 22:423–427.

Figueroa, A., Jaime, S. J., and Alvarez-Alvarado, S. (2016). Whole-body vibration as a potential countermeasure for dynapenia and arterial stiffness. *Integrative Medicine Research.* 5:204–211.

Fox, C. S., et al. (2007). Abdominal visceral and subcutaneous adipose tissue compartments association with metabolic risk factors in the Framingham Heart Study. *Circulation.* 116:39–48.

Fricke, O., et al. (2009). Hormonal and metabolic responses to whole body vibration in healthy adults. *Endocrinologist.* 19:24–30.

Friedman, J., et al. (1990). Exercise-training increases glucose transporter protein GLUT4 in skeletal muscle of obese Zucker (fa/fa) rats. *FEBS Letters.* 268:13–16.

Fujioka, S., et al. (1987). Contribution of intra-abdominal fat accumulation to the impairment of glucose and lipid metabolism in human obesity. *Metabolism.* 36:54–59.

Garvey, W., et al. (1998). Evidence of the defects in trafficking and translocation of GLUT4 glucose transporter in skeletal muscle as a cause of human insulin resistance. *Journal of Clinical Investigation.* 101:2377–2386.

Giunta, M., et al. (2012). Growth hormone-releasing effects of whole body vibration alone or combined with squatting plus external load in severely obese female participants. *Obesity Facts*. 5:567–574.

Giunta, M., et al. (2013). Combination of external load and whole body vibration potentiates the GH-releasing effect of squatting in healthy females. *Hormone and Metabolic Research*. 45:611–616.

Goto, K. and Takamatsu, K. (2005). Hormone and lipolytic responses to whole body vibration in young men. *Japanese Journal of Physiology*. 55:279–284.

Grundy, S. M. (2006). Metabolic syndrome: Connecting and reconciling cardiovascular and diabetes worlds. *Journal of the American College of Cardiology*. 47:1093–1100.

Grundy, S. M., et al. (2005). Diagnosis and management of the metabolic syndrome: A statement for health care professionals: An American Heart Association/National Heart, Lung, and Blood Institute scientific statement. *Circulation*. 112:2735–2752.

Gusi, N., Raimundo, A., and Leal, A. (2006). Low-frequency vibratory exercise reduces the risk of bone fracture more than walking: A randomized controlled trial. *BMC Musculoskeletal Disorders*. 30:87–92.

Hazell, T. J., et al. (2008). Vertical whole-body vibration does not increase cardiovascular stress to static semi-squat exercise. *European Journal of Applied Physiology*. 104:903–908.

Herrero, A. J., et al. (2011a). Whole-body vibration alters blood flow velocity and neuromuscular activity in Friedreich's ataxia. *Clinical Physiology and Functional Imaging*. 31:139–144.

Herrero, A. J., et al. (2011b). Effects of whole-body vibration on blood flow and neuromuscular activity in spinal cord injury. *Spinal Cord*. 49:554–559.

Hitt, H. C., et al. (2007). Comorbidity of obesity and pain in a general population: Results from the Southern Pain Prevalence Study. *Journal of Pain*. 8:430–436.

Hwang, H. J. and Kim, S. H. (2015). The association among three aspects of physical fitness and metabolic syndrome in a Korean elderly population. *Diabetology & Metabolic Syndrome*. 12:112. doi:10.1186/s13098-015-0106-4.

Iemitsu, M., et al. (2008). Arterial stiffness, physical activity, and atrial natriuretic peptide gene polymorphism in older subjects. *Hypertension Research*. 31:767–774.

International Diabetes Federation (2006). The IDF worldwide definition of the metabolic syndrome. www.idf.org/metabolic-syndrome.

Ites, K. I., et al. (2011). Balance interventions for diabetic peripheral neuropathy: A systematic review. *Journal of Geriatric Physical Therapy*. 34:109–116.

Jensen, T. E. and Richter, E. A. (2012). Regulation of glucose and glycogen metabolism during and after exercise. *Journal of Physiology*. 590:1069–1076.

Johnson, P. K., et al. (2014). Effect of whole body vibration on skin blood flow and nitric oxide production. *Journal of Diabetes Science and Technology*. 8:889–894.

Kahn, R., et al. (2005). The metabolic syndrome: Time for a critical appraisal. Joint statement from the American Diabetes Association and the European Association for the Study of Diabetes. *Diabetes Care*. 28:2289–2304.

Kaur, J. (2014) A comprehensive review on metabolic syndrome. *Cardiology Research and Practice*. 2014:943162. doi:10.1155/2014/943162.

Kemmler, W., et al. (2009). Exercise decreases the risk of metabolic syndrome in elderly females. *Medicine & Science in Sports & Exercise*. 41:297–395.

Kerschan-Schindl, K., et al. (2001). Whole-body vibration exercise leads to alterations in muscle blood volume. *Clinical Physiology*. 21:377–382.

Kissebah, A. H., et al. (1982). Relation of body fat distribution to metabolic complications of obesity. *Journal of Clinical Endocrinology and Metabolism*. 54:254–260.

Klein, S., et al. (2007). Waist circumference and cardiometabolic risk: A consensus statement from Shaping America's Health. Association for Weight Management and Obesity

Prevention; NAASO, The Obesity Society; the American Society for Nutrition; and the American Diabetes Association. *Diabetes Care.* 30:1647–1652.

Klip, A. (2009). The many ways to regulate glucose transporter 4. *Applied Physiology, Nutrition, and Metabolism.* 34:481–87. http://dx.doi.org/10.1139/H09-7.

Klip, A., et al. (2009). Regulation of glucose transporter 4 traffic by energy deprivation from mitochondrial compromise. *Acta Physiologica.* 196:27–35. http://dx.doi.org/10.1111/j.1748-1716.2009.01974.x.

Kvorning, T., et al. (2006). Effects of vibration and resistance training on neuromuscular and hormonal measures. *European Journal of Applied Physiology.* 96:615–625.

Kylin, E. (1923). Studien über das Hypertonie-Hyperglyca "mie-Hyperurika" miesyndrom," *Zentralblatt für Innere Medizin.* 44:105–127.

Lewitt, M. S. (2017). The role of the growth hormone/insulin-like growth factor system in visceral adiposity. *Biochemistry Insights.* 10:1178626417703995.

Lohman, E. B., et al. (2007). The effect of whole body vibration on lower extremity skin blood flow in normal subjects. *Medical Science Monitor.* 13:CR71–76.

Lohman, E. B., et al. (2012). A comparison of whole body vibration and moist heat on lower extremity skin temperature and skin blood flow in healthy older individuals. *Medical Science Monitor.* 18:CR415–424.

Lorenzen, C., et al. (2009). Inconsistent use of terminology in whole body vibration exercise research. *Journal of Science and Medicine in Sport.* 12:676–678.

Lythgo, N., et al. (2009). Whole-body vibration dosage alters leg blood flow. *Clinical Physiology and Functional Imaging.* 29:53–59.

Machado, A., et al. (2010). Whole-body vibration training increases muscle strength and mass in older women: A randomized-controlled trial. *Scandinavian Journal of Medicine & Science in Sports.* 20:200–207.

Maloney-Hinds, C., Petrofsky, J. S., and Zimmerman, G. (2008). The effect of 30 Hz vs. 50 Hz passive vibration and duration of vibration on skin blood flow in the arm. *Medical Science Monitor.* 14:CR112–116.

Menéndez, H., et al. (2016). Acute effects of simultaneous electromyostimulation and vibration on leg blood flow in spinal cord injury. *Spinal Cord.* 54:383–389.

Milanese, C., et al. (2012). Effects of whole-body vibration with or without localized radiofrequency on anthropometry, body composition, and motor performance in young nonobese women. *Journal of Alternative and Complementary Medicine.* 18:69–75.

Milanese, C., et al. (2013). Ten-week whole-body vibration training improves body composition and muscle strength in obese women. *International Journal of Medical Sciences.* 10:307–311.

National Cholesterol Education Program. (2001). NCEP Expert panel on the detection, evaluation, and treatment of high blood cholesterol in adults. Executive summary of the third report of the National Cholesterol Education Program expert panel on detection, evaluation, and treatment of high blood cholesterol in adults (Adult treatment panel III). *JAMA.* 285:2486–2497.

Nazarov, V. and Spivak, G. (1987). Development of athlete's strength abilities by means of biomechanical stimulation method. *Theory Prac Phys Culture.* 12:445–450.

Neufer, P., Shinebarger, M., and Dohm, G. (1992). Effect of training and detraining on skeletal muscle glucose transporter (GLUT4) content in rats. *Canadian Journal of Physiology and Pharmacology.* 70:1286–1290.

Niiranen, T. J., et al. (2017). Prevalence, correlates, and prognosis of healthy vascular aging in a western community-dwelling cohort: The Framingham Heart Study. *Hypertension* 30. pii: HYPERTENSIONAHA.117.09026.

Oda, E. (2008). The metabolic syndrome as a concept of adipose tissue disease. *Hypertension Research.* 31:1283–1291.

Oda, E. (2012). Metabolic syndrome: Its history, mechanisms, and limitations. *Acta Diabetologica*. 49:89–95.

Ohkawara, K., et al. (2007). A dose-response relation between aerobic exercise and visceral fat reduction: Systematic review of clinical trials. *International Journal of Obesity*. 31:1786–1797.

Olijhoek, J. K., et al. (2004). The metabolic syndrome is associated with advanced vascular damage in patients with coronary heart disease, stroke, peripheral arterial disease or abdominal aortic aneurysm. *European Heart Journal*. 25:342–348.

Otsuki, T., et al. (2008a). Systemic arterial compliance, systemic vascular resistance, and effective arterial elastance during exercise in endurance-trained men. *American Journal of Physiology-Regulatory, Integrative and Comparative Physiology*. 295:R228–235.

Otsuki, T., et al. (2008b). Arterial stiffness acutely decreases after whole-body vibration in humans. *Acta Physiologica*. 194:189–194.

Ploug, T., et al. (1990). Effect of endurance training on glucose transport capacity and glucose transporter expression in rat skeletal muscle. *American Journal of Physiology-Endocrinology and Metabolism*. 259:E778–786.

Pou, K. M., et al. (2007). Visceral and subcutaneous adipose tissue volumes are cross-sectionally related to markers of inflammation and oxidative stress. The Framingham Heart Study. *Circulation*. 116:1234–1241.

Rauch, F., et al. (2010). Reporting whole-body vibration intervention studies: Recommendations of the International Society of Musculoskeletal and Neuronal Interactions. *Journal of Musculoskeletal and Neuronal Interactions*. 10:193–198.

Reaven, G. M. (2006). The metabolic syndrome: Is this diagnosis necessary? *The American Journal of Clinical Nutrition*. 83:1237–1247.

Rehn, B., et al. (2007). Effects on leg muscular performance from whole-body vibration exercise: A systematic review. *Scandinavian Journal of Medicine & Science in Sports*. 17:2–11.

Ren, J., et al. (1994). Exercise induces rapid increases in GLUT4 expression, glucose transport capacity, and insulinstimulated glycogen storage in muscle. *Journal of Biological Chemistry*. 269:14396–14401.

Ridker, P. M., Wilson, P. W. F., and Grandy, S. M. (2004). Should C-reactive protein be added to metabolic syndrome and to assessment of global cardiovascular risk? *Circulation*. 109:2818–2825.

Rittweger, J. (2010). Vibration as an exercise modality: How it may work, and what its potential might be. *European Journal of Applied Physiology*. 108:877–904.

Rittweger, J., Schiessl, H., and Felsenberg, D. (2001). Oxygen uptake during whole-body vibration exercise: Comparison with squatting as a slow voluntary movement. *European Journal of Applied Physiology*. 86:169–173.

Rittweger, J., et al. (2002). Oxygen uptake in whole-body vibration exercise: Influence of vibration frequency, amplitude, and external load. *International Journal of Sports Medicine*. 23:428–432.

Roelants, M., et al. (2004). Effects of 24 weeks of whole body vibration training on body composition and muscle strength in untrained females. *International Journal Sports Medicine*. 25:1–5.

Rosenberger, A., et al. (2017). Changes in muscle cross-sectional area, muscle force, and jump performance during 6 weeks of progressive whole-body vibration combined with progressive, high intensity resistance training. *Journal of Musculoskeletal and Neuronal Interactions*. 17:38–49.

Rubin, C., et al. (2003). Transmissibility of 15-hertz to 35-hertz vibrations to the human hip and lumbar spine: Determining the physiologic feasibility of delivering low-level anabolic mechanical stimuli to skeletal regions at greatest risk of fracture because of osteoporosis. *Spine*. 28:2621–2627.

Ruderman, N. B., Schneider, S. H., and Berchtold, P. (1981). The "metabolically-obese", normal-weight individual. *The American Journal of Clinical Nutrition.* 34:1617–1621.

Sá-Caputo, D. C., et al. (2014). Whole body vibration exercises and the improvement of the flexibility in patient with metabolic syndrome. *Rehabilitation Research and Practice.* 2014:628518. doi:10.1155/2014/628518.

Sañudo, B., et al. (2013). Whole body vibration training improves leg blood flow and adiposity in patients with type 2 diabetes mellitus. *European Journal of Applied Physiology.* 113:2245–2252.

Sartorio, A., et al. (2010). GH responses to two consecutive bouts of whole body vibration, maximal voluntary contractions or vibration alternated with maximal voluntary contractions administered at 2-h intervals in healthy adults. *Growth Hormone & IGF Research.* 20:416–421.

Sartorio, A., et al. (2011). Growth hormone and lactate responses induced by maximal isometric voluntary contractions and whole-body vibrations in healthy subjects. *Journal of Endocrinological Investigation.* 34:216–221.

Shanb, A. A. E., et al. (2017). Whole body vibration versus magnetic therapy on bone mineral density in elderly osteoporotic individuals. *Journal of Back and Musculoskeletal Rehabilitation.* 14. doi:10.3233/BMR-160607.

Sierra-Guzmán, R., et al. (2017). Effects of synchronous whole body vibration training on a soft, unstable surface in athletes with chronic ankle instability. *International Journal of Sports Medicine.* 38:447–455.

Simmons, R. K., et al. (2010). The metabolic syndrome: Useful concept or clinical tool? Report of a WHO Expert Consultation. *Diabetologia.* 53:600–605.

Slatkovska, L., et al. (2010). Effect of whole-body vibration on BMD: A systematic review and meta-analysis. *Osteoporosis International.* 21:1969–1980.

Sossa, C., et al. (2013). Lifestyle and dietary factors associated with the evolution of cardiometabolic risk over four years in West-African adults: The Benin study. *Journal of Obesity.* 2013: 298024.

Stehno-Bittel, L. (2008). Intricacies of fat. *Physical Therapy.* 88:1265–1278.

The global burden of disease: 2004 update. World Health Organization Geneva. www.who.int/healthinfo/global_burden_disease/GBD_report_2004update_full.pdf/; (accessed 18.02.16).

Thomas, D. E., Elliott, E. J., and Naughton, G. A. (2006). Exercise for type 2 diabetes mellitus. *Cochrane Database of Systematic Reviews.* 3:CD002968.

Tjønna, A., et al. (2008). Aerobic interval training versus continuous moderate exercise as a treatment for the metabolic syndrome: A pilot study. *Circulation.* 118:346–354.

Vague, J. (1947). Sexual differentiation. A factor affecting the forms of obesity. *Presse Medicale.* 30:39–40.

Vague, J. (1956). The degree of masculine differentiation of obesities: A factor determining predisposition to diabetes, atherosclerosis, gout, uric calculous disease. *American Journal of Clinical Nutrition.* 4:20–34.

Verschueren, S. M., et al. (2004). Effect of 6-month whole body vibration training on hip density, muscle strength, and postural control in postmenopausal women: A randomized controlled pilot study. *Journal of Bone and Mineral Research.* 19:352–359.

Veves, A., Backonja, M., and Malik, R. (2009). Painful diabetic neuropathy: Epidemiology, natural history, early diagnosis, and treatment options. *Pain Medicine.* 9: 660–674.

Vissers, D., et al. (2009). The effect of whole body vibration short-term exercises on respiratory gas exchange in overweight and obese women. *Physician and Sportsmedicine.* 37:88–94.

Vissers, D., et al. (2010). Effect of long-term whole body vibration training on visceral adipose tissue: A preliminary report. *Obesity Facts.* 3:93–100.

Wei, Y., et al. (2008). Skeletal muscle insulin resistance: Role of inflammatory cytokines and reactive oxygen species. *American Journal of Physiology-Regulatory, Integrative and Comparative Physiology*. 294:673–680.

Wong, A., et al. (2016a). Combined whole-body vibration training and l-citrulline supplementation improves pressure wave reflection in obese postmenopausal women. *Applied Physiology, Nutrition, and Metabolism*. 41:292–297.

Wong, A., et al. (2016b). Whole-body vibration exercise therapy improves cardiac autonomic function and blood pressure in obese pre- and stage 1 hypertensive postmenopausal women. *Journal of Alternative and Complementary Medicine*. 22:970–976.

World Health Organization (1999). Definition, diagnosis, and classification of diabetes mellitus and its complications: Report of a WHO Consultation. Geneva.

World Health Organization (2014). Global status report on noncommunicable diseases, 2014. Geneva. http://apps.who.int/iris/bitstream/10665/148114/1/9789241564854_eng.pdf?ua=1/ (accessed 18.02.16).

World Health Organization (2016). Cardiovascular death and disability can be reduced more than 50 percent. Geneva. www.who.int/mediacentre/news/releases/pr83/en/ (accessed 18.02.16).

Yoosefinejad, A. K., et al. (2014). The effectiveness of a single session of whole-body vibration in improving the balance and the strength in type 2 diabetic patients with mild to moderate degree of peripheral neuropathy: A pilot study. *Journal of Bodywork and Movement Therapies*. 18:82–86.

Yoosefinejad, A. K., et al. (2015). Short-term effects of the whole-body vibration on the balance and muscle strength of type 2 diabetic patients with peripheral neuropathy: A quasi-randomized-controlled trial study. *Journal of Diabetes & Metabolic Disorders*. 23:45. doi:10.1186/s40200-015-0173-y.

Zaki, M. E. (2014). Effects of whole body vibration and resistance training on bone mineral density and anthropometry in obese postmenopausal women. *Journal of Osteoporosis*. 2014:702589. doi:10.1155/2014/702589.

10 Effects of Mechanical Vibration on Bone Tissue

Christiano Bittencourt Machado, Borja Sañudo, Christina Stark, and Eckhard Schoenau

CONTENTS

10.1 PRINCIPLES OF BONE ANATOMY AND BIOLOGY

The skeleton is mainly constructed by bone tissue, and it presents many functions in the human body: support for soft tissues, a basis for muscle activity (biomechanical levers), protection for vital organs, production of blood cells (bone marrow), an important calcium, phosphate and other ions reservoir. Moreover, three ossicles located in the middle ear act in the mechanical aspect of hearing (malleus, incus and stapes).

More than 200 bones classified as long, short, flat, irregular or sesamoid compose the human adult skeleton. Detailed information about them can be obtained in classic anatomy textbooks (Moore et al., 2013; Paulsen and Waschke, 2011). Long bones present a shaft (diaphysis) and form most bones of the upper and lower limbs. Short bones are cube-shaped, mainly composed by trabecular bone (for example, bones of the wrist and ankle). In flat bones, trabecular bone is surrounded by two parallel layers of compact bone (bones in skull and sternum). The irregular bones present irregular shapes and cannot be classified in another category, like the vertebrae and coxal bone. Finally, sesamoid bones (for example, the patella) are embedded in muscle tendons. These bones are intimately associated with nonosseus tissues during a lifetime: tendons, ligaments and cartilage, the latter participating in the structure of important elements such as costal cartilage, larynx, trachea, bronchi, nose and ears (Carter and Beaupré, 2001).

Because of its complex properties, bone tissue must have a complex structure. Rho et al. (1998) divided it into five different levels: (1) the macrostructure, composed of cortical and cancellous bone; (2) the microstructure, composed by the Haversian systems, osteons and trabeculae (from 10 to 500 μm); (3) the sub-microstructure (from 1 to 10 μm), presenting the lamellae; (4) the nanostructure (from 100 nm to 1 μm), composed by collagen fibers and minerals and (5) the sub-nanostructure, represented by the molecular structure of the bone elements (minerals, collagen and other organic proteins), with an order dimension of less than 100 nm.

Several biomolecules participate in bone structure: collagen, water, hydroxyapatite crystals, proteoglycans like decorin and biglycan, and noncollagenous proteins like osteocalcin and osteopontin. The mineral part of bone consists of hydroxyapatite crystals – $Ca_{10}(PO_4)_6(OH)_2$. Proteoglycans and noncollagenous proteins are believed to control the rate of mineralization and to act as a chemoattractant for bone cells (Marks and Odgren, 2002).

Four types of cells are identified in bone (Figure 10.1) (Machado, 2013). First of all, fully differentiated mononuclear, occurring as a layer of contiguous cuboidal cells called **osteoblasts**, responsible for unmineralized bone matrix production—the osteoid (95% of type I collagen and 5% of noncollagenous proteins). Mechanical stress appears to stimulate mesenchymal cells to originate osteoblasts through a differentiation process. The rate in which osteoblasts lay down osteoid (apposition rate) is approximately 1 μm/day. These cells seem to regulate mineralization of the bone matrix.

Spaces within the bone matrix called lacunae encompass mature osteoblasts named **osteocytes,** flattened cells responsible for the maintenance of bone tissue, matrix synthesis and resorption to a limited extent. Each osteocyte is incorporated by one lacuna, and the canaliculi are tunnels that make the communication among osteocytes. An interesting finding is that an osteocyte may function like an osteoblast or an osteoclast, depending on the organelles inside them.

Bone resorption is provided by large, mobile and multinucleated cells called **osteoclasts**. They are formed by fusion of monocytes from bone marrow. Bone is eroded by a demineralization process, and then by enzymatic action to dissolve collagen (Figure 10.2). During resorption, osteoclasts rest on the bone surface. Two plasma membrane specializations can be identified: an infolded area for bone resorption, the ruffled border, and a clear zone, responsible for creating an environment for resorption, acting as an adhesion point between the osteoclast and bone matrix. Cytokines (signaling molecules) and hormones such as calcitonin and parathormone coordinate osteoclast activity (Majeska, 2009).

Finally, the **bone lining cells** are believed to be precursors of osteoblasts. They are flat, elongated, inactive cells at bone surfaces. Other speculations about their functions is acting in mineral transfer and initiation of bone remodeling process in the presence of a mechanical or chemical stimulus.

An important classification rises if we consider bone porosity degree (the volume fraction of soft tissue): the trabecular (cancellous) and compact (cortical) bone. **Trabecular bone** presents porosity degree ranging from 75% to 95%, showing an

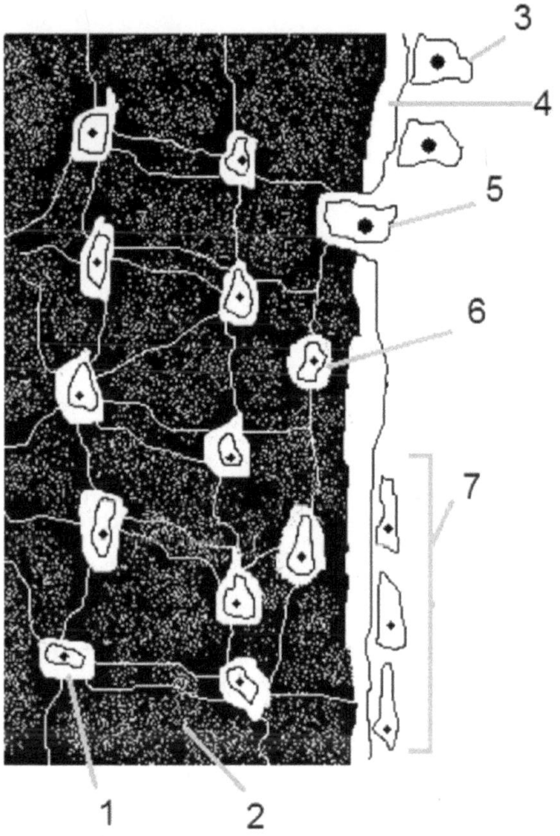

FIGURE 10.1 An illustration showing bone cells in action. It is possible to identify the osteocytes inside the lacunae (1) in the mineralized bone tissue (2). Osteoblasts originated from osteoprogenitor cells (3) lay down osteoid (4) to the formation of new osteocytes (5). Bone lining cells can be visualized in (6) localized in bone surface. (Reprinted from: *Ultrasound in Bone Fractures: from Assessment to Therapy*, p. 7, Christiano B. Machado. Copyright (2013) with permission from Nova Science Publishers, Inc.).

important metabolic role. It can be found in the ends of long bones (epiphysis), flat and cuboidal bones. Non-calcified regions are filled with marrow. **Compact bone** forms the bone cortex, its porosity ranging from 5% to 10%. It forms the shaft of long bones and the shell of cancellous bone. It is mainly responsible for mechanical functions and protection. Concentric lamellae are formed by a network of packed collagen fibrils which run in perpendicular planes in adjacent lamellae (also known as a plywood-like arrangement). The important Haversian canals (with 50 μm of diameter approximately), aligned to the long bones, containing nerves and blood vessels. These canals are connected by transverse Volkmann's canals (Martin et al., 2010; Machado, 2013). Compact bone can still be classified as primary and secondary bones. **Primary bone** forms the osteon and the Haversian canals (circumferential lamellar bone), and it is mineralized tissue laid on the surface of bone. There is

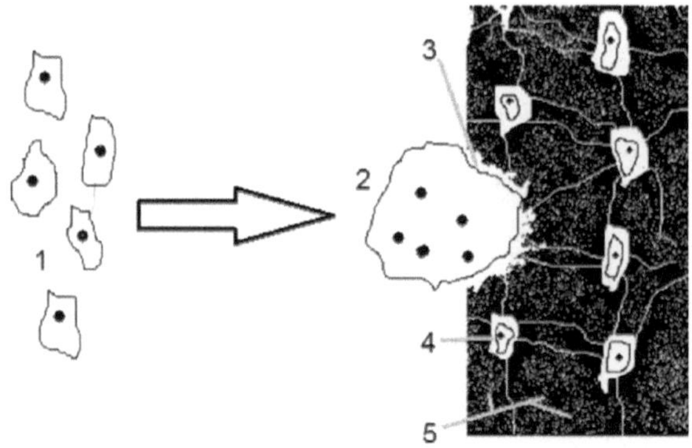

FIGURE 10.2 An illustration showing osteoclasts in action. A fusion of monocytes (1) originates the osteoclast, a multinuclear cell (2). The osteoclasts are responsible for bone resorption through a demineralization process (3). Osteocytes inside the lacunae (4) and bone mineralized matrix (5) can be also identified. (Reprinted from: *Ultrasound in Bone Fractures: from Assessment to Therapy*, p. 8, Christiano B. Machado. Copyright (2013) with permission from Nova Science Publishers, Inc.).

another type of primary bone, the plexiform bone, consisting of woven and lamellar bone producing a "brick wall" semblance. The **secondary bone** is produced by remodeling (secondary osteons appear under the primary bone).

An analysis on bone microstructure would reveal another level of organization: the lamellar bone and the woven bone. **Lamellar bone** is formed by the lamellae, parallel layers of collagen fibers and mineral crystals which provide bone anisotropy. These layers are mineralized using collagen as a "template". A layer of osteoid is found on the surface under the osteoblasts, and the deposit of inorganic matrix is made toward bone surface, or surrounding the osteoblast to produce osteocytes. On the other hand, randomly arranged collagen fibers and crystals form **woven bone**, which is weaker than lamellar bone.

The ossification process takes place by two phenomena: the intramembranous (or direct) ossification and the endochondral (or indirect) ossification. In **intramembranous ossification** (for example, occurring in the bones of calvaria, some facial bones and parts of the mandible and clavicle), mesenchymal cells (MSCs) transform directly into osteoblasts. Initially, MSCs produce types III, V and XI collagen, as well as collagen type I. **Endochondral ossification** is a process of bone development using hyaline cartilage to recruit, proliferate and differentiate embryonic MSCs, being progressively mineralized and replaced by bone.

Cartilage tissue is of utmost importance for bone tissue biology and development. It is present in human body joints, nose, bronchial tubes, ears and intervertebral disks, being also essential for endochondral ossification and fracture regeneration. The flexible **articular cartilage** contains no blood vessels or nerves, composed by so-called **chondroblasts** (cells that produce the extracellular matrix) and the **chondrocytes** (chodroblasts caught in the matrix) laying in spaces called lacunae. Water fills approximately 70% of all cartilage matrices. The extracellular matrix is mainly composed of proteoglycans (15% to 40% of the dry weight) and type II collagen (40% to 70% of the dry weight), however there are other types of collagen in articular cartilage. The **hyaline cartilage** is the most prevalent type of cartilage found in human body. It can be identified in articular surfaces, anterior end of the ribs, tracheal rings and growth plates (nonmineralized region of growth near the end of developing bones). The **elastic cartilage** forms the external ear, epiglottis and Eustachian tubes, presenting great elasticity; and the **fibrocartilage**, forming the pubic symphysis, intervertebral disks and tendon-bone attachments.

Long bones pass through a very specialized process of growth, as briefly explained in Figure 10.3. A region at each end of the bone called **epiphyseal plate** is formed by hyaline cartilage, and endochondral ossification takes place for the initial bone development. These plates are no longer necessary in adults, being totally ossified.

Muscles and bones form the basis of human locomotion. Muscles are responsible for the drive and represent the contractile element of movement. Bones represent the stable framework at which the muscles attach in order to fulfil their function. Muscle activity is controlled by the central and peripheral nervous system. It is well recognized that nerves, muscles and bones represent a functional unit. Already more than 100 years ago the anatomist Julius Wolff described the connection between muscles and skeletal development in his "law of the transformation of the bones" (Wolff, 1892; Frost, 1998; Frost, 2004). This law implies that the skeletal system adapts to the forces acting on it, namely the maximal forces. The decisive osteoanabolic stimulus for the osteoblasts is correspondingly the maximum forces acting on bone through the muscles resulting in shear forces. The deformation of the bone is measured by the "mechanostat", which is formed by the network of osteocytes. In the event of maximal force application to the bone, the synthesis of the bone base substance via the osteoblasts is stimulated by the principle of the "functional muscle-bone unit". Physical activity accordingly promotes bone formation. On the contrary, with less use, e.g. in the case of long-term immobilization, rapid degradation of bone is induced. This relationship between muscle force and bone development is described in the "functional muscle-bone unit" model and is shown in Figure 10.4.

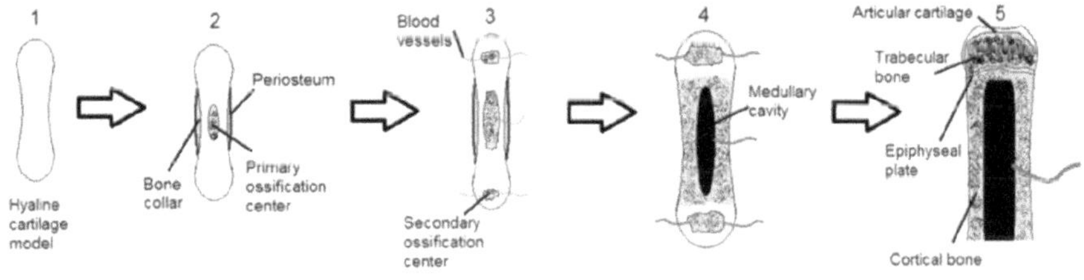

FIGURE 10.3 Growth of long bones: (1) the process initiates with a hyaline cartilage model; (2) a primary ossification center is formed, as well as the bone collar and periosteum; (3) secondary ossification centers are developed in epiphysis regions, and angiogenesis takes place; (4) a medullary cavity containing marrow tissue appears; (5) in children and adolescents, it is possible to observe the epiphyseal plate, as well as the trabecular bone in epiphysis and cortical bone in diaphysis. (Reprinted from: *Ultrasound in Bone Fractures: from Assessment to Therapy*, p. 15, Christiano B. Machado. Copyright (2013) with permission from Nova Science Publishers, Inc.).

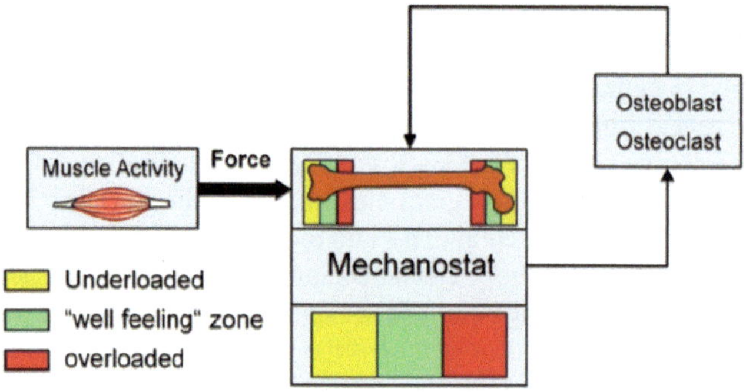

FIGURE 10.4 The "functional muscle-bone unit" model.

10.2 BONE MECHANICS AND VIBRATION

A brief introduction about bone mechanics is important since this chapter aims at exploring the effects of mechanical vibration on bone tissue.

Julius Wolff (1836–1902), a German anatomist, first described the dynamic behavior of bone. The so-called **Wolff's Law** states that bone adapts to the loads it is subjected to during life. Each bone has a specific function and then it is influenced by several events during life. This adaptation process includes increase/decrease in strength according to loading history as well as changes in trabecular and cortical architecture and mineralization.

Considering bone as a solid material, it can undergo five possible deformations: tension, compression, shear, torsion and bending, as seen in Figure 10.5. When the material is loaded in a specific manner and its deformation is recorded, we obtain the **Stress × Strain curve** (or Load × Deformation). It can be observed in Figure 10.6

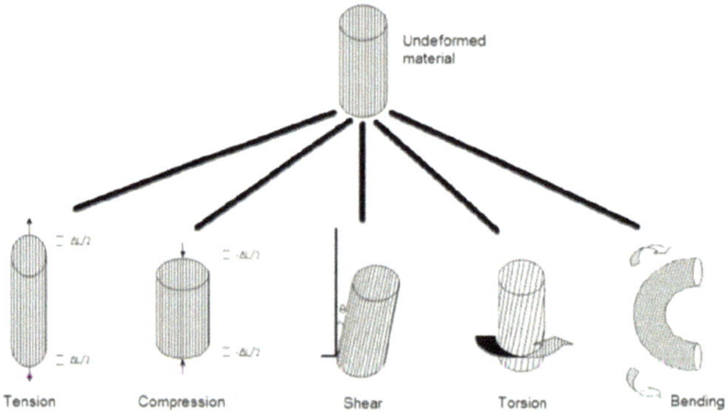

FIGURE 10.5 Possible deformations in a solid material: tension, compression, shear, torsion and bending; ΔL is the variation in length and θ is the angle in relation with the normal. (Reprinted from: *Ultrasound in Bone Fractures: from Assessment to Therapy*, p. 21, Christiano B. Machado. Copyright (2013) with permission from Nova Science Publishers, Inc.).

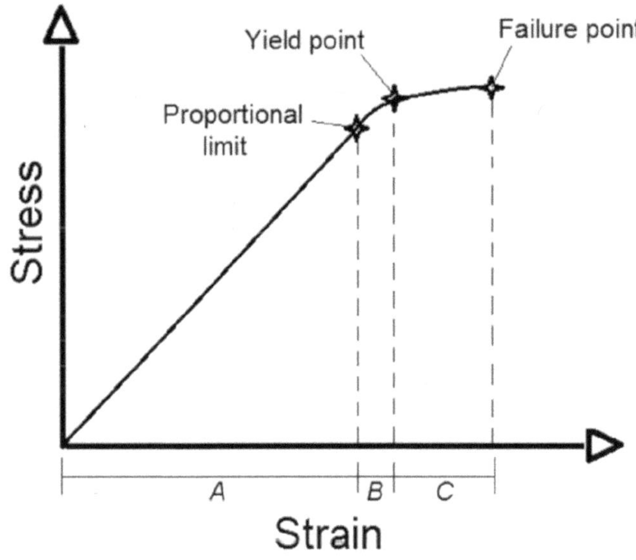

FIGURE 10.6 Stress × Strain curve for a solid material. In A and B, the material is under linear and nonlinear elastic deformation, respectively. Region C represents plastic deformation (the material is permanently deformed).

that a linear elastic deformation takes place (region A), followed by a nonlinear deformation (region B). After the yield point (region C), the material is on a plastic regimen and it may be permanently deformed. With increasing loads, the ultimate load is reached (failure point) and a fracture may occur.

Now it is possible to define some important concepts for solid mechanics: (1) **strength**, which is the load at the yield or fracture point (depending on the material or on the analysis being made); (2) **stiffness** or **rigidity**, which is the necessary load to deform the material; (3) **compliance**, meaning the ease to deform a material, obtained by the reciprocity of stiffness and (4) **stress**, which is the load per unit area (Martin et al., 2010), and can be classified as **normal** or **shear stress**.

Considering an axially loaded body (with a uniformly distributed load), the normal stress σ (in MPa) is obtained as follows (Machado, 2013):

$$\sigma = \frac{F_n}{A} \tag{10.1}$$

where F_n is the normal force, perpendicularly applied to a given area A. The **shear stress** (τ), when the force F_s occurs in shear as shown in Figure 10.7, can be calculated using

$$\tau = \frac{F_s}{A} \tag{10.2}$$

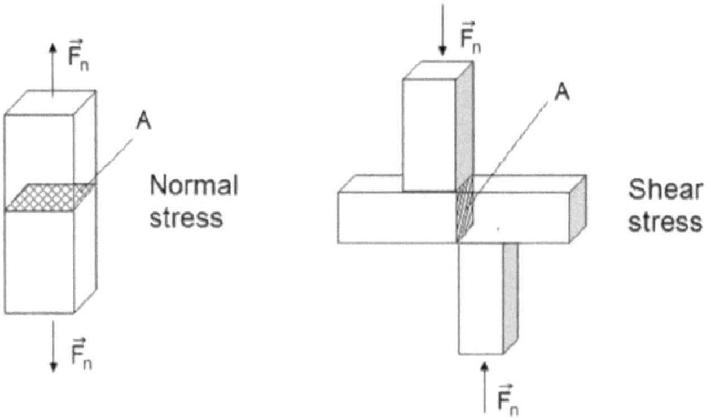

FIGURE 10.7 Normal (σ) and shear (τ) stress applied to area A. (Reprinted from: Ultrasound in Bone Fractures: from Assessment to Therapy, p. 23, Christiano B. Machado. Copyright (2013) with permission from Nova Science Publishers, Inc.).

The **strain** (ε) is a measure of deformation: it is the relative displacement between particles in a body. It can be calculated using the equation (Machado, 2013)

$$\varepsilon = \frac{\Delta l}{l_0} \tag{10.3}$$

where:

Δl is the measured displacement
l_0 is the initial body length

Then the **elasticity** (Young's modulus or modulus of elasticity - E in MPa) can be obtained by the relation:

$$E = \frac{\sigma}{\varepsilon} \tag{10.4}$$

The Young's modulus is the slope of the linear region in Figure 10.5.

Another important parameter is the **Poisson ratio** (υ), which the ratio between the transverse strain component and the longitudinal strain component (Machado, 2013):

$$\vartheta = -\frac{\varepsilon_t}{\varepsilon_l} \tag{10.5}$$

where:

ε_t is the transverse strain
ε_l is the longitudinal strain

Ordinary materials have values of $\upsilon < 0.5$; incompressible materials, $\upsilon = 0.5$.

Strain (ε) can be related to stress (σ) using Hooke's law:

$$\sigma_i = C_{ij}\varepsilon_j \tag{10.6}$$

where C is a matrix with elastic constants C_{ij} (stiffness tensor). Bone is an orthotropic material, and C will present nine independent coefficients:

$$[C] = \begin{bmatrix} C_{11} & C_{12} & C_{13} & 0 & 0 & 0 \\ C_{21} & C_{22} & C_{23} & 0 & 0 & 0 \\ C_{31} & C_{32} & C_{33} & 0 & 0 & 0 \\ 0 & 0 & 0 & C_{44} & 0 & 0 \\ 0 & 0 & 0 & 0 & C_{55} & 0 \\ 0 & 0 & 0 & 0 & 0 & C_{66} \end{bmatrix} \tag{10.7}$$

The elastic constants C_{ij} values for bovine and human cortical bone are shown in Table 10.1.

Bone is considered an anisotropic, linear elastic solid. **Anisotropic** materials present different mechanical properties depending on the direction of analysis. On the contrary, **isotropic** materials are directionally independent. Bone has a special kind of anisotropy, the **orthotropy**, which means that mechanical properties differ according to orthogonal directions (Mitton et al., 2011).

According to Martin et al. (2010), cortical bone has its mechanical properties influenced by osteons. Three mechanisms are responsible for that: (1) continuous replacement of mineralized bone matrix with less calcified material; (2) increase of

TABLE 10.1
Elastic Constants for Bovine (Lasaygues and Pithioux, 2002) and Human (Van Buskirk and Ashman, 1981) Cortical Bone

C_{ij} (in GPa)	Bovine	Human
C_{11}	23.50	20.00
C_{22}	26.00	21.70
C_{33}	34.60	30.00
C_{44}	9.20	6.56
C_{55}	6.00	5.85
C_{66}	6.05	4.74
C_{12}	6.55	10.90
C_{13}	8.35	11.50
C_{23}	8.20	11.50

cortical porosity and (3) introduction of cement line interfaces. Different types of osteons (types L, A and T) can adapt themselves according to loading demands during life (Ascenzi and Bonucci, 1968, 1972).

Bone strength (given by the Young's modulus E) is inversely related to **porosity** (p), i.e., as the former increases, the latter decreases. The following equation proposed by Currey (1988) shows this relation:

$$E = 23.4(1 - p)^{5.74} \tag{10.8}$$

The **apparent density** is the most studied density parameter for bone (values ranging from 1.8 to 2.0 g/cm^3 for cortical bone; 1.0 to 1.4 g/cm^3 for trabecular bone), and it can be calculated by the mass of a volume divided by its total volume. The bone volume fraction (BVF) can be estimated using

$$BVF = \frac{V_b}{V_v} \tag{10.9}$$

where V_b is the sum of the bony matrix volume, and V_v the soft tissue (void) volume. Porosity may be obtained using

$$p = \frac{V_v}{V_T} \tag{10.10}$$

where V_T is the bone total volume. The apparent density ρ_a (in g/cm^3) is finally obtained using

$$\rho_a = \rho_b - (\rho_b - \rho_v)p \tag{10.11}$$

where ρ_b and ρ_V are the density of bone tissue and soft tissue, respectively. Morgan et al. (2003) related E with ρ_a using the equation

$$E = 8920\rho_a^{1.83} \tag{10.12}$$

Trabecular bone presents a smaller mineral content and greater water content than in cortical bone. Lamellar orientation is also different.

10.2.1 Bone Vibration

A paper of Van der Perre and Lowet (1996) summarized a basic theory of vibration analysis with an emphasis on the mechanical characteristics that can be derived from the resonant frequencies of a long bone.

A structure may vibrate at specific frequencies when it is allowed to move freely after an excitation. These frequencies are called the eigenfrequencies, or natural frequencies. Considering a long slender beam to model a long bone, the resonant frequencies are given by the equation

$$f_n = \alpha_n \sqrt{\frac{EI}{A\rho L^4}} \tag{10.13}$$

where:

f_n	is the resonant frequency
n	the mode number
α_n	a coefficient depending on n and boundary conditions
E	the Young's modulus
I	the area moment of inertia
A	the area of cross section
ρ	the density
L	the beam length

The term EI is the bending rigidity; ρA is the mass per unit of length of the beam. The specific bending rigidity (SBR) is given by

$$SBR = \frac{EI}{A\rho} \tag{10.14}$$

This parameter indicates how effectively an amount of material ρA is distributed with respect to the bending rigidity (EI). Considering that a long bone can be modeled as a hollow cylindrical shell filled with marrow inside, Equation (10.13) turns to

$$f_n = \alpha_n \sqrt{\frac{EI}{\left(A_1\rho_1 + A_2\rho_2\right)L^4}} \tag{10.15}$$

And from Equation (10.15) the bending rigidity can be estimated by

$$EI = f^2\left(A_1\rho_1 + A_2\rho_2\right)L^4 \tag{10.16}$$

where f is the natural frequency f_n of the bending mode selected for the measurement. The term $(A_1\rho_1 + A_2\rho_2)L$ can be measured and it corresponds to the total wet bone mass (M). The term $A_2\rho_2 L$ is known as the bone mass. For *in vitro* experiments, Equation (10.16) can be expressed as

$$EI = f^2 ML^3 \tag{10.17}$$

i.e., only the total wet bone mass must be known. Instead, for *in vivo* measurements, a parameter called the total bone mineral content ($TBMC$) can be calculated using

$$EI_{TBMC} = f^2 TBMC.L^3 \tag{10.18}$$

Bone mass has good correlations with $TBMC$. Considering that $A_1\rho_1 + A_2\rho_2$ is linearly related to $A_2\rho_2$, the SBR can be assessed with information about the resonant frequency and bone length, using

$$SBR = f^2 L^4 \tag{10.19}$$

The bending vibration modes of a slender beam occur in two perpendicular planes: a plane of minimum bending rigidity of the beam, and the other being the plane of maximum bending rigidity (Van der Perre and Lowet, 1996).

10.2.2 Bone Mass, Bone Loss and Age

The terms bone mass and bone density are typically used synonymously in the literature. In physics, density has been defined by Archimedes around 300 BC as the mass of a body divided by its volume (called "physical density" here). In clinical practice and research, "bone density" usually has a different meaning: it describes to what extent a radiation beam is attenuated by bone judged from a two dimensional projection image ("areal bone density").

It is important to remember that radiation beam attenuation does not only depend on physical density, but also on bone size (length of the path of the beam across the bone). A small bone therefore has a lower areal bone density than a larger bone, even if the physical density is the same. Consequently, low areal bone density can simply reflect a small size of an otherwise normal bone.

It is therefore crucial to distinguish between bone mass and physical bone density. Bone mass is equivalent to the weight of the bone, which depends on bone size. Physical bone density represents the mass of mineral relative to the outer bone volume and is independent of bone size. Interpretation of areal bone density is often difficult because it is somewhere between these two clear-cut definitions.

In order to use radiological skeletal assessment in childhood adequately, some basic principles have to be considered: knowledge of skeletal development, principles of measurement techniques and potential sources of error. Skeletal development: the macrostructure (geometry) of bone changes continuously during growth. This has considerable influence on the results obtained with densitometry. Another skeletal influence is the relative contribution of cortical and trabecular bone: low bone density may reflect a mineralization defect (i.e., the amount of mineral is not adequate for the volume of mineralized bone tissue), but may also be due to a reduced amount of normally mineralized structural elements (e.g., a thin bone cortex or an abnormally low number of trabeculae). This has important clinical implications. Whereas mineralization defects are mostly due to disorders in calcium and phosphate metabolism (e.g., vitamin D deficiency), a reduced amount of structural elements is often caused by bone inactivation (e.g., low muscle mass or function). In any case, a normal bone density, whether areal or physical, does not necessarily mean that bone structure is normal. Renal osteodystrophy is a good example of this (Schoenau, 1998).

A low bone mass may mean that osteopenia/osteoporosis is present in a child with normal height or even more in a tall child. However, a low bone mass in a child with short stature may simply indicate that the skeleton is adequately adapted to the requirements. Conversely, a high bone mass may constitute osteosclerosis in children with a normal height, but may be normal for a child with tall stature. Thus, the inadequate use of mass and density can result in wrong diagnoses and treatment or inadequate risk prediction.

It is a standard procedure both in pediatrics and in adult medicine to relate diagnostic parameters to age. The World Health Organization recommends to compare individual bone densitometry either with age-related reference data (z-score) or with results of 20–30 year old healthy persons who presumably are at peak bone mass (t-score) (Genant et al., 1999). However, they contrast with the recommendations by Mazess and Cameron (the developers of single photon absorptiometry) in one of the first bone densitometric studies ever performed in children almost 50 years ago (Garn and Wagner, 1969; Mazess and Cameron, 1971). They referred bone mineral content (i.e. the mass of bone mineral) to bone width in order to eliminate the effects of skeletal size. Apart from relating bone mineral content to age, they provide data relative to height and weight in order to provide indices that are independent of body size. This was necessary as bone mineral content was highly correlated with bone width (r=0.85), height and weight (r=0.83). The partial correlations between age and bone mineral content dropped from r=0.75 to less than r=0.10 after adjustment for body size. These early observations have been confirmed by a number of later studies, which have added some more information regarding the determinants of bone mass development (Figure 10.8) (Molgaard et al., 1998; Schoenau et al., 2002). The strongest predictors of bone mass development are height and muscle mass. There is a moderate correlation between age and bone mineral content during childhood, but it is quite different between boys and girls. In contrast, the relationship between muscle and bone varies with age and sex. Using age-related reference data would only perhaps label "healthy" children with osteopenia (Figure 10.9). These data in

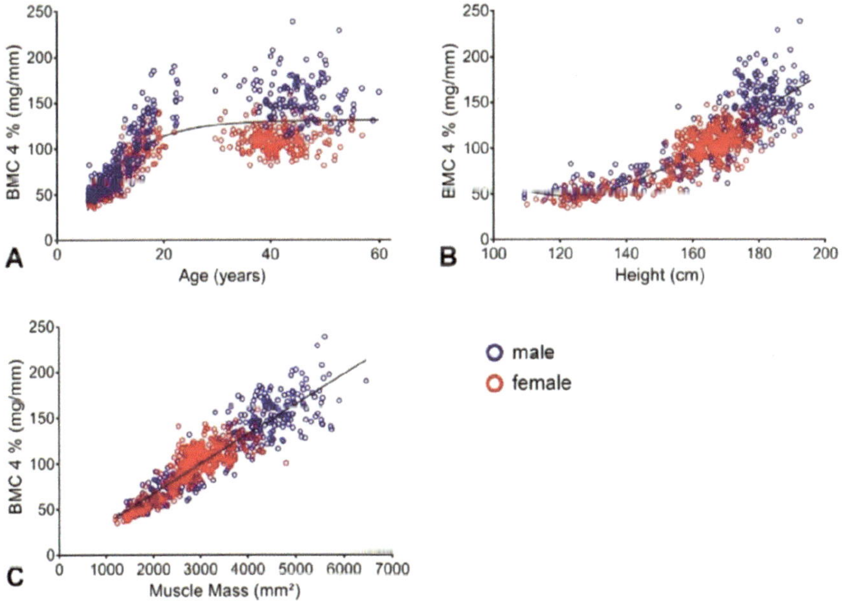

FIGURE 10.8 Relationship between bone mass (BMC=Bone Mineral Content, distal radius) analysed by peripheral computer tomography and age (A), height (B), and local muscle mass (C) studied in healthy children and their parents. Methods and probands are described previously (Schoenau et al., 2002).

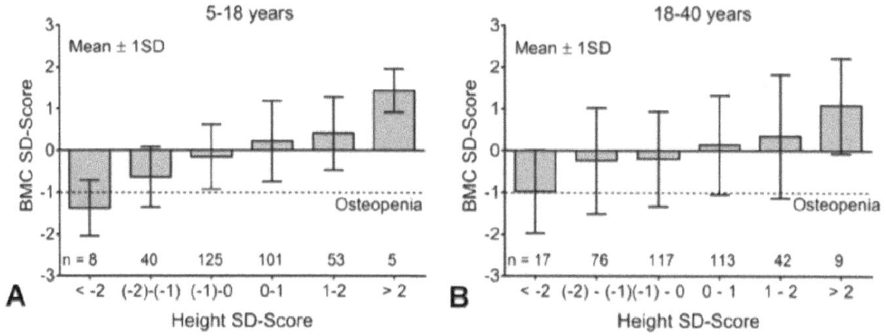

FIGURE 10.9 SD- or Z-score of bone mass (BMC at the distal radius) in relationship to SD- or Z-score of height in healthy children and adults (analyzed by data published previously—Schoenau et al., 2002).

healthy children illustrate that short-normal stature is associated with "osteopenia" using WHO standards.

The clinical relevance of this problem has been demonstrated in patients with constitutional delay of growth and puberty. These patients present a final height in the lower normal range. Using linear absorption techniques in combination with age-related reference data suggested a high risk for osteoporosis (Finkelstein et al., 1992). However, the reported bone mass deficit of these patients disappears when results are corrected for body height or bone size (Bertelloni et al., 1998). Similar effects have been described in children with renal failure and transplantation (Reusz et al., 2000).

A simple example should help understand and summarize this scenario: mice have lower bone mass than elephants. Therefore mice will be diagnosed as osteoporotic, when age-matched elephant standards are used as a reference. However, there is no evidence that mice have more fractures than elephants. As Galileo Galilei commented in the 17th century, small bones perform the same function in small animals as large bones in large animals (Carter et al., 1991). The same should apply to small and tall children and adolescents.

Based on such simple ideas, it can be reasonably suggested that analyses of bone mass (and bone structure) should focus on the question of whether they are adequate for bone function. The main purpose of bones is to provide enough strength (not merely enough mass) to keep voluntary physical loads from causing spontaneous fractures, regardless of whether these loads are chronically subnormal (as in inactivity or muscle disorders), normal or supra-normal. Achieving that "mechanical competence" would be the ultimate test of a bone's health and the main goal of its biological mechanisms (Frost and Schonau, 2000).

Measuring bone mineral density (or mass) has been proposed for many years as a way to identify adults at risk for fracture. However, although bone mineral density measurements can predict fracture risk, the sensitivity and specificity of this prediction is actually quite low 32.50% of all osteoporotic fractures occur at a T-score above −2.5, the cut-off value for osteoporosis according to WHO criteria

(Genant et al., 1999). Therapeutic intervention therefore should not be restricted to patients who fulfil WHO criteria for osteoporosis.

Bone mass does not reflect fracture risk in young people (below 50 years of age) (Hui et al., 1988). Ma and Jones (2002, 2003) showed in a population-based case-control study that the association between bone mass and fractures is weaker in childhood than in adults and that sensitivity and specificity of fracture prediction for a given individual are modest. Goulding et al. (2000) demonstrated in a longitudinal study on children that forearm fractures were mainly predicted by a history of previous forearm fractures. Bone mass and high body weight also predicted fractures, but to a lesser extent. Similar to adults, isolated bone mass analysis is a (weak) predictor of fracture risk in children, but does not appear useful for fracture prediction in the individual.

Figure 10.10 illustrates the generally accepted concept of optimal bone mass acquisition. Based on this concept many studies have analysed the influence of calcium intake on bone mass development. A twin study indicated a positive but modest effect of higher calcium intake on bone mass in children (Johnston et al., 1992). Interestingly, calcium intake had no impact on bone mass during puberty, even though calcium supply is widely believed to be critically important during the pubertal growth period. Further follow-up in these twins showed that the significant differences in bone mineral density between the calcium-supplemented and unsupplemented twins disappeared after withdrawal of calcium tablets. Another study found that 18 months of controlled calcium supplementation led to higher bone mineral mass, but these gains disappeared within 18 months of the trial (Lee et al., 1996). Studies on biochemical markers which reflect bone modeling and remodeling showed that calcium intake reduced remodeling rates but not modeling (Slemenda et al., 1997). These data are compatible with the hypothesis that calcium supplementation decreases remodeling activity. Lower remodeling rates might occur via decreased parathyroid hormone levels and would lead to lower cortical porosity. However, this apparent gain in bone mass is entirely reversible once remodeling rates

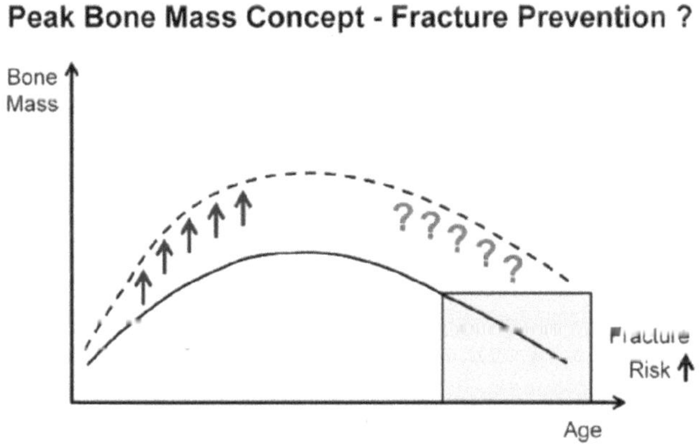

FIGURE 10.10 Peak bone mass concept: fracture prevention by improvement of individual bone mass during childhood and adolescence.

increase. Importantly, calcium supplementation does not seem to stimulate modeling, which is the main process to increase bone strength (cortical thickness, cortical area) during childhood and adolescence.

The long-term effects of physical activity may be similarly reversible. A study on soccer players showed that exercise in youth confers a high and clinically relevant benefit in peak BMD, but that cessation of exercise results in accelerated loss (Karlsson et al., 2000). Thus, those who had stopped playing for over 35 years and who were now at age 60 years or older had no significant residual benefit in BMD. Fracture rate was not lower than expected for age.

On the other hand, animal experiments have shown that a period of disturbed bone mass acquisition in early life may not result in an altered peak bone mass (Schoenau, 1998). The study demonstrated that areas of the juvenile skeleton are not completely remodeled but actually replaced during skeletal growth. Bone formed early in life, regardless of quality, is gradually resorbed and replaced by new bone through modeling. This mechanism "repairs" insufficient bone development after the negative influences have stopped. In summary, short-term effects on bone mass, regardless of positive or negative, do not seem to have any effect in the long run. The skeletal system seems to adapt to current requirements rather than to those of an earlier developmental phase.

For skeletal assessment in childhood we have to distinguish between osteoporosis and mineralization disorders. A mineralization disorder (e.g. rickets) is caused by a loss of bone substance, which can be treated with calcium and vitamin D. In case of osteoporosis we have to stimulate bone growth or decrease bone absorption. Per definition osteoporosis at childhood always includes a minimum of one inadequate fracture additional to reduced bone areal density (Lewiecki et al., 2008).

In the case of osteoporosis, two forms have to be distinguished: primary and secondary osteoporosis. Primary osteoporosis is caused by direct disturbances of bone metabolism, e.g. osteogenesis imperfecta and idiopathic juvenile osteoporosis. Secondary osteoporosis is usually found in immobilizing diseases. These occur particularly in children who have disturbances of their motor abilities due to chronic or congenital diseases, skeletal dysplasia and muscular or neurological causes. In the case of secondary osteoporosis, a sufficient stimulus for bone formation is lacking and therefore the skeletal system remains fragile.

Bone densitometric data often are difficult to interpret in children and adolescents because of large inter- and intra-individual variations in bone size. We propose a functional approach to bone densitometry that addresses two questions: Is bone strength normally adapted to the largest physiological loads, i.e., muscle force? Amd is muscle force adequate for body size? This two-step algorithm has been described before (Schoenau, 2005). Previously published reference data has been used to evaluate results from children with pre-terminal chronic renal failure and renal transplant recipients to show the difference between traditional analysis and utilization of the "functional muscle-bone unit" two-step algorithm. In both groups mean height, muscle cross-sectional area (MCSA), and bone mineral content (BMC) were low for age, but muscle MCSA was normal for height. In the renal transplant recipients, the BMC/muscle MCSA ratio was decreased ($p<0.05$), suggesting that bone cortical strength was not adapted adequately to muscle force. In contrast, chronic renal failure patients had a normal cortical BMC/muscle MCSA ratio, suggesting that their musculoskeletal system was adapted normally to their (decreased) body size.

Another example are patients with cerebral palsy (CP). The prevalence of bone health deficits in children with CP might be often overestimated because age and height-adjusted reference percentiles for BMC and areal bone mineral density (aBMD) assessed by dual energy X-ray absorptiometry (DXA) do not consider reduced muscle activity. The prevalence of positive DXA based findings for bone health deficits in children with CP was compared to the prevalence of positive findings after applying a functional muscle-bone unit diagnostic algorithm (FMBU-A) considering reduced muscle activity in 297 whole body DXA scans of children with CP (Duran et al., 2017). Prevalence of positive age and height-adjusted BMC and aBMD below P3 percentile findings and prevalence after applying the FMBU-A were calculated. It was found a prevalence of positive age adjusted BMC below P3 of 33.3% and aBMD 50.8%. Prevalence of height-adjusted findings for BMC and aBMD below P3 was 16.8% and 36.0% respectively. After applying the FMBU-A BMC and aBMD they were significantly ($P<0.001$) lower: 8.8% and 14.8% respectively. Therefore, we suggest that the prevalence of bone health deficits in children with CP is overestimated when using age and height-adjusted BMC and aBMD. When applying the FMBU-A the prevalence decreases significantly. We recommend applying the FMBU-A when assessing bone health in children with CP.

10.3 EFFECTS OF WHOLE BODY VIBRATION ON BONE: ANIMAL MODELS

As it was previously reported in this chapter, bone formation requires dynamic mechanical loading able to induce tissue responses. It is widely known that bone formation cells (osteoblasts) and bone degrading cells (osteoclasts) can be activated under mechanical stimuli and thus, diverse biomechanical methods have been proposed for preventing bone loss and increasing bone quantity and quality. While different pharmacological approaches (e.g., parathyroid hormone, bisphosphonates, calcitonin or antiresorptive drugs such as alendronate) are commonly used in the treatment of osteoporosis, numerous side effects (nausea, headache, dizziness, among others) have been reported, and the cost must also be considered, and thus, non-adherence to these medications are common. Therefore, new strategies are suggested in the management of this condition. In particular, low-magnitude high-frequency (LMHFV, acceleration <1 g) vibration loading has been shown to promote bone formation and decrease bone resorption. In fact, up-regulated gene expression relate, not just to osteogenesis but also to chondrogenesis, and bone remodeling was reported after LMHFV, which may contribute to fracture healing (Chung et al., 2014).

There is no doubt that the response of the trabecular and cortical bone to vibration is multifactorial and the mechanisms controlling these responses are not clear. Mechanical signals trigger a group of cellular and molecular changes in bone homeostasis that seems to be mediated by the activation of osteoblasts-like cells and different markers for bone formation (e.g., osteocalcin), but also decreasing the formation of osteoclasts and bone resorption (Lau et al., 2011). These responses have also been attributed to an increment in blood flow and angiogenesis (Chow et al.,

2016) as seems to be a relationship between vascularization and bone remodeling. But all these mechanisms are speculations that need a deep investigation to be able to draw solid conclusions.

To date, different regimens of mechanical vibration have been assessed in order to achieve the optimal osteogenic effect. In animal models (rodents, sheep, and turkeys) vibration frequencies ranging from 8 to 90 Hz and acceleration of 0.2–4 g has been shown to have osteogenic effects (see Judex et al., 2015 for review). However, while previous studies conducted in ovariectomized rats have found improvements in bone health (Sehmisch et al., 2009; Tezval et al., 2011; Judex et al., 2007; Gnyubkin et al., 2016), others have reported no alterations (van der Jagt et al., 2012; Brouwers et al., 2010; Qin et al., 2014) or even decreases in mineral content (Pasqualini et al., 2013; Xie et al., 2016) after LMHFV. Therefore, the protocols designed to stimulate bone formation using this approach have not been uniformly successful. Most studies have assessed different frequencies, amplitudes or accelerations, but the number of bouts, duration of exposure or resting times (Xie et al., 2016; Zhang et al., 2014), have also varied between studies. Despite the possible combinations, the effects of high accelerations (above 1 g) have rarely been investigated (Gnyubkin et al., 2016; Nowak et al., 2014) and the majority of experiments have been performed with low magnitudes (<1 g) and frequencies ranging from 30 to 50 Hz.

Identifying the variables that can enhance the osteogenic response to LMHFV would be of interest for the scientific community. In this sense, the study considered as the pioneer in this area (Rubin et al., 2001) showed that LMHFV (30 Hz at 0.3 g) over one year was capable of increasing the trabecular bone volume and density up to 30% and leads to an increment in bone strength and stiffness in sheep. But when we look at the efficacy of these techniques in improving skeletal morphology and biomechanical properties in ovariectomised rodents, the results are inconclusive. Most of the studies applied the vibration in their experiments for 15–20 minutes per day (five days a week); however, frequency and acceleration differ. For example, Xie et al. (2016) suggested that 16 weeks of vibration (30 Hz, 0.3 g, 20 min/day) would improve mechanical properties of bone in ovariectomized rat femurs; however, they observed a detriment in bone mineral density (BMD) and bone stiffness. With a similar protocol (30 Hz, 0.6 g, 20 min/day) conducted over 12 weeks, Sasso et al. (2015) did not observe a response in cortical bone, although they reported a counteracted trabecular bone loss, which used to be more affected by osteoporosis. When the frequency is slightly increased (35 Hz), we see again contradictory results. Numerous studies used the combination 35 Hz–0.25–0.3 g and while some of them suggested the anabolic effect of mechanical stimulation (Li et al., 2015; Chung et al., 2014; Zhang et al., 2014; Leung et al., 2009), others have not (Qin et al., 2014).

Li et al. (2015) highlighted the importance of LMHFV (35 Hz, 0.25 g for 15 minutes per day) for the stimulation of genes related to the proliferation of osteoblasts (formation of bone and cartilage). With the same protocol Zhang et al. (2014) observed improved microarchitecture and bone formation rate but bone mechanical properties were not altered. Leung et al. (2009) also showed enhanced callus formation, mineralization and fracture healing after a comparable protocol. Similarly, improved femoral and tibial cortical area and thickness was showed after five weeks of LMHFV (0.3 g, 45 Hz, 15 minutes, five days per week) in mouse bone (Vanleene and Shefelbine, 2013). By contrast, Qin et al. (2014) suggested that LMHFV promoted bone formation in induced osteoarthritis rat models, showing functional deterioration induced by accelerated cartilage degeneration.

Augmentation of the frequency of stimulation should result in a higher strain, which is needed for bone formation (Ozcivici et al., 2010). Thus, high acceleration levels resulted in osteogenesis for ovariectomized animals, especially for cortical bone not for trabecular bone (Hatori et al., 2015). Nowak et al. (2014) observed that vibration (50 Hz, ≈5 g, five days per week) performed for six months leads to a decrease in the bone resorption marker levels. In a shorter period (three weeks), Falcai et al. (2015) also showed that vibration significantly increased bone formation and decreased bone resorption, although to a lesser extent that swimming or jumping. In an interesting experiment, different frequencies (8, 52 and 90 Hz) at the same acceleration (0.7 g) were compared indicating that only 90 Hz stimulated bone formation in healthy rats (Pasqualini et al., 2013). A recent study (Gnyubkin et al., 2016) suggested that even higher accelerations (2 g) would be more effective in stimulating bone growth but only after short periods of time (three weeks), after nine-week differences in structural cortical or trabecular parameters were not observed. This fact shows that protocol duration may also influence the effect of vibration treatment (Xie et al., 2016).

Another important parameter to bear in mind is the rest between loadings. Several studies have explored the effects of LMHFV on bone quality when the mechanical stimulus was separated by several rest days rather than using daily loading (Li et al., 2015; Zhang et al., 2014; Judex et al., 2015; Ma et al., 2012). There is evidence suggesting that after stimulation the bone cells lose mechanical sensitivity rapidly and therefore, rest intervals between each vibration cycle would be needed (Robling et al., 2002). Thus, osteoblasts could positively respond to vibration initially and lose sensitivity or even respond negatively over time (Quing et al., 2015). In this sense, compared to continuous vibration, intermittent short bouts of vibration followed by resting periods are more efficient in stimulating bone formation in ovariectomized rats (Li et al., 2015; Zhang et al., 2014; Ma et al., 2012). These authors showed that the expression of proteins related to the formation and development of bone and cartilage increased with the duration of resting period, indicating that a seven-day resting period has a greater impact on bone (osteogenesis regulators) than shorter rest periods (Li et al., 2015). Further, when vibration was followed by seven days of rest, bone mechanical properties were improved (Zhang et al., 2014). Even the number of daily vibration bouts (Sen et al., 2011) or its duration could have an impact on bone health. A recent study (Judex et al., 2015) demonstrated that increasing bout duration produced greater bone formation rates than in control mice. With a fixed frequency (45 Hz) and acceleration (0.3 or 0.6 g) authors varied the number of bouts per day (one, two or four), and the duration of each bout (15, 30, or 60 minutes) for three weeks and suggested that increasing the duration of the bout to 30 minutes led to a significant increment in bone formation rates (83%) compared to controls; however, extending the duration (60 minutes) did not provide additional benefits. It is important to note that two bouts per day produced greater benefits than one bout per day but doubling bout number to four bouts per day has no additional effects. Both bout duration (30 minutes) and number (two bouts) were associated with a better bone quantity and architecture. For example, the group subjected to vibration at 0.3 g (two bouts of 30 minutes) had a greater periosteal area (7%) and bone marrow area (13%) than controls of those mice exposed to the same protocol with only one bout of 30 or 60 minutes.

It could be speculated that if the vibration is combined with other therapies we could find additional benefits; nevertheless, when LMHFV was combined with other strategies (i.e., pharmacology or strengthening), some contradictory results have been reported. Some examples illustrate how LMHFV combined with pharmacological treatment may affect the bone's microstructure of ovariectomized-induced rats (Correa et al., 2017; Hoffmann et al., 2016; Hatori et al., 2015). Although the benefits of a combined therapy of alendronate with vibration could not be discerned in the study conducted by Correa et al. (2017), others showed an anabolic effect of vibration combined with this agent on bone (Giro et al., 2008; Chen et al., 2014) enhancing the effects of the parathyroid hormone in trabecular thickness (Hoffmann et al., 2016). Although the effects on bone biomechanical or morphological properties after the combination of vibration therapy with pharmacological treatments cannot be demonstrated (Hoffmann et al., 2016). On the other hand, Li et al. (2012) combined vibration (45 Hz, 0.2 g for 30 minutes per day, five days per week) with resistance training (four sets of 12 repetitions at 65%–75% of 1RM) and reported that this interaction was able at improving cancellous bone. This combination may provide important insights into the potential therapies to target bone disorders.

From the reading of this section we could extract some key messages. For example, it has been suggested that age may decrease the osteogenic potential and therefore exercise programs should begin as soon as possible because they are more effective in the prevention of bone loss than late treatment (Boskey and Coleman, 2010). Thus, a short program seems to be more effective than LMHFV programs with larger durations (Lynch et al., 2011; van der Jagt et al., 2012; Nowak et al., 2014). However, due to high variability of experimental procedures, it is difficult to determine a dose-response relationship with the LMHFV protocols beneficial for bone health. It was reported that mechanical stimulation with frequencies between 35 and 45 Hz with acceleration magnitudes between 0.3 and 0.6 g (ten to 30 minutes per day) could lead to trabecular bone formation rates greater than 66% after only three weeks (Judex and Rubin, 2010). These changes can be accompanied with improvements in bone morphology (cortical area and thickness), which may be acceptable for effective osteoporosis treatment (Hoffmann et al., 2016). Bout duration was reported to be a strong modulator of the bone formation rates, even more than the number of daily bouts and acceleration magnitude. Thus, 30 minute bouts (once or twice a day) seem to be more beneficial than shorter (15 minute) or longer (60 minute) durations.

10.4 EFFECTS OF WHOLE BODY VIBRATION ON BONE: CLINICAL STUDIES

During the last decade whole body vibration (WBV) has been proposed as a strategy for the prevention and treatment of osteoporosis able to enhance bone mass and improve neuromuscular function. If in the previous section it was obvious how in animals studies vibration was effective for increasing bone mass and improving bone architecture and strength, in humans the number of studies is limited and not always in line with the findings observed in animals (Beck, 2015). It seems obvious to think that animal observations cannot be directly applied to the human condition, but it becomes necessary to understand why the results of the few human studies investigating the effect of WBV on bone are heterogeneous and less pronounced than in animal studies. As will be discussed below, there are numerous factors that include the type of device (synchronous or side-alternating), the frequency, the amplitude, the magnitude, the duration or number of repetitions, and even the position or exercise performed on the vibration platform, that must be considered. The methodological differences with respect to these training protocols might contribute to the discrepancies between trials.

Despite this, recent longitudinal studies have demonstrated beneficial effects on bone health after WBV (Verschueren et al., 2004; Gusi et al., 2006; Ruan et al., 2008). Increments in BMD, improvements in bone turnover markers and also decreased bone loss has been reported (Corrie et al., 2014) with hardly any side effects (Beck, 2015); however, other studies have reported limited improvements in these outcomes (Rubin et al., 2004; Iwamoto et al., 2005; von Stengel et al., 2011a; Beck et al., 2010; Slatkovska et al., 2011). Although results highly depend on vibratory parameters used. Most WBV studies, involving both synchronous or side-alternating platforms, have been conducted using low to moderate vibration at frequencies of 15 to 60 Hz. The amplitude ranged from 0.7 to 5 mm but with acceleration magnitudes between 0.1 g and 10 g (mainly intensities ≤1 g but also greater than 1 g in some trials). Vibration sessions used to last ten to 20 minutes, three to seven days a week and with a duration of the intervention ranging from eight to 72 weeks.

The patients' population also included a wide range (e.g., postmenopausal women, children, older adults, clinical populations), which again, complicates the possibility of drawing conclusions, and there is still no clear guidance on the optimal target population (Wysocki, et al. 2011). Numerous studies focused on postmenopausal women diagnosed with osteoporosis (Gómez-Cabello et al., 2014; Verschueren et al., 2004; Gusi et al., 2006; Lai et al., 2013; Ruan et al., 2008; von Stengel et al., 2011a,b; Beck et al., 2010; Slatkovska et al., 2011; Iwamoto et al., 2005; Rubin et al., 2004; Bemben et al., 2010; Stolzenberg et al., 2013), and six focused on children and adolescents (Ward et al., 2004; Gilsanz et al., 2006; Lam et al., 2013; Pitukcheewanont and Safani, 2006; Matute-Llorente et al., 2014; Harrison et al., 2015). The remaining studies evaluated older participants (Santin-Medeiros et al., 2015; Gómez-Cabello et al., 2014; Kiel et al., 2015; Sitjà-Rabert et al., 2012; Verschueren et al., 2011; Zha et al., 2012) or those with clinical conditions (Matute-Llorente et al., 2015; Gusso et al., 2016; Wuermser et al., 2015; Kilebrant et al., 2015). The combination of different therapies (e.g., strengthening or pharmacological agents) with WBV is another factor that must be considered.

As in animal studies, adding WBV to resistance exercise could contribute to an improvement in bone responses (Judex and Rubin, 2010). However, the few studies combining these effects that have investigated reported no additional benefits on bone turnover marker responses in postmenopausal women (Bemben et al., 2010; Bemben et al., 2015; Stolzenberg et al., 2013). While, Stolzenberg et al. (2013) investigated the effectiveness of resistance training combined with WBV (four minutes on a side-alternating vibration platform at a magnitude between 3.9 to 10.9 g for nine months) and showed significant improvements in tibia trabecular BMD, no additional benefits of the WBV therapy were found. In the same line, Bemben et al. (2015) could not find either an advantage to adding WBV to resistance exercise for bone adaptations after five, one min WBV exposures (20 Hz and approximately 2.7 g). By contrast, Humphries et al. (2009) examined the effects of this combination not just on BMD but also on hematological measures for bone remodeling in young women (three, one minute WBV sessions at 50 Hz, two days per week for 16 weeks) and showed positive effects of vibration in improving BMD (increasing bone deposition and decreasing bone resorption). Authors concluded that moderate to high exposures of vibration in combination with resistance training may improve femoral BMD (2%–3%) following a four-month intervention, but it is important to note that this study was performed in healthy active women and that in another population group the results may be different.

10.4.1 Whole Body Vibration and Bone in Healthy Adults

WBV may offer advantages to individuals who are either unable or unwilling to initiate a pharmacologic treatment or perform high impact exercise, known to be osteogenic (Wysocki, et al. 2011). Overall results from a meta-analysis showed that "WBV resulted in small improvements in BMD in postmenopausal women, moderate effects in children and adolescents, but no effect on BMD in young adults" (Slatkovska et al., 2010), although this assumption needs to be contrasted.

10.4.1.1 Effects on Children and Adolescents

This therapy has been recommended in children and adolescents with or without disabling conditions (Matute-Llorente et al., 2014) and that the bone response to vibration may be greater in children than in adults (Harrison et al., 2015) which contrasts with Slatkovska et al. (2010) who suggested that vibration could be more effective for people with low initial BMD. Harrison et al. (2015) assessed the immediate response to vibration in bone turnover markers following very short periods of WBV using different platforms, synchronous (low magnitude; 32–37 Hz, 0.085 mm, 0.3 g) or side-alternating (high magnitude; 20 Hz, 4 mm, 6.4 g), and observed that five consecutive days of WBV increased the bone formation marker and bone resorption marker regardless of the type of vibration used. Authors suggested that in paediatric population the healthy growing skeleton might respond quickly to WBV.

To date, positive results have been reported in four studies assessing bone formation on young participants (Ward et al., 2004; Gilsanz et al., 2006; Lam et al., 2013; Pitukcheewanont et al., 2006). Ward et al. (2004) reported an increase in proximal tibial volumetric trabecular BMD (17.7%) following ten minutes per day, five days per week (0.3 g, 90 Hz) in children with disabling conditions after six months. In the same line, Gilsanz et al. (2006) observed femoral and lumbar vertebrae improvements (2.1% and 3.4% respectively) in women aged between 15 and 20 years old after 12 months of intervention. These results are also consistent with those reported by Lam et al. (2013) in girls with adolescent idiopathic scoliosis after a trial with the same duration. Authors used a low-magnitude protocol (0.3 g, 32–37 Hz) on a synchronous platform for 20 minute per day, five days a week and showed improved BMD at the femoral neck and lumbar spine. On average, studies including children with physical disabilities or low BMD (Pitukcheewanont et al., 2006) have reported increases in both trabecular (~2%–6%) and cortical BMD (~2%–3%). By contrast, in healthy boys (eight to ten years) that were submited to WBV (seven–22 sets of ten–20 repetitions at 30–40 Hz, 1.9 and 6.2 g) three days per week for ten weeks, Erceg et al. (2015) found no differences in osteocalcin following the intervention and while an increase in osteoclastic activity was observed, it could not be ensured that exercise alone was primarily responsible for the observed bone development.

10.4.1.2 Effects on Post-Menopausal Women

In this population group vibration has been considered as effective as strength training for improving bone mass (Gómez-Cabello et al., 2014). But again, although some clinical trials have reported benefits to BMD (Verschueren et al., 2004; Gusi et al., 2006; Lai et al., 2013; Ruan et al., 2008), some contradictory results can also be found (von Stengel et al., 2011a,b; Beck et al., 2010; Slatkovska et al., 2011; Iwamoto et al., 2005; Rubin et al., 2004) showing no significant improvement after WBV. These discrepancies can be attributed to the different protocols performed. Most of these studies had six to eight months of follow-up (Verschueren et al., 2011; Beck et al., 2010; Gusi et al., 2006; Russo et al., 2003) and overall, shorter durations (12 weeks) have not influenced bone turnover (Bemben et al., 2010; Iwamoto et al., 2005; Verschueren et al., 2004). Low frequencies (between 12 and 40 Hz) seems to be ineffective in this group. Slatkovska et al. (2011) assessed the effects of a 20 minute daily program (30–90 Hz, 0.3 g) together with calcium and vitamin D supplementation for 12 months and found no changes in BMD or bone structure. Further, Liphardt et al. (2015) analyzed bone microarchitecture and bone strength in osteopenic postmenopausal women after 12 months WBV (20 Hz, 3–4 mm, ten minutes per day, two to three sessions/week) and showed that vibration was not able to improve bone quality in this population. However, Rubin et al. (2004) showed that those women who complied more than 86% of a program (20 minutes per day, 0.2 g at 30 Hz) over 12 months were able to maintain femoral neck and lumbar spine BMD. Ruan et al. (2008) suggested that this therapy conducted over six months (30 Hz, 5 mm, ten minutes per day, five days per week WBV) lead to improvements in femoral neck (4.9%) and lumbar spine BMD (6.2%). Which agrees in part with the results reflected by Verschueren et al. (2004) who after six months (three times per week, 20 minutes, 35–40 Hz, 2 mm) reported significant improvements in hip BMD (1.5%) but not in lumbar spine. Similarly, five minutes per day (30 Hz, 3.2 g) leads to a significant increase by 2.0% in the lumbar spine BMD (Lai et al., 2013), while Gusi et al. (2006), with a lower magnitude program (six by one minute, 12.6 Hz, 0.7 g) conducted over eight months, showed an increase in hip BMD (4.3%) and suggested that WBV is more effective than walking in postmenopausal women. By contrast, Bemben et al. (2010) indicated that this therapy could be even harmful. In a sample of postmenopausal women authors shown no significant group differences in bone marker variables but they found a decrease at the radius BMD after a WBV program consisted of a high frequency (30–40 Hz) vibration stimulus at mean acceleration magnitudes between 2.2 and 2.8 g, three times per week for 32 weeks. It is important to note that participants combined vibration stimulus with resistance exercise and performed dynamic movements during vibration, which may have altered the results. In another study, participants also performed dynamic exercises on the platform (von Stengel et al., 2011a) and showed that WBV (25–30 Hz, 1.7 mm) was not an additional improvement for the bone parameters. Further, Stolzenberg et al. (2013) evaluated the effect of vibration exercise versus proprioceptive training on bone parameters at the tibia in postmenopausal women. Participants performed three different exercises two times per week on a side-alternating device (22–26 Hz, 3.9 to 10.9 g) for nine months and while no significant differences were found for any of the bone parameters measured, authors suggested that resistance training could increase bone density at the distal tibia and the combination with WBV (four minutes) did not entail additional benefits.

Another important parameter to bear in mind is the type of platform, when two different devices (synchronous vs side-alternating) were compared at a similar magnitud (8 g), non-significant changes were observed for hip BMD but a significant increase in lumbar BMD was observed in the group who trained on the side-alternating devices. Together with the type of platform, the region where this treatment appears to be effective also differs. WBV was effective in the prevention of bone loss in two studies (Lai et al., 2013; von Stengel et al. 2011b) which contrast with some other trials that found no statistical significance (Gusi, 2006; von Stengel, 2011a; Rubin et al., 2004; Leung et al., 2014; Iwamoto et al., 2005). The same situation can be expected for the femoral neck, with possitive results in only two trials (Gusi, 2006; Verschueren et al., 2004). However, when computing only high quality studies, a recent meta-analysis concluded that low-magnitude WBV could reduce bone loss in the lumbar spine in postmenopausal women, with no change in femoral neck BMD (Ma et al., 2016).

It was reported that if WBV starts earlier at the period of menopause it could be more efficacious against bone loss (Sasso et al., 2015). In any case, based on the aforementioned results it is not clear that WBV has positive effects on bone health; however, positive results in balance and muscle performance have been reported what might indirectly impact fracture risk in this population group.

10.4.1.3 Effects on Older Adults

Although it would be desirable, not all older adults can undergo high-intensity or weight-bearing exercise, known to be adequate for bone health and shown to prevent fractures (Wysocki, et al. 2011). Unlike many of the studies we have previously reflected, in this group participants were required to do certain body movements while standing on the vibration platform to achieve the greater benefits.

It has been reported that vibration has significant treatment effect on enhancing certain aspects of leg muscle strength in older adults; however, the effects on hip and lumbar spine BMD are unclear (Lau et al., 2011). These authors suggested that BDM at baseline could modulate the effects of this therapy, but the evidence for WBV effectiveness in older adults with different BMD is limited. In any case, and despite Verschueren et al. (2011) reported that six month WBV (30 Hz, 1.6 g) lead to a decrease in hip BMD when compared with the non-vibrated group, previous studies found positive effects in femoral neck BMD (Sitjà-Rabert et al., 2012). Most studies in the elderly used frequencies ranged between 12.6–40 Hz and at least six months of duration. Santin-Medeiros et al. (2015) suggested that the lack of effect in this group could be due to the use of mechanical loads of low magnitude, more than the trial duration (Kiel et al., 2015). Kiel et al. (2015) found non-significant changes in biochemical markers of bone turnover and volumetric BMD in elderly individuals after two years (ten minutes per day, 37 Hz, 0.3 g). Similarly, Gómez-Cabello et al. (2014) reported that a short-term intervention does not seem to be enough to improve BMD. In this study, participants performed ten repetitions of 45 seconds (40 Hz, 2 mm), three times per week for 11 weeks and although small variation on bone structure were showed, non-significant changes on BMC or BMC were found. Despite this, high frequency (Zha et al., 2013) or magnitudes (Ruan et al., 2008) leads to improvements in lumbar spine (2.5%–4.3%) and femoral neck (3.2%). In view of these results, more research is needed in this population to ascertain the best combination for bone health.

10.4.2 Whole Body Vibration and Bone in Clinical Populations

WBV training is not only used in the prevention of osteoporosis, it can also be considered a therapy for clinical populations that cannot take part in other forms of exercise. Just to show some examples, Matute-Llorente et al. (2015) showed important increases for bone mass in adolescents with and without Down syndrome after 20 weeks (three times per week) of WBV training (ten by 30–60 seconds, 25–30 Hz, 2.5–3.6 g). Similarly, positive effects on bone mass were reported in children with severe motor disabilities after a six-month intervention period (Kilebrant et al., 2015) or in adolescents and young adults with cerebral palsy (Gusso et al., 2016) after 20 weeks (nine minutes per day, four times per week at 12–20 Hz, 1 mm) which achieved improvements in bone mass and density, especially in lumbar spine and lower limbs. Some experiments have also been done in women and men with chronic motor complete paraplegia (Wuermser et al., 2015) analysing bone density and microstructure after 20 minutes per day, five days a week, and for six months (34 Hz, 0.3 g), although most participants did not show an improvement in these outcomes. By contrast, in rheumatoid arthritis 24 sessions of WBV (performed twice a week for 12 weeks at 30 Hz, 3 mm) seems to attenuate the progressive loss of BMD observed in this group. These are only some examples of the importance that WBV could have in diverse pathologies, not exclusively on the musculoskeletal system.

In summary, about half of the human studies showed significant effect of WBV on bone and therefore, we cannot draw solid conclusions with respect to efficacy or dose of this therapy. To date, different types of platforms and different vibration protocols resulted in changes of BMD and while inadequate understanding of optimal WBV dose-response could lead to harmful effects (Wysocki, et al. 2011), these devices seem to be safe and feasible, and might be utilized in different settings to improve bone mass and reduce risk of osteoporotic fracture (Beck et al., 2015). The majority of authors agree on the need for well-designed RCTs, with longer duration and with a larger cohort, in order to be able to recommend this therapy for improving bone health in certain populations. Despite this, WBV was capable of preventing bone loss in human and animal models (Prisby et al., 2008) but should not be considered a replacement for all conventional training. Multimodal training programs including WBV could be more effective than conventional training alone on the musculoskeletal system (von Stengel et al., 2011a).

REFERENCES

Ascenzi, A. and Bonucci, E. (1968). The compressive properties of single osteons. *Anatomical Record*, 161: 377–391.

Ascenzi, A. and Bonucci, E. (1972). The shearing properties of single osteons. *Anatomical Record*, 172: 499–510.

Beck, B. R. and Norling, T. L. (2010). The effect of 8 mos of twice-weekly low- or higher intensity whole body vibration on risk factors for postmenopausal hip fracture. *American Journal of Physical Medicine and Rehabilitation*, 89(12):997–1009.

Beck, B. R. (2015). Vibration therapy to prevent bone loss and falls: Mechanisms and efficacy. *Current Osteoporosis Reports*, 13(6):381–389.

Bemben, D. A., et al. (2010). Effects of combined whole-body vibration and resistance training on muscular strength and bone metabolism in postmenopausal women. *Bone*, 47:650–656.

Bemben, D. A., et al. (2015). Effects of whole-body vibration on acute bone turnover marker responses to resistance exercise in young men. *Journal of Musculoskeletal and Neuronal Interactions*, 15(1):23–31.

Bertelloni, S., et al. (1998). Normal volumetric bone mineral density and bone turnover in young men with histories of constitutional delay of puberty. *Journal of Clinical Endocrinology and Metabolism*, 83(12):4280–4283.

Boskey, A. L., and Coleman, R. (2010). Aging and bone. *Journal of Dental Research*, 89(12):1333–48.

Brouwers, J. E. M., et al. (2010). Effects of vibration treatment on tibial bone of ovariectomized rats analyzed by in vivo micro-CT. *Journal of Orthopaedic Research*, 28(1):62–69.

Carter, D. R. and Beaupré, G. S. (2001). *Skeletal Function and Form: Mechanobiology of Skeletal Development, Aging, and Regeneration*. New York: Cambridge University Press.

Carter, D. R., Wong, M., and Orr, T. E. (1991). Musculoskeletal ontogeny, phylogeny, and functional adaptation. *Journal of Biomechanics*, 24 Suppl 1:3–16.

Chen, G. X., et al. (2014). Effect of low-magnitude whole-body vibration combined with alendronate in ovariectomized rats: A random controlled osteoporosis prevention study. *PLoS One*, 9:e96181.

Chung, S. L., Leung, K. S., and Cheung, W. H. (2014). Low-magnitude high-frequency vibration enhances gene expression related to callus formation, mineralization and remodeling during osteoporotic fracture healing in rats. *Journal of Orthopaedic Research*, 32(12):1572–1579.

Chow, S. K., et al (2016). Mechanical stimulation enhanced estrogen receptor expression and callus formation in diaphyseal long bone fracture healing in ovariectomy-induced osteoporotic rats. *Osteoporosis International*, 27(10):2989–3000.

Correa, C. B., et al. (2017). Can the alendronate dosage be altered when combined with high-frequency loading in osteoporosis treatment? *Osteoporosis International*, 28(4):1287–1293.

Corrie, H., et al. (2014). Effects of vertical and side alternating vibration training on fall risk factors and bone turnover in older people at risk of falls. *Age and Ageing*, 44(1):1–8.

Currey, J. D. (1988). The effect of porosity and mineral content on the Young's modulus of elasticity of compact bone. *Journal of Biomechanics*, 21(2):131–139

Duran, I., et al. (2017). The functional muscle-bone unit in children with cerebral palsy. *Osteoporosis International*, 28(7):2081–2093.

Erceg, D. N., et al. (2015). Changes in bone biomarkers, BMC, and insulin resistance following a 10-week whole body vibration exercise program in overweight latino boys. *International Journal of Medical Sciences*, 12(6):494–501.

Falcai, M. J., et al. (2015). The osteogenic effects of swimming, jumping, and vibration on the protection of bone quality from disuse bone loss. *Scandinavian Journal of Medicine and Science in Sports*, 25(3):390–397.

Finkelstein, J. S., et al. (1992). Osteopenia in men with a history of delayed puberty. *New England Journal of Medicine*, 326(9):600–604.

Frost, H. M. (1998). From Wolff's law to the mechanostat: A new "face of physiology". *Journal of Orthopaedic Science*, 3(5):282–286.

Frost, H. M. (2004). A 2003 update of bone physiology and Wolff's Law for clinicians. *The Angle Orthodontist*, 74(1): 3–15.

Frost, H. M. and Schonau, E. (2000). The "muscle-bone unit" in children and adolescents: A 2000 overview. *Journal of Pediatric Endocrinology and Metabolism*,13(6):571–590.

Garn, S. M. and Wagner, B. (1969). The adolescent growth of the skeletal mass and its implications to mineral requirements. In *Adolescent Nutrition and Growth*, ed. F. P. Heald. New York: Appleton-Century Crofts. 139–161.

Genant, H. K., et al. (1999). Interim report and recommendations of the World Health Organization Task-Force for Osteoporosis. *Osteoporosis International*, 10(4):259–264.

Gilsanz, V., et al. (2006). Low-level, high-frequency mechanical signals enhance musculoskeletal development of young women with low BMD. *Journal of Bone and Mineral Research*, 21(9):1464–1474.

Giro, G., et al. (2008). Influence of estrogen deficiency and its treatment with alendronate and estrogen on bone density around osseointegrated implants: Radiographic study in female rats. *Oral Surgery Oral Medicine Oral Pathology Oral Radiology Endodontics*, 105(2):162–167.

Gnyubkin, V., et al. (2016). High-acceleration whole body vibration stimulates cortical bone accrual and increases bone mineral content in growing mice. *Journal of Biomechanics*, 49(9):1899–1908.

Gómez-Cabello, A., et al. (2014). Effects of a short-term whole body vibration intervention on bone mass and structure in elderly people. *Journal of Science and Medicine in Sport*, 17(2):160–164.

Goulding, A., et al. (2000). More broken bones: A 4-year double cohort study of young girls with and without distal forearm fractures. *Journal of Bone and Mineral Research*, 15(10):2011–2018.

Gusi, N., Raimundo, A., and Leal, A. (2006). Low-frequency vibratory exercise reduces the risk of bone fracture more than walking: A randomized controlled trial. *BMC Musculoskeletal Disorders*, 30(7):92.

Gusso, S., et al. (2016). Effects of whole-body vibration training on physical function, bone and muscle mass in adolescents and young adults with cerebral palsy. *Scientific Reports*, 6:225–18.

Harrison, R., et al. (2015). Acute bone response to whole body vibration in healthy pre-pubertal boys. *Journal of Musculoskeletal and Neuronal Interactions*, 15(2):112–122.

Hatori, K., et al. (2015). Single and combined effect of high-frequency loading and bisphosphonate treatment on the bone micro-architecture of ovariectomized rats. *Osteoporosis International*, 26(1):303–313.

Hoffmann, D. B., et al. (2016). Effects of 8-prenylnaringenin and whole-body vibration therapy on a rat model of osteopenia. *Journal of Nutrition and Metabolism*, 2016:6893137.

Hui, S. L., Slemenda, C. W., and Johnston, C. C. (1988). Age and bone mass as predictors of fracture in a prospective study. *Journal of Clinical Investigation*, 81(6):1804–1809.

Humphries, B., et al. (2009). Whole-body vibration effects on bone mineral density in women with or without resistance training. *Aviation, Space and Environmental Medicine*, 80(12):1025–1031.

Iwamoto, J., et al. (2005). Effect of whole-body vibration exercise on lumbar bone mineral density, bone metabolism, and chronic back pain in post-menopausal osteoporotic women treated with alendronate. *Aging Clinical and Experimental Research*, 17(2):157–163.

Johnston, C. C. Jr., et al. (1992). Calcium supplementation and increases in bone mineral density in children. *New England Journal of Medicine*, 327(2):82–87.

Judex, S., Koh, T. J., and Xie, L. (2015). Modulation of bone's sensitivity to low-intensity vibrations by acceleration magnitude, vibration duration, and number of bouts. *Osteoporosis International*, 26(4):1417–1428.

Judex, S. and Rubin, C. T. (2010). Is bone formation induced by high-frequency mechanical signals modulated by muscle activity? *Journal of Musculoskeletal and Neuronal Interactions*, 10(1):3–11.

Judex, S., et al. (2007). Low-magnitude mechanical signals that stimulate bone formation in the ovariectomized rat are dependent on the applied frequency but not on the strain magnitude. *Journal of Biomechanics*, 40(6):1333–1339.

Karlsson, M. K., et al. (2000). Exercise during growth and bone mineral density and fractures in old age. *Lancet*, 355(9202):469–470.

Kiel, D. P., et al. (2015). Low-magnitude mechanical stimulation to improve bone density in persons of advanced age: A randomized, Placebo-Controlled Trial. *Journal of Bone and Mineral Research*, 30(7):1319–1328.

Kilebrant, S., et al. (2015). Whole-body vibration therapy in children with severe motor disabilities. *Journal of Rehabilitation Medicine*, 47(3):223–228.

Lam, T. P., et al. (2013). Effect of whole body vibration (WBV) therapy on bone density and bone quality in osteopenic girls with adolescent idiopathic scoliosis: A randomized, controlled trial. *Osteoporosis International*, 24(5):1623–1636.

Lai, C. L., et al. (2013). Effect of 6 months of whole body vibration on lumbar spine bone density in postmenopausal women: a randomized controlled trial. *Clinical Interventions in Aging*, 8:1603–1609.

Lasaygues, P. and Pithioux, M. (2002). Ultrasonic characterization of orthotropic elastic bovine bones. *Ultrasonics*, 39:567–573.

Lau, R. W., et al. (2011). The effects of whole body vibration therapy on bone mineral density and leg muscle strength in older adults: A systematic review and meta-analysis. *Clinical Rehabilitation*, 25(11):975–988.

Lee, W. T., et al. (1996). A follow-up study on the effects of calcium-supplement withdrawal and puberty on bone acquisition of children. *American Journal of Clinical Nutrition*, 64(1):71–77.

Lewiecki, E. M., et al. (2008). International Society for Clinical Densitometry 2007 adult and pediatric official positions. *Bone*, 43(6):115–1121.

Leung, K. S., et al. (2014). Effects of 18-month low-magnitude high-frequency vibration on fall rate and fracture risks in 710 community elderly—a cluster-randomized controlled trial. *Osteoporosis International*, 25(6):1785–1795.

Leung, K. S., et al. (2009). Low-magnitude high-frequency vibration accelerates callus formation, mineralization, and fracture healing in rats. *Journal of Orthopaedic Research*, 27(4):458–465.

Li, M., et al. (2015). Low-magnitude mechanical vibration regulates expression of osteogenic proteins in ovariectomized rats. *Biochemical and Biophysical Research Communications*, 465(3):344–8.

Li, Z., et al. (2012). Whole-body vibration and resistance exercise prevent long-term hindlimb unloading-induced bone loss: Independent and interactive effects. *European Journal of Applied Physiology*, 112(11):3743–3753.

Lynch, M. A., et al. (2011). Low-magnitude whole-body vibration does not enhance the anabolic skeletal effects of intermittent PTH in adult mice. *Journal of Orthopaedic Research*, 29(4):465–472.

Ma, D. Q. and Jones, G. (2002). Clinical risk factors but not bone density are associated with prevalent fractures in prepubertal children. *Journal of Paediatric and Child Health*, 38(5):497–500.

Ma, D. and Jones, G. (2003). The association between bone mineral density, metacarpal morphometry, and upper limb fractures in children: A population-based case-control study. *Journal of Clinical Endocrinology and Metabolism*, 88(4):1486–1491.

Ma, R. S., et al. (2012). High-frequency and low-magnitude whole body vibration with rest days is more effective in improving skeletal micro-morphology and biomechanical properties in ovariectomised rodents. *Hip International*, 22(2):218–226.

Machado, C. B. (2013). *Ultrasound in Bone Fractures: from Assessment to Therapy.* New York: Nova Science Publishers, Inc.

Majeska, R. J. (2009). Cell Biology of Bone. In *Bone Mechanics Handbook*, (2nd edition, pp. 2–24). ed. S. C. Cowin New York: Informa Healthcare USA, Inc.

Martin, R. B., Burr, D. B., and Sharkey, N. A. (2010). *Skeletal Tissue Mechanics* (1st edition). New York: Springer-Verlag.

Marks, S. C., and Odgren, P. R. (2002). Structure and development of the skeleton. In: *Principles of Bone Biology,* (2nd edition, pp. 3–15) eds. J. P. Bilezikian L. G. Raisz, and G. A. Rodan Burlington, MA: Elsevier Inc.

Matute-Llorente, A., et al. (2014). Effect of whole-body vibration therapy on health-related physical fitness in children and adolescents with disabilities: A systematic review. *Journal of Adolescent Health*, 54(4):385–96.

Matute-Llorente, A., et al. (2015). Effect of whole body vibration training on bone mineral density and bone quality in adolescents with Down syndrome: A randomized controlled trial. *Osteoporosis International*, 26(10):2449–2459.

Mazess, R. B. and Cameron, J. R. (1971). Skeletal growth in school children: Maturation and bone mass. *American Journal of Physical Anthropology*, 35(3):399–407.

Mitton, D., Roux, C., and Laugier, P. (2011). Bone Overview. In *Bone Quantitative Ultrasound*, eds. P. Laugier and G. Haïat (1st edition, pp. 1–28). New York: Springer.

Molgaard, C., Lykke Thomsen, B., and Fleischer Michaelsen, K. (1998). Influence of weight, age and puberty on bone size and bone mineral content in healthy children and adolescents. *Acta Paediatrica*, 87(5):494–499.

Moore, K., Agur, A. M. R., and Dalley, A. F. (2013). *Clinically Oriented Anatomy* (7th edition). Philadelphia, PA: LWW.

Morgan, E. F., Bayraktar, H. H., and Keaveny, T. M. (2003). Trabecular bone modulus–density relationships depend on anatomic site. *Journal of Biomechanics*, 36(7):897–904.

Nowak, A., et al. (2014). High-magnitude whole-body vibration effects on bone resorption in adult rats. *Aviation, Space and Environmental Medicine*, 85(5):518–521.

Ozcivici, E., et al. (2010). Low-level vibrations retain bone marrow's osteogenic potential and augment recovery of trabecular bone during reambulation. *PLoS One*, 5(6):e11178.

Pasqualini, M., et al. (2013). Skeletal site-specific effects of whole body vibration in mature rats: From deleterious to beneficial frequency-dependent effects. *Bone*, 55(1):69–77.

Paulsen, F. and Waschke, J. (2011). *Sobotta Atlas of Human Anatomy* (15th edition). London: Urban & Fischer.

Pitukcheewanont, P. and Safani, D. (2006). Extremely low-level, short-term mechanical stimulation increases cancellous and cortical bone density and muscle mass of children with low bone density – A pilot study. *Endocrinologist*, 16:128–132.

Prisby, R. D., et al. (2008). Effects of whole body vibration on the skeleton and other organ systems in man and animal models: What we know and what we need to know. *Ageing Research Reviews*, 7(4): 319–329.

Qin, J., et al. (2014). Low magnitude high frequency vibration accelerated cartilage degeneration but improved epiphyseal bone formation in anterior cruciate ligament transect induced osteoarthritis rat model. *Osteoarthritis Cartilage*, 22(7):1061–1067.

Qing, F., et al. (2016). Administration duration influences the effects of low-magnitude, high-frequency vibration on ovariectomized rat bone. *Journal of Orthopaedic Research*, 34(7):1147–1157.

Reusz, G. S., et al. (2000). Bone metabolism and mineral density following renal transplantation. *Archives of Diseases in Childhood*, 83(2):146–151.

Rho, J. Y., Kuhn-Spearing, L., and Zioupos, P. (1998). Mechanical properties and the hierarchical structure of bone. *Medical Engineering & Physics*, 20(2):92–102.

Robling, A. G., et al. (2002). Shorter, more frequent mechanical loading sessions enhance bone mass. *Medicine & Science in Sports & Exercise*, 34(2):196–202.

Ruan, X. Y., et al. (2008). Effects of vibration therapy on bone mineral density in postmenopausal women with osteoporosis. *Chinese Medical Journal*, 121(13):1155–1158.

Rubin, C., et al. (2001). Anabolism. Low mechanical signals strengthen long bones. *Nature*, 412(6847):603–604.

Rubin, C., et al. (2004). Prevention of postmenopausal bone loss by a low-magnitude, high-frequency mechanical stimuli: A clinical trial assessing compliance, efficacy, and safety. *Journal of Bone and Mineral Research*, 19(3):343–351.

Russo, C. R., et al. (2003). High-frequency vibration training increases muscle power in postmenopausal women. *Archives of Physical Medicine and Rehabilitation*, 84(12):1854–1857.

Santin-Medeiros, F., et al. (2015). Effects of eight months of whole body vibration training on hip bone mass in older women. *Nutrición Hospitalaria*, 31(4):1654–1659.

Sasso, G. R., et al. (2015). Effects of early and late treatments of low-intensity, high-frequency mechanical vibration on boneparameters in rats. *Gynecology and Endocrinology*, 31(12):980–986.

Schoenau, E. (1998). Problems of bone analysis in childhood and adolescence. *Pediatric Nephrology*, 12(5):420–429.

Schoenau, E. (2005). From mechanostat theory to development of the "Functional Muscle-Bone-Unit". *Journal of Musculoskeletal and Neuronal Interactions*, 5(3):232–238.

Schoenau, E., et al. (2002). Bone mineral content per muscle cross-sectional area as an index of the functional muscle-bone unit. *Journal of Bone and Mineral Research*, 17(6):1095–1101.

Sehmisch, S., et al. (2009). Effects of low-magnitude, high frequency mechanical stimulation in the rat osteopenia model. *Osteoporosis International*, 20(12):1999–2008.

Sen, B., et al. (2011). Mechanical signal influence on mesenchymal stem cell fate is enhanced by incorporation of refractory periods into the loading regimen. *Journal of Biomechanics*, 44(4):593–599.

Sitjà-Rabert, M., et al. (2012). Efficacy of whole body vibration exercise in older people: A systematic review. *Disability and Rehabilitation*, 34(11):883–893.

Slatkovska, L., et al. (2011). Effect of 12 months of whole-body vibration therapy on bone density and structure in postmenopausal women: a randomized trial. *Annals of Internal Medicine*, 155(10):668–679.

Slemenda, C. W., et al. (1997). Reduced rates of skeletal remodeling are associated with increased bone mineral density during the development of peak skeletal mass. *Journal of Bone and Mineral Research*, 12(4): 676–682.

Stolzenberg, N., et al. (2013). Bone strength and density via pQCT in post-menopausal osteopenic women after 9 months resistive exercise with whole body vibration or proprioceptive exercise. *Journal of Musculoskeletal and Neuronal Interactions*, 13(1):66–76.

Tezval, M., et al. (2011). Improvement of femoral bone quality after low-magnitude, high-frequency mechanical stimulation in the ovariectomized rat as an osteopenia model. *Calcified Tissue International*, 88(1):33–40.

Van Buskirk, W. C. and Ashman, R. B. (1981). The elastic moduli of bone. In *Mechanical Properties of Bone AMD*, ed. S. C. Cowin (45, pp. 131–143). New York: American Society of Mechanical Engineers.

Van der Jagt, O. P., et al. (2012). Low-magnitude whole body vibration does not affect bone mass but does affect weight in ovariectomized rats. *Journal of Bone and Mineral Metabolism*, 30(1):40–46.

Van der Perre, G. and Lowet, G. (1996). In vivo assessment of bone mechanical properties by vibration and ultrasonic wave propagation analysis. *Bone*, 18(1):29S–35S.

Vanleene, M. and Shefelbine, S. J. (2013). Therapeutic impact of low amplitude high frequency whole body vibrations on the osteogenesis imperfecta mouse bone. *Bone*, 53(2):507–514.

Verschueren, S. M., et al. (2004). Effect of 6-month whole body vibration training on hip density, muscle strength, and postural control in postmenopausal women: A randomized controlled pilot study. *Journal of Bone and Mineral Research*, 19(3):352–359.

Verschueren, S. M., et al. (2011). The effects of whole-body vibration training and vitamin D supplementation on muscle strength, muscle mass, and bone density in institutionalized elderly women: A 6-month randomized, controlled trial. *Journal of Bone and Mineral Research*, 26(1):42–49.

Von Stengel, S., et al. (2011a). Effects of whole body vibration on bone mineral density and falls: Results of the randomized controlled ELVIS study with postmenopausal women. *Osteoporosis International*, 22(1):317–325.

Von Stengel, S., et al. (2011b). Effects of whole-body vibration training on different devices on bone mineral density. *Medicine and Science in Sports and Exercise*, 43(6):1071–1079.

Ward, K., et al. (2004). Low magnitude mechanical loading is osteogenic in children with disabling conditions. *Journal of Bone and Mineral Research*, 19(3):360–369.

Wolff, J. (1892). Das Gesetz der Transformation der Knochen. *Berlin, Hirschwald: Julius Wolff Institut.*

Wuermser, L. A., et al. (2015). The effect of low-magnitude whole body vibration on bone density and microstructure in men and women with chronic motor complete paraplegia. *Journal of Spinal Cord Medicine*, 38(2):178–186.

Wysocki, A., et al. (2011). Whole-body vibration therapy for osteoporosis: State of the science. *Annals of Internal Medicine*, 155(10):680–686.

Xie, P., et al. (2016). Bone mineral density, microarchitectural and mechanical alterations of osteoporotic rat bone under long-term whole body vibration therapy. *Journal of the Mechanical Behavior of Biomedical Materials*, 53:341–349.

Zha, D. S., et al. (2012). Does whole-body vibration with alternative tilting increase bone mineral density and change bone metabolism in senior people? *Aging Clinical and Experimental Research*, 24(1):28–36.

Zhang, R., et al. (2014). Seven day insertion rest in whole body vibration improves multi-level bone quality in tail suspension rats. *Plos One*, 9(3):e92312.

11 Undesirable and Unpleasant Adverse Side Effects of the Whole Body Vibration Exercises

D. da Cunha de Sá-Caputo, Christiano Bittencourt Machado, Redha Taiar, and Mario Bernardo-Filho

CONTENTS

11.1 INTRODUCTION

Vertical platforms and side alternating platforms are widely used to produce mechanical vibrations that will be transmitted to the human body (Rauch et al., 2010). Biomechanical parameters related to the mechanical vibration must be considered when a subject is submitted to the whole body vibration (WBV) exercise, such as frequency, peak-to-peak displacement and peak acceleration. The spatio-temporal context and biophysical parameters, as the work and rest time, total time of the exposition, number of bouts, periodicity in the week, position of the subject in the platform must be also controlled.

Several important effects of the WBV exercises have been reported, as:

(i) Increase of muscle strength and power (Marín-Cascales et al., 2017; Jacobson et al., 2017) and bone mineral density (El-Shamy, 2017; Tan et al., 2016);

(ii) Decrease in the level of the pain and the risk of falls (Alev et al., 2017; Maddalozzo et al., 2016; Yang et al., 2017; Yang et al., 2015);
(iii) Improvement in quality of life (Carvalho-Lima et al., 2017; Pessoa et al., 2017), cognition (Fuermaier et al., 2014; Regterschot et al., 2014), balance and posture (Uhm and Yang, 2017; Wilson et al., 2017), blood flow (Menéndez et al., 2015; Johnson et al., 2014), gait (Goudarzian et al., 2017; Choi et al., 2016) and flexibility (Sá-Caputo et al., 2014; Annino et al., 2017).

In addition, several populations have been managed with the impact of the WBV exercise in:

(a) Clinical disorders, such as fibromyalgia (Alev et al., 2017; Sañudo et al., 2013), diabetes (Manimmanakorn et al., 2017; Lee, 2017), obesity (Alvarez-Alvarado et al., 2017; Wong et al., 2016), chronic obstructive pulmonary disease (Gloeckl et al., 2017; Spielmanns et al., 2017), osteoporosis (Bowtell et al., 2016; Luo et al., 2017), cerebral palsy (Krause et al., 2017; Ko et al., 2016), Parkinson's disease (Chouza et al., 2011; Arias et al., 2009), multiple sclerosis (Yang et al., 2016; Uszynski et al., 2016), osteogenesis imperfecta (Högler et al., 2017; Hoyer-Kuhn et al., 2014) and osteoarthritis (Lai et al., 2017; Bokaeian et al., 2016);
(b) Promotion of health and human welfare to the elderly (Goudarzian et al., 2017; Ko et al., 2017);
(c) Fitness and sports (Maeda et al., 2016; Martínez-Pardo et al., 2015; Cloak et al., 2016; Dallas et al., 2015).

The action mechanism implicated with the effect of WBV exercises would be also associated with an enhancement of the muscle activity during the exposition to the WBV. Probably, tonic vibration reflex responses (Hand et al., 2009; Issurin et al., 1994) and increase of the motor unit recruitment might be involved. These responses would be induced by changes in the length of the muscle spindles. Moreover, presumably, an action of the mechanical vibration in the central nervous system may lead a neuroendocrine response. This could be justified due to the observed alteration on the plasma concentration of the various hormonal and non-hormonal biomarkers (Bosco et al., 2000; Cardinale et al., 2010).

11.2 UNDESIRABLE AND UNPLEASANT EFFECTS OF WHOLE BODY VIBRATION EXERCISES

Besides of the positive effects of whole body vibration exercise, adverse (undesirable and unpleasant) side effects of these exercises can occur and have been reported. The causes of these side effects are related to various factors.

Rittweger et al. (2000) have assessed the physiological mechanisms of fatigue by vibration exercise in young healthy subjects. Vibration exercise and cardiovascular data were compared to progressive bicycle ergometry until exhaustion. Vibration exercise was performed in two sessions, with a 26 Hz vibration on a ground plate, in combination with squatting plus additional load (40% of body weight). After vibration exercise, perceived exertion on Borg's scale was 18, and thus as high as after bicycle ergometry. The authors have signalized that itching erythema and an increase in cutaneous blood flow were observed in about half of the individuals exposed to the vibration exercise. It follows that exhaustive whole-body vibration would elicit a mild cardiovascular exertion and that neural as well as muscular mechanisms of fatigue may play a role.

Crewther et al. (2004) have verified the effect of the gravitational forces (*g*-forces) associated with different postures (standing single leg, standing double leg, semi-squat), amplitudes (1.25, 3.0 and 5.25 mm), frequencies (10, 20 and 30 Hz) and at some anatomical sites (tibial tuberosity, greater trochanter, jaw). It was observed that untrained subjects exposed to mechanical vibration suffered from side effects that depend on the frequency used. They have varied from 10 up to 30 Hz. Adverse side effects such as hot feet, itching of the lower limbs, vertigo and severe hip discomfort were reported in this study associated to the protocols using 30 Hz frequency.

Cronin and Crewther (2004) investigated whether triceps surae stiffness alters in response to WBV. The stiffness of the right and left leg of eleven relatively untrained subjects was measured prior to a warm-up, post-warm-up and post-vibration using a damped oscillation technique. After warm-up, one leg was vibrated (frequency 26 Hz, amplitude 6 mm) whilst the other leg acted was the control. The subjects were exposed to the mechanical vibration five times for a duration of 60 seconds (60 seconds rest between each repetition). Considering the adverse side effects, after the exposition to the mechanical vibration a number of subjects experienced pain to the lower leg muscles and a loss of function. In some subjects, the pain was experienced at the jaw and the neck. These adverse side effects required treatment in six subjects but the pain had resolved within seven to ten days in all subjects. Other individuals during the course of the exposition to the vibratory stimulation reported an itching sensation in the lower limb. Moreover, the authors recommend that, considering the injury potential associated with WBV further research into safe dose-response relationships.

Satou et al. (2007) investigated the relation between WBV and changes in wakefulness in ten healthy male university students. The subjects were exposed to WBV (frequency 10 Hz, acceleration level 0.6 m/s^2) for 12 minutes in a seated position. The questionnaire of Visual Analog Sleepiness Scale (VASS) and the Kwansei Gakuin Sleepiness Scale (KSS) and the electroencephalogram (EEG) measurement by Alpha Attenuation Test (AAT) (repeated three times each opened and closed eye for one minute) were used in the study. Wakefulness levels were defined as the ratio of mean alpha-wave power during eyes closed versus eyes opened. VASS and KSS increased and subjective levels of wakefulness decreased from pre- to post-exposure in all subjects, regardless of vibration exposure. The objective wakefulness levels of AAT were reduced at the post-exposure test in all subjects. In the case with exposure to whole-body vibration was a significant difference from the case without exposure to whole-body vibration. The authors suggest that a short-term exposure to whole-body vibration may cause a reduction of wakefulness level. Moreover, this could help to better understand some adverse side effects.

Monteleone et al. (2007), reported a case of significant morbidity following one session of five minutes of WBV exercise in an athletic adult individual in apparent good health, but with asymptomatic nephrolithiasis. A female (40 years old) amateur athletic (runner) with a past medical history of renal calculi was exposed to a single session of WBV. The individual stood on the base of the WBV platform with the 90° angle knee holding a support bar for balance. She underwent a series of five 30-second repetition of WBV (frequency of 30 Hz, amplitude of 4.5 mm) (repetitions separated by a one-minute rest). This sequence was performed at five-minute intervals until a total effective WBV exposition time of five minutes. Considering the adverse side effects of the WBV, primarily, 12 hours after the session, the individual reported sudden right flank pain and the onset of fever to 40°. Moreover, she presented positive Giordano maneuver. Further clinical evaluations were performed: (i) Ultrasound examination revealed the presence of a stone in the right renal pelvis in association with hydronephrosis; (ii) Laboratory examination indicated abnormities, as hemoglobin 9.7 g/dL, creatinine clearance 48 mL/min, hematuria, proteinuria; (iii) Urine culture demonstrated colony-forming units of Escherichia coli > 40 million per field.

Therapy with antibiotics (intramuscular) eliminated the pain and fever. The authors suggest that mechanical vibration applied during a session of the WBV training can, by mobilizing the stone, lead a bacteremia, causing the onset of symptoms. In consequence, it is suggested that nephrolithiasis not diagnosed prior to the exposition to WBV can lead the subject to unnecessary dangerous consequences, and this clinical condition can contribute to the appearance of unexpected adverse side effects.

Bertschinger and Dosso (2008) described a case of a 43-year-old man with strong myope (−13 D in the right eye and −18 D in the left eye), without prior medical history or drug use, who was consulted urgently due to a sudden decrease in the visual acuity on the left eye. The man presented in two weeks a spontaneous vitreous hemorrhage after starting exposition to WBV exercise for 20 minutes, twice per week. After a two-week period of medical intervention, the intravitreous hemorrhage was partially resolved. The authors report that WBV exercise could cause this adverse side effect in the eye and that to their knowledge, it was suggested that this was the first report of a possible ocular side effect due to the WBV exercise.

Vela et al. (2010) presented cases of intraocular lens (IOL) dislocation that appeared shortly after the individuals exercised on an oscillating/vibratory platform. Spontaneous late dislocation of an IOL in the capsular bag has been described as a serious complication of cataract surgery. Vela et al. (2010) have reported clinical cases of IOL dislocation that appeared immediately after the exposition to whole-body vibration. A 71-year-old woman who presented with lens subluxation in the right eye and complete posterior IOL dislocation in the left eye after exercising on an oscillating/vibratory platform for ten minutes. The corrected distance visual acuity (CDVA) was 20/32 in the right eye and counting fingers (CF) in the left eye. A 62-year-old woman who presented with unilateral IOL dislocation within the capsular bag in the right eye one day after using an oscillating/vibratory platform.

The CDVA was CF in the right eye and 20/20 in the left eye. It is necessary to consider that the patients had no history of ocular or systemic disorders. Timing from IOL implantation to dislocation was approximately six and four years, respectively. Pars plana vitrectomy with removal of the dislocated IOL was performed in both subjects. It is marked that the movement produced in whole-body vibration would be sufficient to induce an in-the-bag dislocation. However, in predisposed eyes presenting weakness or damage of the zonular fibers, vibration may facilitate IOL dislocation. Considering the adverse side effects, it is signalized that subjects with predisposition to IOL dislocation may be at increased risk when using vibration devices. In addition, it is suggested that cataract surgeons should be aware of this potential complication with IOL dislocation related to the exposition to whole body mechanical vibration.

Amir et al. (2010) have described a case of benign paroxysmal positional vertigo (BPPV) which occurred after use of a WBV training plate. A 44-year-old woman was referred with classic symptoms of BPPV following the exposition to WBV training. The condition resolved spontaneously after several days. It was noted negative adverse side effects in users of this equipment, such as dizziness, headache and a sensation of imbalance. However, there have been no reported cases involving vertigo. It is suggested that WBV exercises may cause side effects, including vertigo, by generating forces that can influence the internal organs, which may potentially cause labyrinthine trauma or dislocation of otoconia, leading to BPPV.

Gillan et al. (2011) have reported a case of spontaneous vitreous hemorrhage (VH) in a healthy individual exposed to a session of WBV training. A 52-year-old man attended the clinic with a sudden drop in his right visual acuity after WBV training. The patient performed a 20-minute session of WBV for the first time in two weeks. No other exercise had been performed. Within two minutes of completing WBV, he developed a new floater in his right eye, and within five minutes he had an important drop in vision. His vision was 6/18 in the affected eye and he was found to have a central VH. The fellow eye was asymptomatic with acuity 6/6 and normal fundus. Four weeks later, his vision had improved to 6/9, but he was noted to have superior and inferior vitreous condensations with resolving VH. No retinal breaks were detected. By eight weeks, the hemorrhage associated with the vitreous changes continued to resolve but acuity was unchanged. B-scan ultrasound showed localized, inferior, posterior vitreous detachments (PVD) corresponding to the inferior vitreous condensations. The PVD was seen to be inferior to the disk. Considering the adverse side effects, it is relevant to point out that Gillan et al. (2011) have suggested that the VH and localized PVD seen in the clinical case would have an association with WBV: (a) vibrations directly caused the VH, which led to contraction in the vitreous or along the posterior hyaloid face, resulting in a localized PVD subsequently; (b) vibrations themselves caused a localized PVD, which caused hemorrhage from the retinal vasculature at the moment of PVD.

Franchignoni et al. (2013) have reported that a healthy elite athlete (steeplechase runner) suffered two episodes of gross hematuria 72 hours after exposition to

WBV exercise. A 34-year-old male national-level steeplechase runner with 20 years experience came to the doctor following two episodes of reddish-colored urine (one and eight days prior). He said that he ran 10–15 km daily and performed sessions to enhance muscle strength and power once a week. Three weeks earlier, he had added a weekly session of WBV training (five repetitions of one min, 30 Hz, 4 mm amplitude, a semi-squat position with 100° knee flexion) with feet positioned at a distance of about 20 cm from the sagittal axis of the plate. Shortly after the third WBV exposition, he had an episode of bright red urine. He continued his usual running schedule without any other symptom. The urine became macroscopically normal the day after. Seven days later, following the next WBV session (before the daily run), a reddish-colored urine reappeared. At that point, he stopped any physical activity. He stated that he had no history of other hematuria (HT) episodes or abdominal injuries, and did not recall any pain (particularly abdominal pain) or bruising during or after activities in the previous month. The patient stated that he had voided the bladder before WBV session. He had no fever or arthralgia. The next morning (about 40 hours after the HT onset), blood screening analysis was unremarkable, and the urinalysis revealed an important number of red blood cells. Abdominal ultrasonographic evaluation showed no abnormalities. Blood and urine analyses, performed two days later, were normal. The patient was advised to stop WBV training and to take fluid before and during exertion. He did not experience any episode of HT during a one-year follow-up with periodic check-ups. Due to the history and clinical course of sudden HT and the analyses performed, it is suggested that the most probable explanation seems to be a bladder injury, which was produced by repeated impact of the posterior bladder wall against the bladder base, in turn, causing focal mucosal and vascular lesions. Considering the adverse side effects, it is relevant to remark the suggestion of Franchignoni et al., 2013, that the concomitance of the two types of trauma, daily running and the subsequent WBV could have been critical in an individual.

Sonza et al. (2015) studied the acceleration magnitudes at different body segments for different frequencies of WBV. Additionally, vibration sensation ratings by subjects served to create perception vibration magnitude and discomfort maps of the human body. Measurements were performed at 12 different frequencies, two intensities (3 and 5 mm amplitudes) of rotational mode WBV. Vertical accelerometry of the head, hip and lower leg with the same WBV settings was performed. Considering the adverse side effects, at 5 mm amplitude, 61.5% of the subjects (healthy sport students) reported discomfort in the foot region (21–25 Hz), 46.2% for the lower back (17, 19 and 21 Hz) and 23% for the abdominal region (9–13 Hz). Furthermore, it is suggested that the subject feels discomfort or pain, this could be a sign that the chosen settings (frequency, amplitude, vibration mode) are not appropriate for this individual.

Hwang and Ryu (2016) submitted chronic stroke individuals to WBV exercise at different vibration frequencies and examined its immediate effect on their postural sway. The 10 Hz WBV exercise did not affect the postural sway of stroke individuals. The 40 Hz WBV exercise increased postural sway in the latero-medial direction. It is suggested that WBV exercise application to stroke patients in the clinical field may have adverse side effects and therefore caution is necessary.

It is highlighted that oscillating/vibratory platforms providing side-alternating vibration may pose some health risks with high amplitudes (the feet are positioned too far from the axis of rotation in this kind of oscillating/vibratory platform). Cochrane (2011) has pointed out that some of the related side-effects of the use of WBV exercise would be due to lack of familiarity of the participants with the WBV exercise.

11.3 ABSOLUTE AND RELATIVE CONTRAINDICATIONS TO THE EXPOSITION OF INDIVIDUALS TO THE WHOLE BODY VIBRATION EXERCISES

To try to minimize the undesirable and unpleasant effects of the WBV exercises, absolute and relative contraindications to the exposition of individuals to the WBV exercises must be considered. A condition that must be highlighted to try to avoid the undesirable unpleasant effects is to perform a complete and careful clinical evaluation of the individual with professional criteria.

WBV training would be not recommended for pregnancy, epilepsy, headache, blurred vision or serious ocular disease, acute hernia, discopathy, hip and knee implants and individuals with pacemakers (see Table 11.1).

Pregnancy is considered as an absolute contraindication. Authors have pointed out that vibration exposure may cause the reduction of uterine blood flow, menstrual disturbances and abnormal pregnancy such as abortions or stillbirths (Penkov, 2007).

TABLE 11.1
Absolute and Relative Contraindications to the Exposition of an Individual to the Whole Body Vibration Exercises

Absolute Contraindications	Relative Contraindications
Cardiovascular problem (without control)	Hypertension (controlled)
Pacemaker	Diabetes (controlled)
Recent surgery	Implants (care)
Prosthesis of pelvis and knee	Osteopenia (care)
Epilepsy	Initial osteoporosis (care)
Severe diabetes	Headache
Recent intrauterine device (IUD)	
Metal pins	
Advanced osteoporosis	
Serious ocular disease	
Deep vein thrombosis	
Pregnancy	
Acute hernia	
Phobia of the platform movement	

Epilepsy is also considered as contraindication for WBV training. The potential serious injury from falling may happen if a WBV trainee has a seizure during standing on the platform. But, it is important to consider that there is no research in this case.

The hip and knee implants and individuals with pacemaker have been concerned that the vibration may cause migration of the implants and pacemaker.

Vibration exposure may cause vertebral discopathy by increasing internal pressure, increasing anteroposterior shear flexibility, and decreasing resistance to buckling instability (Wilder, 1993).

Vibration in individuals with vertebral disc disorder should be prohibited. Severe headache and blurred vision or serious ocular diseases may lead to falls and serious injury during WBV training (Moseley and Griffin, 1986).

Recent fractures and wounds should be considered for developing non-union fractures and nonhealing wounds after WBV training.

Vela et al. (2010) pointed out that WBV exercise is not advised for pregnant women or for individuals with pacemakers, tumors, epilepsy, acute hernia, headaches or surgical implants (especially hip and knee prostheses). It is suggested that, people who have had cataract surgery may not remember or may not be aware of the fact that IOLs are intraocular prostheses and are therefore a contraindication to exercise of this type.

11.4 RECOMMENDATIONS OF THE INTERNATIONAL SOCIETY OF MUSCULOSKELETAL AND NEURONAL INTERACTIONS TO IMPROVE THE QUALITY OF REPORTS ABOUT WHOLE BODY VIBRATION EXERCISES

Rauch et al. (2010), have reported recommendations of the International Society of Musculoskeletal and Neuronal Interactions to improve the quality of publications about the WBV exercises to try to establish criteria for WBV interventions. Moreover, these recommendations might also contribute to reduce the possibility of undesirable effects. Naturally, a correct clinical evaluation of the individual that will be exposed to the mechanical vibration generated in the oscillating/vibratory platform is desirable.

The recommendations are also related to:

(i) Characteristics of the commercial oscillating/vibratory platform and of the mechanical vibration produced in these platforms;
(ii) Biomechanical parameters related to the mechanical vibration;
(iii) Position of the individual on platform and about the footwear;
(iv) The various steps of the temporal concerns of the protocol.

Figure 11.1 indicates parameters related to the commercial device of the oscillating/vibratory platform.

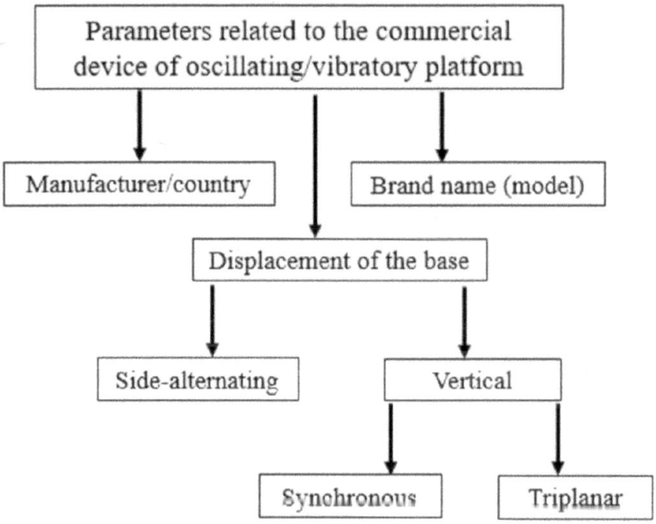

FIGURE 11.1 Parameters related to the commercial device of oscillating/vibratory platform.

Figure 11.2 indicates biomechanical parameters related to the mechanical vibration that is generated in the oscillating/vibratory platform.

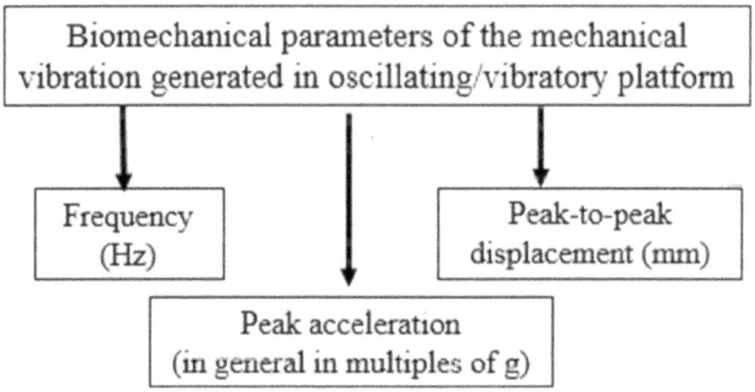

FIGURE 11.2 Biomechanical parameters of the mechanical vibration generated in oscillating/vibratory platform. g is a symbol that represents the Earth's gravity. It is a constant (9.81 m/s^2) and corresponds the nominal acceleration due to gravity at the Earth's surface at sea level (Rauch et al., 2010).

Other considerations are related to the position of the individual that is exposed to the mechanical vibration in the oscillating/vibratory platform and about footwear, as shown in Figure 11.3.

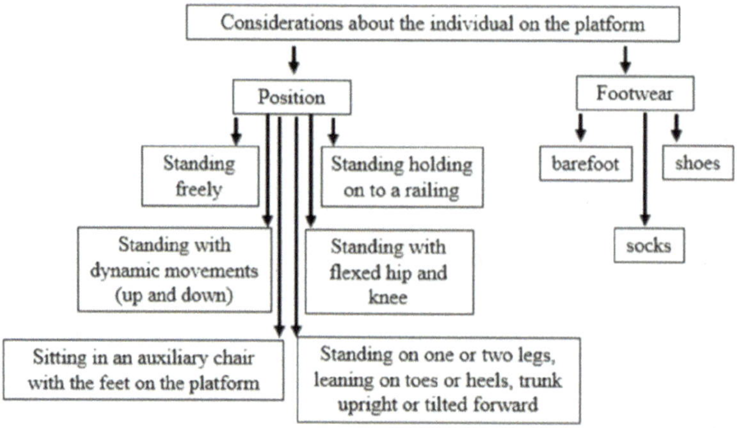

FIGURE 11.3 Considerations related to the individual exposed to the whole body vibration exercise.

The various steps of the temporal concerns of the protocol must be also considered, as it is shown in Figure 11.4.

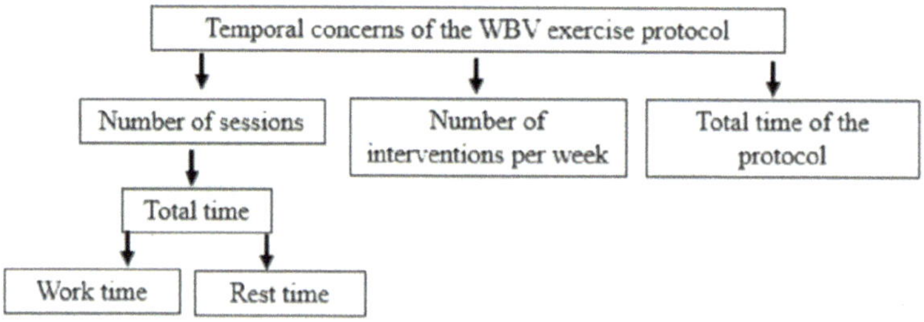

FIGURE 11.4 Temporal concerns of the whole body vibration protocol.

11.5 APPROACHES TO THE SAFETY AND CARE ASPECTS OF WBV EXERCISES

The International Organization for Standardization (ISO) has provided several approaches (requirements, specifications, guidelines or characteristics) that can be used to ensure that various available materials, products, processes and services are fit for their purpose.

Ittianuwat et al. (2017) reported that the ISO 2631-1 standard (1997) defines the three limits of the mechanical vibration that people can be expected to tolerate in industrial exposure, that are for comfort, performance proficiency and safety. Combinations of various parameters are necessary, as the frequency, direction and time of exposure differ. However, Rittweger (2000) reported that the use of this ISO in relation to the WBV exercise is not well understood.

Lienhard et al. (2014) have pointed out that the ISO has published guidelines for occupational vibrations, and daily exposure to WBV has been found to exceed the limit determined by the ISO. It is relevant to discuss if the guidelines are, in general, applicable to training or medical devices, as the guidelines focus on very high vibration frequencies (up to 80 Hz), induced over long periods of time (four to eight hours), and transmitted through the bottom and not through the feet. Moreover, although less than 11% of the platform acceleration was transmitted to the head, it is unknown if such acceleration magnitudes are detrimental.

Pel et al. (2009) have investigated three-dimensional platform accelerations of three different WBV devices without and with three volunteers of different weight in squat position (150° knee flexion). The devices tested were two professional devices, the Power Plate and the Galileo-Fitness, and one home-use device, the Power-Maxx. The transmission of vertical platform accelerations of each device to the lower limbs in eight healthy volunteers in squat position (100° knee flexion). It was shown that the platforms of two professional devices vibrated in an almost perfect vertical sine wave at frequencies between 25–50 and 5–40 Hz, respectively. It is reported that the platform accelerations were slightly influenced by body weight. The Power-Maxx platform mainly vibrated in the horizontal plane at frequencies between 22 and 32 Hz, with minimal accelerations in the vertical direction. The weight of the volunteers reduced the platform accelerations in the horizontal plane but amplified those in the vertical direction about eight times. The vertical accelerations were highest in the Galileo (~15 units of g) and the Power Plate (~8 units of g), and lowest in the Power-Maxx (~2 units of g). The authors have reported that the transmission of vertical accelerations at a common preset vibration frequency of 25 Hz were largest in the ankle and that transmission of acceleration reduced about ten times at the knee and hip. It was concluded that large variation in three-dimensional accelerations exist in commercially available devices. The results suggest that these differences in mechanical behavior induce variations in transmissibility of vertical vibrations to the (lower) body.

Caryn et al. (2014) have pointed out that excessive mechanical energy transferred to the head and eye can cause injury. The purpose of this study was to evaluate how changes in platform frequency and knee flexion angle affect acceleration transmission to the head. Participants stood on a WBV platform with knee flexion angles of 0, 20, and 40° to evaluate how changes in knee flexion affected head acceleration. Seven specific platform frequencies were tested between 20-50 Hz at two peak-to-peak displacement settings (1 and 2 mm nominal). Accelerations were measured with triaxial accelerometers at the platform and head to generate transmissibility ratios. It is reported that platform-to-head transmissibility was not significantly different between the two platform peak-to-peak amplitudes. As it was expected and Rauch et al. (2010) have indicated, the transmissibility measures varied depending on platform frequency and knee angle. Flexing the knees resulted in reduced head transmissibility at all frequencies, as Rauch et al. (2010) have pointed out. Platform-to-head transmissibility values exceeded 1.0 at both 20 and 25 Hz platform vibration frequencies with the knees in full extension. To reduce the risk of injury to structures of the head during vibration exercise, using platforms frequencies below 30 Hz with small knee flexion angles (<40°) should be avoided.

Sonza et al. (2015) studied the acceleration magnitudes at different body segments for different frequencies of WBV. Additionally, vibration sensation ratings by subjects served to create perception vibration magnitude and discomfort maps of the human body. In an experiment, young adults, participated in WBV severity perception ratings, based on a Borg scale. Measurements were performed at different frequencies, two intensities (3 and 5 mm amplitudes) of rotational mode WBV. On a separate day, a second experiment included vertical accelerometry of the head, hip and lower leg with the same WBV settings. It is reported that the highest lower limb vibration magnitude perception based on the Borg scale was extremely intense for the frequencies between 21 and 25 Hz; somewhat hard for the trunk region (11–25 Hz) and fairly light for the head (13–25 Hz). The highest vertical accelerations were found at a frequency of 23 Hz at the tibia, 9 Hz at the hip and 13 Hz at the head. At 5 mm amplitude, 61.5% of the subjects (healthy sport students) reported discomfort in the foot region (21–25 Hz), 46.2% for the lower back (17, 19 and 21 Hz) and 23% for the abdominal region (9–13 Hz). The range of 3–7 Hz represents the safest frequency range with magnitudes less than 1 g*sec for all studied regions. It is also pointed out that, in general, the number of occurrences for discomfort, related to the WBV in our experiments was not very high. However, one should consider that the subjects of this study were healthy sports students and discomfort ratings are likely to be different for patients and/or elderly subjects. It is suggested that the discomfort can be related to resonance phenomena within the body and must be considered as a possible source of undesirable side effects of using WBV. It is reported that perceptions of highest vibration in the body and most discomfort can differ. However, for the regions of head, lower back, lower legs and foot, the frequencies from the responses "feeling most" and "discomfort" are close or coincide. In agreement with Sonza et al. (2015) and with the recommendations of Rauch et al., 2010, it is highly important to consider in the protocols involving the exposition to WBV; if the subject feels discomfort or pain, this could be a sign that the chosen settings (frequency, amplitude, vibration mode) are not appropriate for this individual. Moreover, Sonza et al. (2015) suggest that the WBV frequency and amplitude settings should be based on an individual's perceptions.

11.6 CONCLUSIONS AND PERSPECTIVES

The complexity of the interactions of the mechanical vibration with the body is high, however, important several clinical applications have been described. Benefits for different populations have been reported, since trained and untrained healthy individuals, young and elderly people and professional athletes. However, undesirable and unpleasant adverse effects of WBV exercise have been presented. Putting together all this information, it is clear that more research is needed in order to better understand the potential of WBV exercise. However, with the appropriate and necessary care, WBV exercise seems to be reasonably safe and suitable intervention to be used in various populations of different ages and clinical conditions.

11.7 KEYWORDS FOR INDEX

Whole body vibration, mechanical vibration, biological effects, clinical applications, undesirable effects, biomechanics.

11.8 ACKNOWLEDGMENTS

The authors thank to Conselho Nacional de Desenvolvimento Científico e Tecnológico (CNPq), Fundação Carlos Chagas Filho de Amparo à Pesquisa do Estado do Rio de Janeiro (FAPERJ) and Universidade do Estado do Rio de Janeiro (UERJ), UNESA, Université de Reims

REFERENCES

Alev, A., et al. (2017). Effects of whole body vibration therapy in pain, function and depression of the patients with fibromyalgia. *Complementary Therapies in Clinical Practice* 28:200–203.

Alvarez-Alvarado, S., et al. (2017). Benefits of whole-body vibration training on arterial function and muscle strength in young overweight/obese women. *Hypertension Research* 40:487–492.

Amir, I., Young, E., and Belloso, A. (2010). Self-limiting benign paroxysmal positional vertigo following use of whole-body vibration training plate. *Journal of Laryngology & Otology* 124:796–798.

Annino, G., et al. (2017). Acute changes in neuromuscular activity in vertical jump and flexibility after exposure to whole body vibration. *Medicine (Baltimore)* 96:e7629.

Arias, P., et al. (2009). Effect of whole body vibration in Parkinson's disease: A controlled study. *Movement Disorders* 24:891–8.

Bertschinger, D. R. and Dosso, A. (2008). Vitreous hemorrhage and whole-body vibration training--is there an association? *Journal Francais D'Ophtalmologie* 31(8):e17.

Bokaeian, H. R., et al. (2016). The effect of adding whole body vibration training to strengthening training in the treatment of knee osteoarthritis: A randomized clinical trial. *Journal of Bodywork and Movement Therapies* 20:334–340.

Bosco, C., et al. (2000). Hormonal responses to whole-body vibration in men. *European Journal of Applied Physiology* 81:449–454.

Bowtell, J. L., et al. (2016). Short duration small sided football and to a lesser extent whole body vibration exercise induce acute changes in markers of bone turnover. *BioMed Research International* (2016):3574258.

Cardinale, M., et al. (2010). Hormonal responses to a single session of whole body vibration exercise in older individuals. *British Journal of Sports Medicine* 44:284–288.

Carvalho-Lima, R. P., et al. (2017). Quality of life of patients with metabolic syndrome is improved after whole body vibration exercises. *African Journal of Traditional, Complementary and Alternative Medicines* 14:59–65.

Caryn, R. C., Hazell, T. J., and Dickey, J. P. (2014). Transmission of acceleration from a synchronous vibration exercise platform to the head. *International Journal of Sports Medicine* 35:330–338.

Choi, E. T., et al. (2016). The effects of visual control whole body vibration exercise on balance and gait function of stroke patients. *Journal of Physical Therapy Science* 28:3149–3152.

Chouza, M., et al. (2011). Acute effects of whole-body vibration at 3, 6, and 9 Hz on balance and gait in patients with Parkinson's disease. *Movement Disorder* 26:920–1.

Cloak, R., Nevill, A., and Wyon, M. (2016). The acute effects of vibration training on balance and stability amongst soccer players. *European Journal of Sport Science* 16: 20–26.

Cochrane, D. J. (2011). The potential neural mechanisms of acute indirect vibration. *Journal of Sports Science and Medicine* 10:19–30.

Crewther, B., Cronin, J., and Keogh, J. (2004). Gravitational forces and whole body vibration: Implications for prescription of vibratory stimulation. *Physical Therapy in Sport* 5:37–43.

Cronin, J. and Crewther, B. (2004). Training volume and strength and power development. *Journal of Science and Medicine in Sport* 7:144–155.

Dallas, G., et al. (2015). The acute effects of different training loads of whole body vibration on flexibility and explosive strength of lower limbs in divers. *Biology of Sport.* 32: 235–241.

El-Shamy, S. (2017). Effect of whole body vibration training on quadriceps strength, bone mineral density, and functional capacity in children with hemophilia: A randomized clinical trial. *Journal of Musculoskeletal and Neuronal Interactions* 17:19–26.

Franchignoni, F., Vercelli, S., and Ozcakar, L. (2013). Hematuria in a runner after treatment with whole body vibration: A case report. *Scandinavian Journal of Medicine & Science in Sports* 23:383–385.

Fuermaier, A. B., et al. (2014). Whole-body vibration improves cognitive functions of an adult with ADHD. *Attention Deficit and Hyperactivity Disorders* 6:211–220.

Gillan, S. N., Sutherland, S., and Cormack, T. G. (2011). Vitreous hemorrhage after whole-body vibration training. *Retinal Cases and Brief Reports* 5:130–131.

Gloeckl, R., et al. (2017). What's the secret behind the benefits of whole-body vibration training in patients with COPD? A randomized, controlled trial. *Respiratory Medicine* 126:17–24.

Goudarzian, M., et al. (2017). Effects of whole body vibration training and mental training on mobility, neuromuscular performance, and muscle strength in older men. *Journal of Exercise Rehabilitation* 13:573–580.

Hand, J., Verscheure, S., and Osternig L. (2009). A comparison of whole-body vibration and resistance training on total work in the rotator cuff. *Journal of Athletic Training* 44:469–474.

Högler, W., et al. (2017). The effect of whole body vibration training on bone and muscle function in children with osteogenesis imperfecta. *Journal of Clinical Endocrinology and Metabolism* 102:2734–2743.

Hoyer-Kuhn, H., et al. (2014). A specialized rehabilitation approach improves mobility in children with osteogenesis imperfecta. *Journal of Musculoskeletal and Neuronal Interactions* 14:445–53.

Hwang, K. J. and Ryu, Y. U. (2016). Whole body vibration may have immediate adverse effects on the postural sway of stroke patients. *Journal of Physical Therapy Science* 28:473–477.

Issurin, V. B., Liebermann, D. G., and Tenenbaum, G. (1994). Effect of vibratory stimulation training on maximal force and flexibility. *Journal of Sports Sciences* 12:561–566.

Ittianuwat, R., Fard, M., and Kato, K. (2017). Evaluation of seatback vibration based on ISO 2631-1 (1997) standard method: The influence of vehicle seat structural resonance. *Ergonomics* 60:82–92.

Johnson, P. K., et al. (2014). Effect of whole body vibration on skin blood flow and nitric oxide production. *Journal of Diabetes Science and Technology* 8:889–894.

Jacobson, B. H., et al. (2017). Acute effect of biomechanical muscle stimulation on the counter-movement vertical jump power and velocity in division I football players. *The Journal of Strength and Conditioning Research* 31:1259–1264.

Ko, M. S., et al. (2016). Effects of three weeks of whole-body vibration training on joint-position sense, balance, and gait in children with cerebral palsy: A randomized controlled study. *Physiotherapy Canada* 68:99–105.

Ko, M. C., et al. (2017). Whole-body vibration training improves balance control and sit-to-stand performance among middle-aged and older adults: A pilot randomized controlled trial. *European Review of Aging and Physical Activity* 18;14:11.

Krause, A., et al. (2017). Alleviation of motor impairments in patients with cerebral palsy: Acute effects of whole-body vibration on stretch reflex response, voluntary muscle activation and mobility. *Frontiers in Neurology* 16;8:416.

Lai, Z., et al. (2017). Effects of whole body vibration exercise on neuromuscular function for individuals with knee osteoarthritis: Study protocol for a randomized controlled trial. *Trials* 18:437.

Lee, K. (2017). Effects of whole-body vibration therapy on perception thresholds of type 2 diabetic patients with peripheral neuropathy: A randomized controlled trial. *Journal of Physical Therapy Science* 9:1684–1688.

Lienhard, K., et al. (2014). Determination of the optimal parameters maximizing muscle activity of the lower limbs during vertical synchronous whole-body vibration. *European Journal of Applied Physiology* 114:1493–1501.

Luo, X., et al. (2017). The effect of whole-body vibration therapy on bone metabolism, motor function, and anthropometric parameters in women with postmenopausal osteoporosis. *Disability and Rehabilitation* 39:2315–2323.

Maddalozzo, G. F., et al. (2016). Comparison of 2 multimodal interventions with and without whole body vibration therapy plus traction on pain and disability in patients with non-specific chronic low back pain. *Journal of Chiropractic Medicine* 15:243–251.

Maeda, N., et al. (2016). Effect of whole-body-vibration training on trunk-muscle strength and physical performance in healthy adults: Preliminary results of a randomized controlled trial. *Journal of Sport Rehabilitation* 25:357–363.

Manimmanakorn, N., et al. (2017). Effects of whole body vibration on glycemic indices and peripheral blood flow in type ii diabetic patients. *Malaysian Journal of Medical Sciences* 24:55–63.

Marín-Cascales, E., Alcaraz, P. E., and Rubio-Arias, J. A. (2017). Effects of 24 weeks of whole body vibration versus multicomponent training on muscle strength and body composition in postmenopausal women: A randomized controlled trial. *Rejuvenation Research* 20:193–201.

Martínez-Pardo, E., et al. (2015). Effects of whole-body vibration training on body composition and physical fitness in recreationally active young adults. *Nutrición Hospitalaria* 32:1949–1959.

Menéndez, H., et al. (2015). Influence of isolated or simultaneous application of electromyostimulation and vibration on leg blood flow. *European Journal of Applied Physiology* 115:1747–1755.

Monteleone, G., et al. (2007). Contraindications for whole body vibration training: A case of nephrolitiasis. *Journal of Sports Medicine and Physical Fitness* 47:443–445.

Moseley, M. J. and Griffin, M. J. (1986). Effects of display vibration and whole-body vibration on visual performance. *Ergonomics* 29:977–983.

Pel, J. J., et al. (2009). Platform accelerations of three different whole-body vibration devices and the transmission of vertical vibrations to the lower limbs. *Medical Engineering & Physics* 31:937–944.

Penkov, A. (2007). Influence of occupational vibration on the female reproductive system and function. *Akush Ginekol (Sofiia)* 46:44–48.

Pessoa, M. F., et al. (2017). Vibrating platform training improves respiratory muscle strength, quality of life, and inspiratory capacity in the elderly adults: A randomized controlled trial. *Journals of Gerontology. Series A, Biological Sciences and Medical Sciences* 72:683–688.

Rauch, F., et al. (2010). Reporting whole-body vibration intervention studies: Recommendations of the International Society of Musculoskeletal and Neuronal Interactions. *JMNI* 10:193–198.

Regterschot, G. R., et al. (2014). Whole body vibration improves cognition in healthy young adults. *PLoS One* 9(6):e100506.

Rittweger, J., Beller, G., and Felsenberg, D. (2000). Acute physiological effects of exhaustive whole-body vibration exercise in man. *Clinical Physiology* 20:134–142.

Sá-Caputo D. C., et al. (2014). Whole body vibration exercises and the improvement of the flexibility in patient with metabolic syndrome. *Rehabilitation Research and Practice* (2014):628518.

Sañudo, B., et al. (2013). Changes in body balance and functional performance following whole-body vibration training in patients with fibromyalgia syndrome: A randomized controlled trial. *Journal of Rehabilitation Medicine* 45:678–684.

Satou, Y., et al. (2007). Effects of short-term exposure to whole-body vibration on wakefulness level. *Industrial Health* 45:217–223.

Sonza, A., et al. (2015). A whole body vibration perception map and associated acceleration loads at the lower leg, hip and head. *Medical Engineering & Physics* 37:642–649.

Spielmanns, M., et al. (2017). Low-volume whole-body vibration training improves exercise capacity in subjects with mild to severe COPD. *Respiratory Care* 62:315–323.

Tan, L., et al. (2016). Effect of 4-week whole body vibration on distal radius density. *Chinese Medical Sciences Journal* 31:95–99.

Uhm, Y. H. and Yang, D. J. (2017). The effects of whole body vibration combined biofeedback postural control training on the balance ability and gait ability in stroke patients. *Journal of Physical Therapy Science* 29:2022–2025.

Uszynski, M. K., et al. (2016). Comparing the effects of whole-body vibration to standard exercise in ambulatory people with Multiple Sclerosis: A randomised controlled feasibility study. *Clinical Rehabilitation* 30:657–68.

Vela, J. I., et al. (2010). Intraocular lens dislocation after whole-body vibration. *Journal of Cataract & Refractive Surgery* 36:1790–1791.

Wilder, D. G. (1993). The biomechanics of vibration and low back pain. *American Journal of Industrial Medicine.* 23:577–588.

Wilson, S. J., et al. (2017). The influence of an acute bout of whole body vibration on human postural control responses. *Journal Motor Behavior* 23:1–8.

Wong, A., et al. (2016). Whole-body vibration exercise therapy improves cardiac autonomic function and blood pressure in obese pre- and stage 1 hypertensive postmenopausal women. *Journal of Alternative and Complementary Medicine* 22:970–976.

Yang, F. et al. (2015). Controlled whole-body vibration training reduces risk of falls among community-dwelling older adults. *Journal of Biomechanics* 48:3206–12.

Yang, F., et al. (2016). Effects of controlled whole-body vibration training in improving fall risk factors among individuals with multiple sclerosis: A pilot study. *Disability and Rehabilitation* 15:1–8.

Yang, F., et al. (2017). Effects of vibration training in reducing risk of slip-related falls among young adults with obesity. *Journal of Biomechanics* 57:87–93.

Index

A

B

C

F

G

H

I

W

Y